AF864433

Selected Titles in This Series

203 **Hanna Nencka and Jean-Pierre Bourguignon, Editors,** Geometry and nature: In memory of W. K. Clifford, 1997

202 **Jean-Louis Loday, James D. Stasheff, and Alexander A. Voronov, Editors,** Operads: Proceedings of Renaissance Conferences, 1997

201 **J. R. Quine and Peter Sarnak, Editors,** Extremal Riemann surfaces, 1997

200 **F. Dias, J.-M. Ghidaglia, and J.-C. Saut, Editors,** Mathematical problems in the theory of water waves, 1996

199 **G. Banaszak, W. Gajda, and P. Krasoń, Editors,** Algebraic K-theory, 1996

198 **Donald G. Saari and Zhihong Xia, Editors,** Hamiltonian dynamics and celestial mechanics, 1996

197 **J. E. Bonin, J. G. Oxley, and B. Servatius, Editors,** Matroid theory, 1996

196 **David Bao, Shiing-shen Chern, and Zhongmin Shen, Editors,** Finsler geometry, 1996

195 **Warren Dicks and Enric Ventura,** The group fixed by a family of injective endomorphisms of a free group, 1996

194 **Seok-Jin Kang, Myung-Hwan Kim, and Insok Lee, Editors,** Lie algebras and their representations, 1996

193 **Chongying Dong and Geoffrey Mason, Editors,** Moonshine, the Monster, and related topics, 1996

192 **Tomek Bartoszyński and Marion Scheepers, Editors,** Set theory, 1995

191 **Tuong Ton-That, Kenneth I. Gross, Donald St. P. Richards, and Paul J. Sally, Jr., Editors,** Representation theory and harmonic analysis, 1995

190 **Mourad E. H. Ismail, M. Zuhair Nashed, Ahmed I. Zayed, and Ahmed F. Ghaleb, Editors,** Mathematical analysis, wavelets, and signal processing, 1995

189 **S. A. M. Marcantognini, G. A. Mendoza, M. D. Morán, A. Octavio, and W. O. Urbina, Editors,** Harmonic analysis and operator theory, 1995

188 **Alejandro Adem, R. James Milgram, and Douglas C. Ravenel, Editors,** Homotopy theory and its applications, 1995

187 **G. W. Brumfiel and H. M. Hilden,** $SL(2)$ representations of finitely presented groups, 1995

186 **Shreeram S. Abhyankar, Walter Feit, Michael D. Fried, Yasutaka Ihara, and Helmut Voelklein, Editors,** Recent developments in the inverse Galois problem, 1995

185 **Raúl E. Curto, Ronald G. Douglas, Joel D. Pincus, and Norberto Salinas, Editors,** Multivariable operator theory, 1995

184 **L. A. Bokut′, A. I. Kostrikin, and S. S. Kutateladze, Editors,** Second International Conference on Algebra, 1995

183 **William C. Connett, Marc-Olivier Gebuhrer, and Alan L. Schwartz, Editors,** Applications of hypergroups and related measure algebras, 1995

182 **Selman Akbulut, Editor,** Real algebraic geometry and topology, 1995

181 **Mila Cenkl and Haynes Miller, Editors,** The Čech Centennial, 1995

180 **David E. Keyes and Jinchao Xu, Editors,** Domain decomposition methods in scientific and engineering computing, 1994

179 **Yoshiaki Maeda, Hideki Omoro, and Alan Weinstein, Editors,** Symplectic geometry and quantization, 1994

178 **Hélène Barcelo and Gil Kalai, Editors,** Jerusalem Combinatorics '93, 1994

177 **Simon Gindikin, Roe Goodman, Frederick P. Greenleaf, and Paul J. Sally, Jr., Editors,** Representation theory and analysis on homogeneous spaces, 1994

176 **David Ballard,** Foundational aspects of "non"standard mathematics, 1994

(See the AMS catalog for earlier titles)

CONTEMPORARY MATHEMATICS

203

Geometry and Nature

In Memory of W. K. Clifford

A Conference on New Trends in
Geometrical and Topological Methods
in Memory of William Kingdon Clifford
July 30–August 5, 1995
Madeira, Portugal

Hanna Nencka
Jean-Pierre Bourguignon
Editors

American Mathematical Society
Providence, Rhode Island

The conference on New Trends in Geometrical and Topological Methods in Memory of W. K. Clifford was held at the University of Madeira, Madeira, Portugal, from July 30–August 5, 1995. Support was provided by Centro de Ciencia e Tecnologia da Madeira, Junta Nacional de Investigacao Cientifica e Tecnologica, Fundacao Luso-Americana para o Desenvolvimento, Fundacao Jose Berardo, British Council, Camara Municipal do Funchal, Banco Internacional do Funchal, Centro de Ciencias Matematicas, Caixa Geral de Depositos, and Secretaria Regional de Turismo e Cultura.

1991 *Mathematics Subject Classification*. Primary 15Axx, 16Axx, 30Fxx, 53Cxx, 83Cxx.

Library of Congress Cataloging-in-Publication Data

Conference on New Trends in Geometrical and Topological Methods (1995 : São João da Madeira, Portugal)

Geometry and nature : in memory of W. K. Clifford : a Conference on New Trends in Geometrical and Topological Methods in memory of William Kingdon Clifford, July 30–August 5, 1995, Madeira, Portugal / Hanna Nencka, Jean-Pierre Bourguignon, editors.

p. cm.—(Contemporary mathematics, ISSN 0271-4132 ; 203)

Includes bibliographical references.

ISBN 0-8218-0607-6 (pbk. : alk. paper)

1. Geometry—Congresses. 2. Topology—Congresses. 3. Mathematical physics—Congresses. 4. Clifford, William Kingdon, 1845–1879. I. Clifford, William Kingdon, 1845–1879. II. Nencka, Hanna, 1956– . III. Bourguignon, Jean-Pierre, 1947– . IV. Title. V. Series: Contemporary mathematics (American Mathematical Society) ; v. 203.

QC20.7.G44C66 1995
530.15′6—dc21

96-46141
CIP

---

10 9 8 7 6 5 4 3 2 1   02 01 00 99 98 97

Contents

LIST OF PARTICIPANTS

WILLIAM ABIKOFF, University of Connecticut U-9, 196 Auditorium Road, Storrs, CT 06269-3009, USA
abikoff@math.uconn.edu, abikoff@matisse.math.uconn.edu

ANA ABREU, Universidade da Madeira, Colégio dos jesuítas, Largo do colégio, PT-9000 Funchal, Madeira, Portugal

RUI ALMEIDA, Universidade da Madeira, Colégio dos jesuítas, Largo do colégio, PT-9000 Funchal, Madeira, Portugal
ruialm@dragoeiro.uma.pt

PAULO DE ALMEIDA, Instituto Superior Tecnico, Departamento de Matematica, Av. Rovisco PT- Lisboa, PT- Lisboa, Portugal
palmeida@ist.utd.pt

SHUN-ICHI AMARI, University of Tokyo, Riken Frontier Research Program, Japan
amari@sat.t.u-tokyo.ac.jp

ABHAY ASHTEKAR, Center for Gravitational Physics and Geometry, 104 Davey Lab, Penn. State University, University Park, PA 16802-6300 USA
ashtekar@phys.psu.edu

JEAN-PIERRE BOURGUIGNON, IHES, F-91440 Bures-sur-Yvette, France
jpb@orphee.polytechnique.fr

GUY BURDET, Centre de Physique Théorique, CNRS Luminy, Case 907, F-13288 Marseille Cedex 9, France
burdet@cptsu2.univ-mrs.fr

JAROLIM BUREŠ, Charles University, Sokolovska 83, 18600 Prague 8, Czech Republic
jbures@karlin.mff.cuni.cz

ALICIDIO CAETANO, Groupo de matematica, complexo II, Unibersidade e Lisboa, Av. Gama pinto 21, PT-1699 Lisboa cedex. Portugal

MICHEL CAHEN, U.L.B. Campus Plaine CP218, Bvd. du Triomphe, B-1050, Brussels, Belgium
sgutt@ulb.ac.be

BRANDON CARTER, D.A.R.C. Observatoire de Paris, F-92 Meudon, France
carter@mesiob.obspm.fr

YVONNE CHOQUET-BRUHAT, Mécanique Relativiste, Université Paris VI, Tours 66-3e étage, 4 place Jussieu, F-75252 Paris cedex 05, France
choquet@cicrp.jussieu.fr

PHILIPPE COMBE, Centre de Physique Théorique, CNRS Luminy, Case 907, F-13288 Marseille Cedex 9, France
combe@cptsu2.univ-mrs.fr

JOSE MANUEL CASTANHEIRA DA COSTA, Universidade da Madeira, Colégio dos jesuítas, Largo do colégio, PT-9000 Funchal, Madeira, Portugal

J.S.R. CHISHOLM, University of Kent, Cornwallis Building, Canterbury CT2 7NF, U.K.

MONTY CHISHOLM, University of Kent, Cornwallis Building, Canterbury CT2 7NF, U.K.

MARIO CUNHA, Universidade da Madeira, Colégio dos jesuítas, Largo do colégio, PT-9000 Funchal, Madeira, Portugal
mario@dragoeiro.uma.pt

GIANFAUSTO DELL'ANTONIO, Dipartamento di Matematica, Universita di Roma 1, piazzale Aldo Moro 2, I-00185 Roma, Italy
gianfa@tsmi19.sissa.it, dellantonio@sci.uniroma1.it

MAREK DEMIAŃSKI, N. Copernicus Astronomical Center, Polish Academy of Sciences, Bartycka 18, PL-00 716 Warsaw, Poland
mde@camk.edu.pl

SERGIO DOPLICHER, Universita degli Studi di Roma "La Sapienza", Dipartimento di Matematica, Instituto "Guido Castelnuovo", I-00185 Roma, Piazzale Aldo Moro, 2, Italy
serdopli@itcaspur.caspur.it

MICHEL DUBOIS-VIOLETTE, Laboratoire de Physique Théorique, Bâtiment 211, Université de Paris XI, F-91405 Orsay Cedex, France
flad@qcd.th.u-psud.fr, flad@psisun.u-psud.fr

J. TABARDA DUARTE, Universidade da Madeira, Colégio dos jesuítas, Largo do colégio, PT-9000 Funchal, Madeira, Portugal

JEAN-CHRISTOPHE DUCOM, Centre de Physique Théorique, CNRS Luminy, Case 907, F-13288 Marseille Cedex 9, France
ducom@cpt.univ-mrs.fr

CHRISTIAN DUVAL, Centre de Physique Théorique, CNRS Luminy, Case 907, F-13288 Marseille Cedex 9, France
duval@cpt.univ-mrs.fr

THOMAS FRIEDRICH, Institut für Reine Mathematik, Humboldt-Universität zu Berlin, Unter den Linden 6, Pf 1297, O-1086 Berlin, Germany
friedric@mathematik.hu-berlin.de

MARIBEL GONÇALVES, Universidade da Madeira, Colégio dos jesuítas, Largo do colégio, PT-9000 Funchal, Madeira, Portugal
maribel@dragoeiro.uma.pt

ALFRED GRAY, University of Maryland, College Park, MD 20742-0001, USA
gray@hypatia.umd.edu

MARTIN GROTHAUS, Fakultät für physik, Universitat Bielefeld, D-4800 Bielefeld, Germany

SIMONE GUTT, Université de Metz, Département de Mathématiques et Informatique, Ile de Saulcy, F-57045 Metz cedex 1, France
sgutt@ulb.ac.be

PETR HAJICEK, Institute for Theoretical Physics, University of Berne, Sidlerstrasse 5, CH-3012 Berne, Switzerland
hajicek@butp.unibe.ch

JOHN HARNAD, C.R.M, Université de Montréal, C.P. 6128, succ centre ville, Montréal, Québec, Canada H3C 3J7, and Department of Mathematics and Statistics, Concordia University, 7141 Sherbrook W., Montréal, Québec, Canada H4B 1R6
harnad@alcor.concordia.ca, harnad@mathcn.umontreal.ca

JENNY HARRISON, Department of Mathematics, University of California, 1000 Centennial Dr., Berkeley, CA 94720, USA
harrison@math.berkeley.edu

PETER A. HORVÁTHY, Département de Mathématiques, Université de Tours, F-37200 Tours, France
horvathy@balzac.univ-tours.fr

BIRGER IVERSEN, Aarhus University, Aarhus, Denmark
birger@mi.aau.dk

DANIEL KASTLER, Centre de Physique Théorique, CNRS Luminy, Case 907, F-13288 Marseille Cedex 9, France
kastler@cpt.univ-mrs.fr

JERZY KIJOWSKI, Center for Theoretical Physics, Polish Academy of Sciences, 02-668 Warsaw, Al. Lotników 32/46, Poland
kijowski@theta1.ifpan.edu.pl

DIETER KOTSCHICK, Mathematical Institute, University of Basel, Rheinsprung 21, CH-4051 Basel, Switzerland
kotschick@urz.unibas.ch

IRWIN KRA, SUNY at Stony Brook, Stony Brook, NY 11794-3651, USA
irwin@math.sunysb.edu

JERZY LEWANDOWSKI, Institute for Theoretical Physics, University of Warsaw, 00-681 Warsaw, ul. Hoża 69, Poland
lewand@fuw.edu.pl

JOHN MADORE, LPTHE, Université Paris XI, F-91405 Orsay cedex
madore@qcd.th.u-psud.fr

SANDRA MENDOÇAL, Universidade da Madeira, Colégio dos jesuítas, Largo do colégio, PT-9000 Funchal, Madeira, Portugal
sandra@dragoeiro.uma.pt

HANNA NENCKA, Universidade da Madeira, Colégio dos jesuítas, Largo do colégio, PT-9000 Funchal, Madeira, Portugal
nencka@dragoeiro.uma.pt, nencka@cptsu2.univ-mrs.fr

MARIO NOVELLO, Centro Brasileiro de Pasquisas Fisicas, rua Dr. Xavier Sigaud 150, URCA, 22290 Rio de Janeiro
novello@lafexsu1.lafex.cbpf.br

TAXIARCHIS PAPACOSTAS, Physics Department, University of Crete, 714-09 Iraklion, Crete, Greece
taxiar@iesl.forth.gr

MARTINE PERRIN, Centre de Physique Théorique, CNRS Luminy, Case 907, F-13288 Marseille Cedex 9, France
perrin@cptsu2.univ-mrs.fr

ANTONIO PIRES, Universidade da Madeira, Colégio dos jesuítas, Largo do colégio, PT-9000 Funchal, Madeira, Portugal
pires@dragoeiro.uma.pt

CHARLES J. READ, Trinity College, Cambridge, CB2 1TQ, England
cr25@phx.cam.ac.uk

IVOR ROBINSON, Mathematics Program, The University of Texas at Dallas, Jo 4-2, PO Box 830688, Richardson, TX 75083-0688, USA
robinson@utdallas.edu

JOSE-FRANCESCO RODRIGUES, CMAF, Universidade de Lisboa, Av. Prof. Gama Pinto 2, 1699 Lisboa Codex, Portugal

EUGENIA DA SA, Departamento de Matematica, Universidade da Porto, Porto, Portugal

THOMAS SCHÜCKER, Centre de Physique Théorique, CNRS Luminy, Case 907, F-13288 Marseille Cedex 9, France
schucker@cpt.univ-mrs.fr

GEOFFREY SEWELL, Department of Physics, Queen Mary and Westfield College, Mile End Road, London, E1 4NS, England
sewell@v1.ph.qmw.ac.uk

JOSE LUIS DA SILVA, Universidade da Madeira, Colégio dos jesuítas, Largo do colégio, PT-9000 Funchal, Madeira, Portugal
luis@dragoeiro.uma.pt

JOSE LAURINDO SOBRINHO, Ribeira dos Pretêtes, PT-9125 Caniço, Madeira, Portugal

JEAN-MARIE SOURIAU, Centre de Physique Théorique, CNRS Luminy, Case 907, F-13288 Marseille Cedex 9, France
souriau@cpt.univ-mrs.fr

RAYMOND F. STREATER, Department of Mathematics, King's College, Strand, London WC2R 2LS, England
udah110@kcl.ac.uk

LUDWIG STREIT, CCM, Universidade da Madeira, Colégio dos jesuítas, Largo do colégio, PT-9000 Funchal, Madeira, Portugal
streit@dragoeiro.uma.pt, streit@phisik.uni-bielefeld.de

ANDREZEJ TRAUTMAN, Institute for Theoretical Physics, University of Warsaw, ul. Hoża 69, 00-681 Warsaw, Poland
amt@fuw.edu.pl, trautman@tsmi19.sissa.it

ROLAND TRIAY, Centre de Physique Théorique, CNRS Luminy, Case 907, F-13288 Marseille Cedex 9, France
triay@cpt.univ-mrs.fr

ROBIN W. TUCKER, School of Physics, University of Lancaster, Lancaster, LA1 4YB, UK
rwt@vax1.physics.lancaster.ac.uk

RITA VASCONCELOS, Departamento de Matematica, Universidade da Madeira, Colégio dos jesuítas, Largo do colégio, PT-9000 Funchal, Madeira, Portugal
rita@dragoeiro.uma.pt

F. WIEGEL, Physics Department, Twente University, Netherlands

STANISLAW L. WORONOWICZ, Department of Mathematical Methods in Physics, Faculty of Physics, University of Warsaw, Hoża 74, 00-682 Warsaw, Poland
slworono@fuw.edu.pl

Preface

This issue of Contemporary Mathematics series is an outgrowth of a Conference "New Trends in Geometrical and Topological Methods" devoted to the memory of William Kingdon Clifford (Exeter 1845-Madeira 1879). The conference took place at the University of Madeira (Portugal) in July/August 1995. It's aim was to: - bring together and to put in close and informal contact active workers in fields linked to the Clifford's work, - foster interchanges of ideas and discussions between mathematicians and theoretical physicists. The meeting was divided into six one-day sessions, each one devoted to a specific aspect of Clifford's work. This volume is an attempt to present to a larger community of mathematicians and physicists, the Clifford legacy in new perspectives. The new concepts, ideas and results provided by Clifford's work are presented in this issue. This volume contains papers presented or submitted to the Conference. Each contribution is self-contained. The volume is divided in the following chapters: the first one is devoted to Clifford Algebras, the second one to Riemannian Surfaces, the third one to Information Geometry, the fourth one to Non Commutative Geometry, the fifth one to Symplectic Geometry and Self-similar Structures, the sixth one to Cosmology and General Relativity, the seventh one to Field Theory.

It is obvious that this ordering should not to be taken too seriously since in many cases there is an interplay of subjects.

This conference was made possible thanks to the support of several organizations and associations, and to the help of many friends. I am grateful to the Madeira' s Government, Dr. Francisco Santos (regional secretary of education), and L. Mendonca (President of the Madeira's Parliament). I would like to express my gratitude to the members of the Scientific and Organizing Committees, to the members of the Centre de Physique Theorique CNRS, Marseille, Luminy (France). I am grateful to the University of Madeira for all the help I had during the organization of the conference. Specially, I would like to thank the President of the Installation Committee of University of Madeira Prof. Joao David Pinto Correia, Prof. Jose Manuel Castanhera, (Vowel of I.C.U.Ma), Dr. Ana Isabel Spranger (advisor of the President of I.C.U.Ma), and Rui Almeida (Department of Physics). I wish to acknowledge the support from CITMA and its President Dr. Maximiano Martins and all the members of CITMA's Installation Committee. I am grateful to JNICT, FLAD, Fundacao Jose Berardo, British Council, Camara Municipal do Funchal, BANIF, Centro de Ciencias Matematicas, Caixa Geral de Depositos and Secretaria Regional de Turismo e Cultura for their financial support. I would like to express my deepest and sincere gratitude to Profs. Irwin Kra and William Abikoff for the advice and to Prof. Philippe Combe and Dr. Jean-Christophe Ducom for all their help during the conference and for editing this volume.

It was with great sadness that we learned of the untimely death of Prof. Birger Iversen shortly after this conference. His contributions to the field were highly valued, and he will be greatly missed by friends and colleagues.

Hanna Nencka, Madeira , July 1996

Contemporary Mathematics
Volume **203**, 1997

W.K. Clifford as a geometer

William Kingdon Clifford was born on May 4th, 1845 in Exeter, England. He received his early education at Mr.Templeton's school in Exeter. From this school he proceeded, in 1860, to King's College, London. Having in 1863 obtained a minor scholarship, he entered the Trinity College, Cambridge. In 1868 he was elected a fellow of the Trinity College and in 1871 he became Professor of Applied Mathematics and Mechanics at the University College, London, a position he occupied until his death.

In June, 1874, Clifford was elected a Fellow of the Royal Society. On June 18th, 1866, he became a very active member of the London Mathematical Society, and served on its Council for every session from 1868-1869 to 1876-1877.

Clifford was above all a geometer, his mathematical writings can be for a large part classified as geometry[1]. In the treatment of geometrical questions he showed a notorious preference for symbolical methods, and as might be expected, a marvellous command over analytical expression. He took much pleasure in speculative constructions of complex or non-Euclidian systems of space-relations. He had a very strong geometrical imagination. The properties of space remained a perpetual subject of contemplation for him. Purely analytical considerations undertaken by W.K. Clifford without any impulse from reference to geometry are few. His most important works are devoted to Riemann's surfaces, biquaternions, and classification of loci.

Clifford attributed to geometry the widest imaginable scope. He was a mathematician as well as a metaphysician; and geometry was to him an important factor in the problem of "describing the universe". In his lecture *"On the postulates of the Science of Space"* he stated his own views on this question. The pages in which he expressed them are likely to be remembered, as marking an important moment in the controversy concerning the nature of space and the origin of our knowledge of it, which is likely to last as long as metaphysical enquiries have any interest for mankind. In his lectures he enumerated four fundamental postulates on which the ordinary conception of space is founded: 1 - its

[1]see e.g. Mathematical Papers by William Kingdon Clifford, Edited by Robert Tucker, Chelsea Publishing Compagny, Bronx, New York.

continuity, 2 - its flatness in its smallest part, 3 - its similarity to itself at every point, and 4 - the possibility of the existence of figures similar to one another, but on different scales of magnitude.

The second postulate consists of one of his favourite hypothesis and has been described in the communication *"On the Space-Theory of Matter"* at the Cambridge Philosophical Society, in 1870, where he developed the idea that space may not be independent of time. This idea was explicitly stated in a book: *"Common Sense of the Exact Sciences"* published by Rowe after his death, where he wrote:

We may conceive our space to have everywhere a nearly uniform curvature, but that slight variations of the curvature may occur from point to point, and themselves vary with time. These variations of the curvature with the time may produce effects which we not unnaturally attribute to physical causes independant of the geometry of our space. We may even go as far as to assign to this variation of the curvature what really happens in that phenomenon which we term the motion of matter.

The study of the geometrical methods of Grassmann so long neglected, made a great and enduring impression on Clifford. In his paper (1876) *"Applications of Grassmann's Extensive Algebra"*, speaking about the work of Grassmann he had the prophetic sentence:

I may be permitted to express my profund admiration of that extraordinary work, and my conviction that its principles will exercise a vast influence upon the future of mathematical science.

It is one more time through geometry that Clifford approached a problem in algebra to provide a newer view, and his name is perpetuated today in the so-called Clifford algebras, of which octonions or biquaternions are special cases. These noncommutative algebras were used by Clifford to study motions in non-Euclidean spaces, certain manifolds which are now known as spaces of Clifford and Klein.

In the spring of 1876, distinct and grave indications of pulmonary disease were noted, but to be careful about himself never occurred to him. In the early months of 1878 there came a sudden change for the worst. Since medicine had no new thing to recommend, and almost nothing to forbid, a last experiment could only be tried and Clifford sailed for Madeira. The change from the bitterness of recent English winters to the fair and temperate air of Madeira had unfortunately no power to restore his waning forces; but it enabled him to spend his last days in ease and comparative enjoyment. Some weeks were added to his life but on March 3, 1879, the end came.

Part I

CLIFFORD ALGEBRAS

Contemporary Mathematics
Volume **203**, 1997

Clifford and the 'Square Root' Ideas

A. TRAUTMAN

ABSTRACT. This review article begins with a short history of the notions associated with spinors; it describes several distinct 'square root' ideas occurring in connection with Clifford algebras, spin groups and pure spinors. Applications of pure spinors to geometry and physics are briefly presented. An appendix contains a simple derivation of the Vahlen-Ahlfors, fractional-linear form of Möbius transformations.

1. Introduction

Spinors—and the structures underlying this notion, such as Clifford algebras, spin groups and their representations, and spin structures on manifolds—are a good example of the subtle relations and mutual influences between mathematics and physics. Their origins can be traced to work of mathematicians, but they owe their name and fame to physicists. One of the great achievements of 20th century physics is the elucidation of the role of fermions—particles with half integer spin, such as electrons and protons, requiring spinors for their quantum-mechanical description—in the observed stability of matter and the chemical properties of atoms. An essential physical law used to explain these properties is the *Pauli exclusion principle* ('no more than one fermion in any state') which, in turn, follows from the applicability of the Fermi-Dirac statistics to fermions: the quantum state of k fermions of the same kind belongs to the kth exterior power of the vector space of one-particle states. A fundamental result of relativistic quantum theory is the theorem on the *connection between spin and statistics* 'explaining' why fermions obey that statistics and, therefore, satisfy the exclusion principle; see, e. g. [**SW**] and the references given there. One is tempted to say that the world around us, and life in particular, are so rich because Nature found

---

1991 *Mathematics Subject Classification.* Primary 15A66; Secondary 83C60.

Work on this paper was supported in part by the Interdisciplinary Laboratory of the International School for Advanced Studies (SISSA) in Trieste and the Polish Committee on Scientific Research (KBN) under grant 2 P302 112 7.

it convenient to use, among its building blocks, entities requiring spinors in their description.

At first, spinors baffled physicists and mathematicians alike; in the words of Darwin (1928)[1]: *The relativity theory is based on nothing but the idea of invariance and develops from it the conception of tensors as a matter of necessity; and it is rather disconcerting to find that apparently something has slipped through the net, so that physical quantities exist, which it would be, to say the least, very artificial and inconvenient to express as tensors.*

Some people who had not accepted Darwin's wisdom expressed in the last sentence were criticized by É. Cartan; see the footnotes in the last section of [**C4**]. Since spinors are difficult to visualize, even though they cannot be expressed as tensors, it is natural to relate their properties to those of the more familiar vectors, tensors, quadratic forms and orthogonal transformations. Many of such heuristic considerations emphasize the appearance of the idea of a 'square' or of a 'square root' in the relations between vectors and spinors.

This article reviews some of the (distinct) square root ideas as well as a few applications of spinors. Proofs are replaced by references to the literature, whenever possible. Extensive information and bibliographies on the subject of Clifford algebras, spin structures, the Dirac operator, and on applications of spinors in geometry and physics can be found in recent books and reviews, such as [**BGV, Bo, Ha, LM, PR**]. The next section contains a rather cursory account of the history of notions related to spinors; this is intended to put in perspective the very significant contribution made by W. K. Clifford. Following Cartan [**C4**], and unlike in most contemporary expositions of the subject, spinor representations are treated here first for *odd*-dimensional spaces. This is justified by the fact that, even though the Clifford algebra Cl_{2m} is isomorphic to the even subalgebra Cl^{+}_{2m+1}, the group $\mathrm{Spin}_{2m}(\mathbb{C})$ is properly contained in $\mathrm{Spin}_{2m+!}(\mathbb{C})$; any results established for the latter hold, by restriction, for the former, but not conversely. The Appendix outlines a simple, but apparently little known, derivation of the fractional-linear form of Möbius transformations of projective quadrics (in particular, of the n-sphere $\mathbb{S}_n$).

2. On the history of Clifford algebras and spinors

There is a prehistory of spinors: for example, in the papers by Euler [**E**] and Rodrigues [**Ro**] on the rational representation of rotations in $\mathbb{R}^3$ one can find—with a little effort and some good will—the map $\mathrm{SU}_2 \to \mathrm{SO}_3$. Hamilton represented rotations in terms of quaternions: every rotation in $\mathbb{R}^3$ is of the form $q \mapsto aqa^{-1}$, where $a \in \mathrm{Sp}_1$ and $q \in \mathbb{R}^3 \subset \mathbb{H}$ are a unit and a pure quaternion, respectively (this shows that the groups Spin_3 and Sp_1 are isomorphic). Cayley (1855) extended Hamilton's observation to $\mathbb{R}^4$; he proved, in essence, the isomorphism of the groups Spin_4 and $\mathrm{Sp}_1 \times \mathrm{Sp}_1$. More information on that early period can be found in [**C1**].

---

[1]In this style, I indicate references, mainly to the physics literature, that are listed in [**BuT**].

Clifford [**Cl1,2**] introduced the algebras that now bear his name by considering the underlying vector space of a Grassmann algebra and endowing it with a new product, thus generalizing the algebras of complex numbers and of quaternions. In his language, the 'geometric algebras' over $\mathbb{R}$ are generated by n 'units' $e_1, \ldots, e_n$, such that

$$e_\mu e_\nu + e_\nu e_\mu = -2\delta_{\mu\nu}, \tag{1}$$

where $\mu, \nu = 1, \ldots, n$. The minus sign in (1) was motivated by the algebras $\mathbb{C}$ and $\mathbb{H}$ over $\mathbb{R}$. Clifford knew most of the basic properties of his algebras, described now in modern texts: their $\mathbb{Z}_2$-grading and (semi-) simplicity, the structure of the center, their periodicity and relations between algebras associated with vector spaces of adjacent dimensions, etc. Using Clifford algebras, Lipschitz [**Lp**] introduced the groups that are now called Spin and Clifford; see, in this connection, the anonymous note [**Corr**]. In 1902 Vahlen found, in terms of 'Clifford numbers', the fractional-linear representation of the Möbius group in n dimension; see the Appendix to this paper.

The modern period in the history of spinors begins probably with the papers by Cartan [**C2, C3**] containing a description of the spin representations of the Lie algebras of orthogonal groups. The discovery of the spin of the electron by Uhlenbeck and Goudsmit (1925) forced physicists to find mathematical tools to describe, within the framework of quantum mechanics, this new degree of freedom. When doing this, Pauli (1927) and Dirac (1928) rediscovered the spin representations and the Clifford algebras associated with three- and four-dimensional vector spaces, respectively. Dirac introduced, at that time, in the context of Minkowski space, the differential operator that now bears his name; the Dirac equation has been very successful in describing the quantum-relativistic behavior of electrons and other particles with spin $\frac{1}{2}$. Soon afterwards, Weyl [**W**] and Fock [**F**] developed a *local* theory of the Dirac operator in curved, (Lorentzian) spacetimes. Shortly after the appearance of the paper by Dirac, Ivanenko and Landau (1928) published another approach to the theory of the 'magnetic electron'. They proposed to use an equation based on the operator $d + \delta$, acting on (inhomogeneous) differential forms, where δ is the formal adjoint of the exterior derivative d. In view of the successes of the Dirac theory, their equation was ignored for a long time. The operator $d + \delta$, whose square is the Laplacian, was considered by Kähler [**K**] and, under the influence of his work, physicists used it to describe fermions on a lattice.

Brauer and Weyl (1935) gave a general construction, based on Clifford algebras, of the spin representations in any number of dimensions and showed how the tensor product of two such representations decomposes into irreducibles. The connections between spinors, totally isotropic spaces and projective geometry seem to have been clearly stated, for the first time, by Veblen [**V1, V2**] and developed in seminar lectures at Princeton, given jointly with Givens [**VG**]. These notes contain remarks that may have influenced Cartan in his work on pure spinors[2] [**C4**]. Chevalley based his *Algebraic theory of spinors* [**Ch**] on the notion of minimal, one-sided ideals of Clifford algebras, an idea considered earlier by Riesz [**Ri**] in the context of the Dirac equation in the theory of general relativity and, less explicitly, by several physicists; see, e. g., [**Sa**]. Chevalley developed the theory of spinors over an arbitrary field and provided complete proofs of many new and old results; in particular, on pure spinors and triality. Haefliger's note [**H**] seems to be the first publication containing the definition of a spin structure on a Riemannian space and the derivation of the obstruction to the existence of such a structure. On the basis of this definition, Atiyah and Singer [**AS**] developed a global theory of the Dirac operator $\mathcal{D}$ on a Riemannian space; they proved the index theorem for the operators $d+\delta$ and $\mathcal{D}$. This paper, together with [**ABS**] and the note by Lichnerowicz [**Li**] on harmonic spinors, begins the modern period of research on global aspects of spinor analysis and its applications to differential geometry and topology. This research is surveyed in [**BGV, Bo, Gi, LM**].

Penrose and, under his influence, many physicists (see [**PR**] for a review and references) have used spinors in the context of general relativity; the Newman-Penrose (1962) formalism, used to study solutions of Einstein's equations, is an extension of the method of moving frames to the principal spin bundle over a (Lorentzian) spacetime $\mathcal{M}$. Cartan (1922) noticed that, in such a spacetime, the Weyl tensor (of conformal curvature), at a point where it is $\neq 0$, defines 4 isotropic (optical, null) directions in the tangent space; sometimes these directions coincide; an enumeration of the possible coincidences leads to an algebraic classification of the Weyl tensors of Lorentzian spacetimes; a spacetime is *algebraically degenerate* if at least two of those directions coincide; it is said to be of type D if there are two distinct pairs of coinciding directions (example: the Schwarzschild metric). Since the manifold of isotropic directions at a point of $\mathcal{M}$ can be identified with the projective space of directions of Weyl (semi-) spinors, the structure group $\mathrm{SL}_2(\mathbb{C})$ of the spin bundle of $\mathcal{M}$ reduces to its Abelian subgroup $\mathbb{C}^\times = \mathbb{C} \setminus \{0\}$; the Newman-Penrose method has been very effective in finding solutions of Einstein's equations in this case.

3. The universal linearization of quadratic forms

The idea of *linearization* pervades much of pure and applied mathematics: it suffices to mention the notions of a differential, of tensor and exterior products or the inverse scattering method. The construction of Clifford algebras solves the following universal problem: given a quadratic space (V, h), where V is a (finite-dimensional) vector space over a field K and $h : V \to K$ is a quadratic form, find all Clifford maps, i. e. linear maps $f : V \to \mathcal{A}$, where $\mathcal{A}$ is an algebra

[2] In fact, Cartan used the expression *spineur simple*; the name 'pure spinor' is due to Chevalley.

over K, with a unit element $1_{\mathcal{A}}$, such that

$$f(v)^2 = h(v)1_{\mathcal{A}}$$

for every $v \in V$. The Clifford algebra of the quadratic space (V, h) is defined as $\mathrm{Cl}(V, h) = \mathrm{T}(V)/\mathrm{J}(V, h)$, where $\mathrm{J}(V, h)$ is the ideal generated by all elements of the tensor algebra $\mathrm{T}(V) = \oplus_{p=0}^{\infty} \otimes^p V$ of the form $v \otimes v - h(v)$. The Clifford algebra of (V, h) contains $K \oplus V$ as a vector subspace, the injection $V \to \mathrm{Cl}(V, h)$ is a Clifford map and every Clifford map $f : V \to \mathcal{A}$ extends to a homomorphism $\tilde{f} : \mathrm{Cl}(V, h) \to \mathcal{A}$ of algebras with units. In particular, if $f : V \to V \subset \mathrm{Cl}(V, h)$ is an isometry, $f \in \mathrm{O}(V, h)$, then $\mathrm{Cl}(f) = \tilde{f}$ is an automorphism of the algebra $\mathrm{Cl}(V, h)$; in other words, Cl is a functor from the category of quadratic spaces to that of algebras with unit elements. The isometry $v \mapsto -v$ extends to the *main automorphism* α of $\mathrm{Cl}(V, h)$. Since this automorphism is involutive, it defines a $\mathbb{Z}_2$-grading of the algebra, $\mathrm{Cl}(V, h) = \mathrm{Cl}^+(V, h) \oplus \mathrm{Cl}^-(V, h)$, where $\mathrm{Cl}^{\pm}(V, h) = \{a \in \mathrm{Cl}(V, h) | \alpha(a) = \pm a\}$. The *transposition* is an antiautomorphism $a \mapsto a^t$ of $\mathrm{Cl}(V, h)$ characterized by being a linear automorphism of the underlying vector space such that $1^t = 1$, $v^t = v$ for every $v \in V$ and $(ab)^t = b^t a^t$ for every $a, b \in \mathrm{Cl}(V, h)$.

Assuming, from now on, that the characteristic of K is not 2, one can associate with the quadratic form h the symmetric linear isomorphism $V \to V^*$ given by $v \mapsto v^h$, where $v^h \in V^*$ is the linear form on V such that $2\langle v', v^h \rangle = h(v + v') - h(v) - h(v')$ for every $v, v' \in V$. The linear map v^h extends to a derivation $i(v)$ of degree -1 of the $\mathbb{Z}$-graded (exterior) algebra $\wedge V$. Denoting by $e(v) \in \mathrm{End} \wedge V$ the exterior multiplication by v, one has $i(v)e(v) + e(v)i(v) = h(v)1_{\mathrm{End} \wedge V}$. Therefore, if $F(v) = e(v) + i(v)$, then $F : V \to \mathrm{End} \wedge V$ is a Clifford map and the linear map

(2) $$c : \mathrm{Cl}(V, h) \to \wedge V, \quad \text{defined by} \quad c(a) = \tilde{F}(a)1_{\wedge V},$$

is an isomorphism of vector spaces over K, natural with respect to isometries: if $f \in \mathrm{O}(V, h)$, then

(3) $$c \circ \mathrm{Cl}(f) = \wedge f \circ c.$$

The map c reduces to the identity on $K \oplus V$ and satisfies

(4) $$c(va) = (e(v) + i(v))c(a)$$

for every $v \in V$ and $a \in \mathrm{Cl}(V, h)$ [**Ch**]. The algebra $\wedge V$ has a natural transposition antiautomorphism, $(v_1 \wedge \cdots \wedge v_p)^t = v_p \wedge \cdots \wedge v_1$, and $c(a^t) = c(a)^t$ for every $a \in \mathrm{Cl}(V, h)$. There is also the main automorphism α_V of $\wedge V$ such that $c \circ \alpha = \alpha_V \circ c$.

Let V_0 be a one-dimensional vector space with a generator e_0 and a quadratic form h_0 such that $h_0(e_0) = -1$. The Clifford map $V \to \mathrm{Cl}^+(V \oplus V_0, h \oplus h_0)$ given by $v \mapsto ve_0$ extends to an isomorphism of algebras $\mathrm{Cl}(V, h) \to \mathrm{Cl}^+(V \oplus V_0, h \oplus h_0)$.

In applications, one considers a *representation* γ of $\mathrm{Cl}(V,h)$ in a finite-dimensional vector space S of 'spinors'. Let (e_μ) be a linear basis in V and let $h_{\mu\nu} = \langle e_\mu, e^h_\nu \rangle$, where $\mu,\nu = 1,\dots,n$ and n is the dimension of V. Using Einstein's summation convention over pairs of repeated indices, every $v \in V$ can be written, in terms of its components, as $v^\mu e_\mu$. The representation γ, restricted to V, is a Clifford map and

$$(\gamma_\mu v^\mu)^2 = h_{\mu\nu} v^\mu v^\nu \mathrm{id}_S.$$

If (e_μ) is an orthogonal basis, i. e. if $h_{\mu\nu} = 0$ for $\mu \neq \nu$, then $\mu_1 < \mu_2 < \cdots < \mu_p$ implies $c(e_{\mu_1} e_{\mu_2} \dots e_{\mu_p}) = e_{\mu_1} \wedge e_{\mu_2} \wedge \cdots \wedge e_{\mu_p}$. Let $c_p(a)$ denote the component in $\wedge^p V$ of $c(a)$, $a \in \mathrm{Cl}(V,h)$. If $c(a) = c_p(a)$, then one says that $a \in \mathrm{Cl}(V,h)$ is of *degree* p. If a is of degree p, then $a^t = (-1)^{\frac{1}{2}p(p-1)} a$.

4. Double-valuedness of spinor representations of orthogonal groups

In this section it is assumed that $K = \mathbb{R}$ and the quadratic form h is non-degenerate. One says that v is a *unit vector* if either $h(v) = 1$ or $h(v) = -1$. The *spin group* associated with (V,h) is the set $\mathrm{Spin}(V,h)$ of products of the elements of all sequences consisting of an even number of unit vectors, with multiplication induced from $\mathrm{Cl}(V,h)$. The adjoint representation ρ of $\mathrm{Spin}(V,h)$ in V, defined by $\rho(a)v = ava^{-1}$ for $a \in \mathrm{Spin}(V,h)$ and $v \in V$, gives the exact sequence of group homomorphisms,

$$1 \to \mathbb{Z}_2 \to \mathrm{Spin}(V,h) \xrightarrow{\rho} \mathrm{SO}(V,h) \to 1.$$

This shows that, with respect to a natural topology and differential structure, the group $\mathrm{Spin}(V,h)$ is a Lie group doubly covering the Lie group $\mathrm{SO}(V,h)$. A representation of the spin group lifts to a representation of the rotation group $\mathrm{SO}(V,h)$ if, and only if, it is trivial on the kernel of ρ; the *spin representations*, which are defined here as restrictions to $\mathrm{Spin}(V,h)$ of non-trivial representations of the algebra $\mathrm{Cl}(V,h)$, never lift; they are sometimes said to define double-valued representations of $\mathrm{SO}(V,h)$.

If $V = \mathbb{R}^{k+l}$ and the signature of h is (k,l), then one writes $\mathrm{Cl}_{k,l}$ instead of $\mathrm{Cl}(V,h)$; a similar notation is used for the orthogonal and spin groups. The number $k-l$ is the *index* of h. If k or $l > 1$, then there are two unit vectors u and v orthogonal one to another and such that $h(u) = h(v)$. The function $t \mapsto \omega(t) = \cos t + uv \sin t$ maps the interval $[0,\pi]$ into a curve in $\mathrm{Spin}_{k,l}$ connecting 1 and -1; the image of this curve by ρ is a non-contractible loop in $\mathrm{SO}_{k,l}$. Let $\pi_1(G^\mathrm{o})$ denote the fundamental group of the connected component G^o of a Lie group G; one has

PROPOSITION 1. *If k and l are non-negative integers and at least one of them is > 1, then the sequence*

$$1 \to \pi_1(\mathrm{Spin}^\mathrm{o}_{k,l}) \to \pi_1(\mathrm{SO}^\mathrm{o}_{k,l}) \to \mathbb{Z}_2 \to 1$$

is exact: the group $\pi_1(\mathrm{SO}^\circ_{k,l})$ has a non-trivial $\mathbb{Z}_2$-grading.

Note that $\mathrm{Spin}_{1,0} = \mathbb{Z}_2$ and $\mathrm{Spin}^\circ_{1,1}$ is isomorphic to $\mathbb{R}$. The groups $\mathrm{Spin}_{k,0}$ and $\mathrm{Spin}_{0,k}$, which are isomorphic, are denoted Spin_k; they are connected for $k > 1$ and simply connected for $k > 2$. The groups $\mathrm{Spin}^\circ_{k,l}$, $k > 2$, are simply connected for $l = 1$, but not for $l > 1$. If the groups Spin_2 and SO_2 are both identified with $\mathrm{U}_1 \subset \mathbb{C}$, then ρ becomes literally the map of taking the square.

Let $\mathcal{M}$ be an oriented manifold with a metric tensor of signature (k,l). Let $\pi : \mathcal{P} \to \mathcal{M}$ be its $\mathrm{SO}_{k,l}$-bundle of orthonormal frames of coherent orientation. A *spin structure* on $\mathcal{M}$ is a principal $\mathrm{Spin}_{k,l}$-bundle $\varpi : \mathcal{Q} \to \mathcal{M}$, together with a morphism $\chi : \mathcal{Q} \to \mathcal{P}$ of principal bundles over $\mathcal{M}$ associated with ρ: for every $a \in \mathrm{Spin}_{k,l}$ and $q \in \mathcal{Q}$ one has $\chi(qa) = \chi(q)\rho(a)$ and $\pi \circ \chi = \varpi$. The topological obstruction to the existence of a spin structure on a pseudo-Riemannian space has been found by Karoubi [**Ka**]. Given a spin structure $\mathcal{Q} \to \mathcal{P} \to \mathcal{M}$ and a spin representation $\gamma : \mathrm{Spin}_{k,l} \to \mathrm{End}\, S$, one defines a *spinor field* of type γ on $\mathcal{M}$ to be a (smooth) map $\psi : \mathcal{Q} \to S$ such that, for every $a \in \mathrm{Spin}_{k,l}$ and $q \in \mathcal{Q}$ one has $\psi(qa) = \gamma(a^{-1})\psi(q)$. Let ω be the map defined in the preceding paragraph; since $\gamma(-1) = -\mathrm{id}_S$, one has, for a spinor field, $\psi(q\omega(\pi)) = -\psi(q\omega(0))$; the frame $\chi(q)\rho(\omega(t))$ results from the frame $\chi(q)$ by a rotation by the angle $2t$; for this reason one sometimes says that 'a rotation by 2π induces a change of the sign of a spinor'.

5. Tensor squares of spinors are multivectors

The tensor product of two spin representations of $\mathrm{Spin}(V,h)$ is a representation of $\mathrm{SO}(V,h)$; this simple fact underlies the physicists' construction of (real) multivectors from spinors.

5. 1. Complex vector spaces. Consider the complex vector space $W = \mathbb{C}^n$ with a quadratic form h defined by $h(z) = (z^1)^2 - (z^2)^2 + \cdots + (-1)^{n+1}(z^n)^2$, where $z = (z^\mu e_\mu) \in \mathbb{C}^n$. Denote by Cl_n the Clifford algebra $\mathrm{Cl}(W,h)$. Let $\eta = e_1 \cdots e_n \in \mathrm{Cl}_n$ be an (oriented) volume element in W. Clearly, $\eta v = (-1)^{n+1} v\eta$ for every $v \in W$. The Kähler definition of the Hodge map $\star : \wedge W \to \wedge W$ reads

$$\star c(a) = c(\eta a), \quad \text{where} \quad a \in \mathrm{Cl}_m,$$

and leads to

$$\star\star = \mathrm{id}_{\wedge W} \quad \text{and} \quad i(v) \circ \star = (-1)^{n+1} \star \circ\, e(v)$$

for every $v \in W$.

The complex spin group $\mathrm{Spin}_n(\mathbb{C})$ is defined as the subset of Cl_n^+ consisting of products of all sequences of an even number of vectors with squares $= 1$; if $a \in \mathrm{Spin}_n(\mathbb{C})$, then $a^t a = 1$. A representation of the algebra Cl_n in a vector space defines, by restriction, a representation of the group $\mathrm{Spin}_n(\mathbb{C})$; one uses the same name and letter for a representation of the algebra and of its restriction to the group. If $a \in \mathrm{Spin}_n(\mathbb{C})$ and $b \in \mathrm{Cl}_n$, then $\mathrm{Cl}(\rho(a))b = aba^{-1}$ so that (3) gives

$$c(aba^{-1}) = \wedge\rho(a) \circ c(b). \tag{5}$$

5. 1. 1. Consider first an *odd*-dimensional complex vector space $W = \mathbb{C}^{2m+1}$ ($m = 1, 2, \ldots$). The even subalgebra Cl^+_{2m+1} is simple and, since it is complex 2^{2m}-dimensional, it has one, up to equivalence, irreducible and faithful 'Pauli' representation σ in a 2^m-dimensional complex space S of spinors. By putting $\sigma_\pm(\eta) = \pm\mathrm{id}_S$ and $\sigma_\pm|\mathrm{Cl}^+_{2m+1} = \sigma$, one extends σ to the representations σ_+ and $\sigma_- = \sigma_+ \circ \alpha$ of Cl_{2m+1} in S; these representations are irreducible and inequivalent, but not faithful: the kernel of σ_- (resp., σ_+) is the vector space of self-dual (resp., antiself-dual) elements of Cl_{2m+1}.

The representations $\sigma_\pm$ can be described explicitly as follows. Consider a Witt decomposition $W = N \oplus P \oplus \mathbb{C}e_{2m+1}$, where N and P are m-dimensional—therefore maximal—totally isotropic subspaces of W and e_{2m+1} is a unit vector orthogonal to $V = N \oplus P$. One takes $S = \wedge N$ and, writing an element of W as $n + p + ze_{2m+1}$, where $n \in N$, $p \in P$ and $z \in \mathbb{C}$, one puts $\sigma_\pm(n + p + ze_{2m+1}) = \pm(\sqrt{2}(e(n) + i(p)) + z\alpha_N)$.

If $f \in \mathrm{End}\, S$, then $f^* \in \mathrm{End}\, S^*$ is defined by $\langle \varphi, f^*(\varphi')\rangle = \langle f(\varphi), \varphi'\rangle$ for every $\varphi \in S$ and $\varphi' \in S^*$. Let β be the antiautomorphism of Cl_{2m+1} defined by $\beta(a) = a^t$ for m even and $\beta(a) = \alpha(a)^t$ for m odd so that $\beta(\eta) = \eta$ for every m. The two representations $a \mapsto \sigma_\pm(\beta(a))^*$ of Cl_{2m+1} in S^* are equivalent to the corresponding representations $\sigma_\pm$: there exists an isomorphism $B : S \to S^*$ such that $\sigma_\pm(\beta(a))^* = B\sigma_\pm(a)B^{-1}$ for every $a \in \mathrm{Cl}_{2m+1}$. Iterating and using Schur's lemma one obtains $B^* = \varepsilon B$, where either $\varepsilon = 1$ or $\varepsilon = -1$. To determine ε, note that $(B\sigma(a))^* = \varepsilon B\sigma(a^t)$ for every $a \in \mathrm{Cl}^+_{2m+1}$. Since $\dim\{f : S \to S^* | f^* = f\} > \dim\{f : S \to S^* | f^* = -f\}$, one has $\varepsilon = \mathrm{sgn}\,(\dim A^+_m - \dim A^-_m)$, where $A^\pm_m = \{a \in \mathrm{Cl}^+_{2m+1} | a^t = \pm a\}$. Moreover, $\dim A^+_m - \dim A^-_m = \sum_{p=0}^n (-1)^p \binom{2m+1}{2p} = 2^m\sqrt{2}\cos(2m+1)\frac{\pi}{4}$ so that **[C4]**

$$B^* = (-1)^{\frac{1}{2}m(m+1)} B. \tag{6}$$

For every $a \in \mathrm{Cl}^+_{2m+1}$ one has

$$\sigma(a)^* = B\sigma(a^t)B^{-1}. \tag{7}$$

The isomorphism B defines also non-degenerate quadratic forms $B \otimes B^{-1}$ and $B \otimes B$ on $\mathrm{End}\, S$ and $S \otimes S$, respectively. Namely, if $f \in \mathrm{End}\, S$, then $(B \otimes B^{-1})(f) = \mathrm{tr}\,(B^{-1} \circ f^* \circ B \circ f)$; if $\varphi, \psi \in S$, then $(B \otimes B)(\varphi \otimes \psi) = \langle \varphi, B\varphi\rangle\langle \psi, B\psi\rangle$; the linear map $S \otimes S \to \mathrm{End}\, S$ defined by $\varphi \otimes \psi \mapsto \varphi \otimes B\psi$ is an isometry for these quadratic forms. Similarly, the algebra Cl^+_{2m+1} has a quadratic form H, $H(a) = 2^{-m}\mathrm{tr}\,\sigma(a^t a)$ and σ is an isometry of (Cl_{2m+1}, H) onto $(\mathrm{End}\, S, 2^{-m} B \otimes B^{-1})$. The even exterior algebra $\wedge^+ W$ has a quadratic form $\wedge^+ h$ obtained by extension of h; the isomorphism (2) restricted to Cl^+_{2m+1} is an isometry equivariant with respect to the action of the spin group. This leads to

PROPOSITION 2. *Let σ be a faithful and irreducible representation of the even Clifford algebra* Cl^+_{2m+1}, *associated with* $W = \mathbb{C}^{2m+1}$, *in a* 2^m*-dimensional complex vector space* S. *There then exists an isomorphism* $B : S \to S^*$ *such that* (6) *and* (7) *hold. The bilinear map*

$$E : S \times S \to \wedge^+ W, \quad E(\varphi, \psi) = c \circ \sigma^{-1}(\varphi \otimes B\psi),$$

(i) satisfies

$$E(\psi, \varphi) = (-1)^{\frac{1}{2}m(m+1)} E(\varphi, \psi)^t;$$

(ii) if $v \in W$, *then*

$$E(\sigma_+(v)\varphi, \psi) = (i(v) + e(v)) \star E(\varphi, \psi);$$

(iii) if $a \in \mathrm{Spin}_{2m+1}(\mathbb{C})$, *then*

$$E(\sigma(a)\varphi, \sigma(a)\psi) = \wedge\rho(a) \circ E(\varphi, \psi);$$

(iv) the linear map $S \otimes S \to \wedge^+ W$, *associated with* E, *is an isometry of the quadratic space* $(S \otimes S, 2^{-m} B \otimes B)$ *onto* $(\wedge^+ W, \wedge^+ h)$ *which is equivariant with respect to the action of the group* $\mathrm{Spin}_{2m+1}(\mathbb{C})$.

To prove (i), put $b = \sigma^{-1}(\varphi \otimes B\psi)$; then $\psi \otimes B\varphi = B^{-1} \circ (\varphi \otimes B\psi)^* \circ B^* = (-1)^{\frac{1}{2}m(m+1)}\sigma(b^t)$. The proof of (ii) is based on $\sigma_+(v) = \sigma(\eta v)$, on the definition of the Hodge dual and on (4). Part (iii) follows from $\sigma(a)\varphi \otimes B\sigma(a)\psi = \sigma(a) \circ (\varphi \otimes B\psi) \circ \sigma(a^t)$, $a^t a = 1$ and and (5). Part (iv) follows from the preceding remarks; see also the proof of Prop. 4. 2. in [**BuT**].

The component of E in $\wedge^{2p} W$ is denoted by E_{2p}. In particular, $E_0(\varphi, \psi) = 2^{-m}\langle \varphi, B\psi \rangle$. The bilinear form E_0 is invariant with respect to the action of $\mathrm{Spin}_{2m+1}(\mathbb{C})$. More generally, if $a \in \mathrm{Cl}^+_{2m+1}$ and $a^t a = 1$, then $\langle \sigma(a)\varphi, B\sigma(a)\psi \rangle = \langle \varphi, B\psi \rangle$. According to part (i) of Prop. 2, one has

$$E_{2p}(\psi, \varphi) = (-1)^{\frac{1}{2}m(m+1)+p} E_{2p}(\varphi, \psi).$$

Putting ν equal to the integer part of $\frac{1}{2}(m+1)$, one obtains that

(8) $$\text{if } \nu - p \text{ is odd, then } E_{2p}(\varphi, \varphi) = 0.$$

5. 1. 2. Consider now the *even*-dimensional subspace $V = \mathbb{C}^{2m}$ of W orthogonal to the unit vector e_{2m+1}. The algebra Cl_{2m} can be identified with a subalgebra of Cl_{2m+1}; the Clifford map $V \to \mathrm{Cl}^+_{2m+1}$, $v \mapsto \eta v$, extends to an isomorphism of algebras $j : \mathrm{Cl}_{2m} \to \mathrm{Cl}^+_{2m+1}$. Since $j(v)^t = (-1)^m j(v)$, one has $j(a)^t = j(\beta(a))$ for every $a \in \mathrm{Cl}_{2m}$. The element $\eta e_{2m+1} = e_1 \cdots e_{2m}$ is a volume in V. The composition $\gamma = \sigma \circ j$ is the 'Dirac' representation of the algebra Cl_{2m} in S. The automorphisms $\gamma_\mu = \gamma(e_\mu)$ $(\mu = 1, \dots, 2m)$ of S generalize the classical Dirac matrices; one has $\sigma_\pm(e_\mu) = \pm\gamma_\mu$ for $\mu = 1, \dots, 2m$ and $\sigma_\pm(e_{2m+1}) = \pm\varGamma$, where $\varGamma = \gamma_1 \cdots \gamma_{2m}$ is the 'chirality' automorphism of S. It follows from these definitions that

(9) $$\gamma^*_\mu = (-1)^m B\gamma_\mu B^{-1} \quad \text{and} \quad \varGamma^* = (-1)^m B\varGamma B^{-1}.$$

Let $k : \wedge V \to \wedge^+ W$ be the isomorphism of vector spaces such that $k \circ c = c \circ j$; explicitly, it is given by $k(w) = w$ for $w \in \wedge^+ V$ and $k(w) = \star w$ for w odd, $w \in \wedge^- V$. Let the Hodge dual in V be denoted by $*$, i. e. $c(\eta e_{2m+1} a) = *c(a)$ for $a \in \mathrm{Cl}_{2m}$. Consider the bilinear map

$$F = k^{-1} \circ E : S \times S \to \wedge V.$$

Denoting by $F_p(\varphi, \psi)$ the component of $F(\varphi, \psi)$ in $\wedge^p V$, one obtains, as a corollary of part (i) of Prop. 2,

$$F_p(\psi, \varphi) = (-1)^{\frac{1}{2}(m-p)(m-p+1)} F_p(\varphi, \psi). \tag{10}$$

Putting $v = e_{2m+1}$ in part (ii) and using $\sigma_+(e_{2m+1}) = \sigma(\eta e_{2m+1})$ leads to

$$F(\Gamma\varphi, \psi) = *F(\varphi, \psi) \quad \text{and} \quad F(\varphi, \Gamma\psi) = (-1)^m \alpha \circ *F(\varphi, \psi). \tag{11}$$

If $\wedge V$ is given the quadratic form $\wedge h$ such that

$$\wedge h(v_1 \wedge \cdots \wedge v_p) = h(v_1) \cdots h(v_p),$$

where $v_1, \ldots, v_p \in V$, then k becomes an isometry of $(\wedge V, \wedge h)$ onto $(\wedge^+ W, \wedge^+ h)$. As a corollary from Prop. 2 one obtains that the map

$$S \otimes S \to \wedge V, \ \text{ given by } \ \varphi \otimes \psi \mapsto c \circ \gamma^{-1}(\varphi \otimes B\psi), \tag{12}$$

is an isometry of the corresponding quadratic spaces, equivariant with respect to the action of $\mathrm{Spin}_{2m}(\mathbb{C})$.

Restricted to the even subalgebra Cl^+_{2m}, the representation γ decomposes into the direct sum of two irreducible and inequivalent, but not faithful, representations γ_+ and γ_- in 2^{m-1}-dimensional spaces S_+ and S_- of 'Weyl' or 'chiral' spinors. One has $S_\pm = \{\varphi \in S : \Gamma\varphi = \pm\varphi\}$ and $S = S_+ \oplus S_-$. If φ and ψ are both Weyl spinors with respect to γ, then (9) and (10) give

$$\text{if} \quad \chi(\varphi) + \chi(\psi) + m + p \equiv 1 \mod 2, \quad \text{then} \quad F_p(\varphi, \psi) = 0. \tag{13}$$

In particular, if φ is a Weyl spinor, then $F_p(\varphi, \varphi) = 0$ unless $p \equiv m \mod 4$.

If the representation γ comes from a representation σ constructed in terms of a Witt decomposition, so that $S = \wedge N$, then $\Gamma = \alpha_N$ and $S_\pm = \wedge^\pm N$.

5. 2. Real vector spaces. Consider now the real vector space $\mathbb{R}^n$ with a quadratic form h of signature (k, l), $k + l = n$. If $(e_1, \ldots, e_n)$ is a frame orthonormal with respect to h, then the volume element $\eta = e_1 \cdots e_n$ satisfies $\eta^2 = (-1)^{\frac{1}{2}(l-k)(l-k+1)}$. The complexification of the Clifford algebra $\mathrm{Cl}_{k,l}$ is isomorphic with Cl_n.

5. 2. 1. Consider first the case of n odd, $n = 2m+1$. The Pauli representation of Cl^+_{2m+1}, restricted to $\mathrm{Cl}^+_{k,l}$, yields a representation σ of this real algebra in a 2^m-dimensional, complex vector space S. The representation σ can be extended to the representations σ_+ and σ_- of $\mathrm{Cl}_{k,l}$ in S by putting $\sigma_\pm(\eta) = \pm\mathrm{id}_S$ when $\eta^2 = 1$ and $\sigma_\pm(\eta) = \pm\sqrt{-1}\,\mathrm{id}_S$ when $\eta^2 = -1$. Recall also that with every complex vector space S one can associate its 'complex conjugate' space $\bar{S}$: the space $\bar{S}$ has the same set of elements as S, but the product of $z \in \mathbb{C}$ by a vector in $\bar{S}$ is equal to the product of $\bar{z}$ by the same vector in S. Denoting by $\bar{\varphi}$ the vector φ, considered as an element of $\bar{S}$, one has $\overline{z\varphi} = \bar{z}\bar{\varphi}$ for every $\varphi \in S$ and $z \in \mathbb{C}$. If $f : S_1 \to S_2$ is a linear map of complex vector spaces, then the linear map $\bar{f} : \bar{S_1} \to \bar{S_2}$ is defined by $\overline{f}(\overline{\varphi}) = \overline{f(\varphi)}$; the maps $\varphi \mapsto \bar{\varphi}$ and $f \mapsto \bar{f}$ are semi-linear, etc. With every representation τ of a *real* algebra $\mathcal{A}$ in a *complex* vector space S one can associate the complex conjugate representation $\bar{\tau}$ in $\bar{S}$, given by $\overline{\tau}(a) = \overline{\tau(a)}$ for every $a \in \mathcal{A}$. Since $\mathrm{Cl}^+_{k,l}$ is central simple for $k+l$ odd, its representations σ and $\bar{\sigma}$ are complex-equivalent; if $\eta^2 = 1$, then the representations $\overline{\sigma_\pm}$ are equivalent to the corresponding representations $\sigma_\pm$; if $\eta^2 = -1$, then $\overline{\sigma_+}$ is equivalent to σ_-. In every case there is a linear isomorphism

$$C : S \to \overline{S} \text{ such that } \overline{\sigma_\mu} = (-1)^{\frac{1}{2}(l-k)(l-k+1)} C\sigma_\mu C^{-1}, \quad \text{where } \sigma_\mu = \sigma_+(e_\mu)$$

for $\mu = 1, \ldots, k+l = 2m+1$. An argument similar to the one used with respect to B in Sec. 5. 1 shows that C can be rescaled so that either $\bar{C}C = \mathrm{id}_S$ or $\bar{C}C = -\mathrm{id}_S$. Moreover, one obtains from (7) and

$$\overline{\sigma(a)} = C\sigma(a)C^{-1}, \quad a \in \mathrm{Cl}^+_{k,l}, \tag{14}$$

that $C^{-1}\bar{B}^{-1}\bar{C}^* B^*$ is in the commutant of σ; therefore, one can rescale B so that

$$B = C^* \bar{B} C \tag{15}$$

and then the sesquilinear form

$$A : S \times S \to \mathbb{C}, \quad \text{given by} \quad A(\varphi, \psi) = \langle \bar{\varphi}, \bar{B}C\psi \rangle$$

is either Hermitean or anti-Hermitean.

According to a terminology used in physics, one says that $\varphi_{\mathrm{c}} = C^{-1}\bar{\varphi}$ is the *charge conjugate* of the spinor $\varphi \in S$. The charge conjugate φ'_{c} of $\varphi' \in S^*$ is defined so that $\langle \varphi_{\mathrm{c}}, \varphi'_{\mathrm{c}} \rangle = \overline{\langle \varphi, \varphi' \rangle}$; by virtue of (14), if $\varphi' = B\psi$, then $\varphi'_{\mathrm{c}} = B\psi_{\mathrm{c}}$. There are two cases to consider **[C1, BuT]**:
(i) The *real* case: if $l - k \equiv 1$ or $7 \mod 8$, then $\bar{C}C = \mathrm{id}_S$; there is then the 2^m-dimensional real vector space

$$S_{\mathbb{R}} = \{\varphi \in S \varphi_{\mathrm{c}} = \varphi\}.$$

and a decomposition of S into complementary subspaces of 'Majorana' spinors, $S = S_{\mathbb{R}} \oplus \sqrt{-1} S_{\mathbb{R}}$. The representation σ is real: $\sigma(a)S_{\mathbb{R}} \subset S_{\mathbb{R}}$ for every $a \in \mathrm{Cl}^+_{k,l}$. The automorphisms $\sigma_\mu = \sigma_+(e_\mu)$ are real (resp., pure imaginary) for $l - k \equiv 7 \mod 8$ (resp., $l - k \equiv 1 \mod 8$). The algebra $\mathrm{Cl}^+_{k,l}$ is isomorphic to the matrix algebra $\mathbb{R}(2^m)$. The algebra $\mathrm{Cl}_{k,l}$ is isomorphic to $\mathbb{R}(2^m) \oplus \mathbb{R}(2^m)$ (resp., $\mathbb{C}(2^m)$) for $l - k \equiv 7 \mod 8$ (resp., $l - k \equiv 1 \mod 8$). The form A restricted $S_{\mathbb{R}}$ is real and has the same symmetry as B.

(ii) The *quaternionic* case: if $l - k \equiv 3$ or $5 \mod 8$, then $\bar{C}C = -\mathrm{id}_S$ and S can be given the structure of a right module over $\mathbb{H}$. Explicitly, denoting by i, j and $\mathrm{k} = \mathrm{ij}$ the quaternionic units, one puts $\varphi\mathrm{i} = \sqrt{-1}\varphi$ and $\varphi\mathrm{j} = \varphi_{\mathrm{c}}$. The algebra $\mathrm{Cl}^+_{k,l}$ is isomorphic to the matrix algebra $\mathbb{H}(2^{m-1})$. The algebra $\mathrm{Cl}_{k,l}$ is isomorphic to $\mathbb{H}(2^{m-1}) \oplus \mathbb{H}(2^{m-1})$ (resp., $\mathbb{C}(2^m)$) for $l - k \equiv 3 \mod 8$ (resp., $l - k \equiv 5 \mod 8$).

The vector space $\mathcal{A} = \{f \in \mathrm{End}\, S \bar{f}C = Cf\}$ is a real algebra spanned by all elements of the form $\varphi \otimes \varphi' + \varphi_{\mathrm{c}} \otimes \varphi'_{\mathrm{c}}$, where $\varphi \in S$ and $\varphi' \in S^*$. The representation σ factors through the injection $\mathcal{A} \to \mathrm{End}\, S$. Moreover, in the real case, the algebra $\mathcal{A}$ is isomorphic to $S_{\mathbb{R}} \otimes_{\mathbb{R}} S_{\mathbb{R}}$. In the quaternionic case, it is isomorphic to the tensor product over $\mathbb{H}$ of the right $\mathbb{H}$-module S by the left $\mathbb{H}$-module S^*. In each case one has

$$\overline{E(\varphi, \psi)} = E(\varphi_{\mathrm{c}}, \psi_{\mathrm{c}}).$$

The homogeneous components of the multivector $E(\varphi_{\mathrm{c}}, \varphi)$ are either real or imaginary, as can be seen from part (i) of Prop. 2 and $\varphi_{\mathrm{cc}} = C\bar{C}\varphi$.

The quadratic form H on $\mathrm{Cl}^+_{k,l}$, $k + l = 2m + 1$, $H(a) = 2^{-m}\mathrm{tr}\,\sigma(a^t a)$, is real; its signature can be evaluated as follows. Consider the polynomial $\varsigma(\xi, \eta) = \frac{1}{2}(1+\xi)^k(1+\eta)^l + \frac{1}{2}(1-\xi)^k(1-\eta)^l = \sum_{p,q;\, p+q \text{ even}} \binom{k}{p}\binom{l}{q}\xi^p\eta^q$. The index of H equals $\varsigma(1, -1)$. Therefore, H is positive-definite if, and only if, either $k = 0$ or $l = 0$; if both k and l are positive, then H is neutral.[3]

5. 2. 2. Consider now the *even*-dimensional subspace V of $\mathbb{R}^{k+l}$, orthogonal to a unit vector u. Depending on whether $u^2 = 1$ or -1, the signature of the restriction h_V of h to V is $(k-1, l)$ or $(k, l-1)$. The map $V \to \mathrm{Cl}^+_{k,l}$, $v \mapsto v\eta$, extends to an isomorphism of algebras $j : \mathrm{Cl}(V, \eta^2 h_V) \to \mathrm{Cl}^+_{k,l}$: if $\eta^2 = 1$ (resp., $\eta^2 = -1$), then $\mathrm{Cl}^+_{k,l}$ is isomorphic to both $\mathrm{Cl}_{k-1,l}$ and $\mathrm{Cl}_{k,l-1}$ (resp., $\mathrm{Cl}_{l-1,k}$ and $\mathrm{Cl}_{l,k-1}$).

If $k + l = 2m$ and $\gamma : \mathrm{Cl}_{k,l} \to \mathrm{End}\, S$ is a Dirac representation in a complex space S of dimension 2^m, then there is $C : S \to \bar{S}$ such that $\overline{\gamma(a)} = C\gamma(a)C^{-1}$ for every $a \in \mathrm{Cl}_{k,l}$. According to previous remarks, one can rescale C so that either $\bar{C}C = \mathrm{id}_S$ (for $l - k \equiv 0$ or $6 \mod 8$) or $\bar{C}C = -\mathrm{id}_S$ (for $l - k \equiv 2$ or $4 \mod 8$). It is convenient to define the chirality automorphism as $\Gamma = (-1)^{\frac{1}{4}(l-k)(l-k+1)}\gamma_1 \cdots \gamma_{2m}$ so that

$$\Gamma^2 = \mathrm{id}_S \quad \text{and} \quad \bar{\Gamma} = (-1)^{\frac{1}{2}(l-k)(l-k+1)} C\Gamma C^{-1}. \tag{16}$$

The representation γ, restricted to $\mathrm{Cl}^+_{k,l}$ decomposes as in the complex case, $\gamma = \gamma_+ \oplus \gamma_-$. The representation $\check{\gamma}$ of $\mathrm{Cl}^+_{k,l}$ in S^*, contragredient to γ, is defined by $\check{\gamma}(a) = \gamma(a^t)^*$, $a \in \mathrm{Cl}^+_{k,l}$. It also decomposes, $\check{\gamma} = \check{\gamma}_+ \oplus \check{\gamma}_-$, where $\check{\gamma}_\pm : \mathrm{Cl}^+_{k,l} \to \mathrm{End}\, S^*_\pm$, $S^*_\pm = \{\varphi \in S\Gamma^*\varphi = \pm\varphi\}$. There is also a similar

[3] I am indebted to P. Lounesto for having pointed out to me a mistake in the formulation of this result given in [**BuT**].

decomposition $\bar{\gamma} = \bar{\gamma}_+ \oplus \bar{\gamma}_-$ of $\bar{\gamma}$ restricted to $\mathrm{Cl}^+_{k,l}$. The representations γ_+ and γ_- are not complex-equivalent; each of the representations $\check{\gamma}_+$, $\check{\gamma}_-$, $\bar{\gamma}_+$ and $\bar{\gamma}_-$ is equivalent to either γ_+ or γ_-. In particular, denoting by $\sim$ the equivalence relation of representations, one obtains from (9)

$$\check{\gamma}_\pm \sim \gamma_\pm \quad \text{for } m \text{ even and} \quad \check{\gamma}_\pm \sim \gamma_\mp \quad \text{for } m \text{ odd}$$

and from (16)

$$\bar{\gamma}_\pm \sim \gamma_\pm \quad \text{for } l - k \equiv 0 \text{ or } 4 \mod 8 \text{ and} \quad \bar{\gamma}_\pm \sim \gamma_\mp \quad \text{for } l - k \equiv 2 \text{ or } 6 \mod 8.$$

5. 2. 3. Many authors of papers on spinors use a notation with indices similar to that codified for tensors by Schouten. This van der Waerden-Penrose notation for Weyl spinors associated with a $2m$-dimensional real vector space can be briefly described as follows. Consider linear bases: (e_A) in S_+, $(e_{A'})$ in S_-, $(e_{\dot{A}})$ in $\bar{S}_+$, (e^A) in S_+^*, etc., where $e_{\dot{A}} = \overline{e_A}$, (e^A) is the basis dual to (e_A) and the indices range from 1 to 2^{m-1}. Instead of saying 'let $\varphi \in S_+^* \otimes S_+ \otimes S_-$' one says 'consider a spinor ${\varphi_A}^{BC'}$'. This notation is convenient when one works with several spinor fields and considers their tensor products and contractions: such is the case of applications of spinor analysis in the theory of relativity [**PR**]. The Dirac 'matrices' $\gamma^\mu = h^{\mu\nu}\gamma_\nu$ change chirality of Weyl spinors; Penrose writes $\gamma^\mu(e_A) = e_{B'}{\sigma^{\mu B'}}_A$. If $\mathcal{M}$ is a Riemannian manifold with a spin structure $\mathcal{Q} \to \mathcal{P} \to \mathcal{M}$ and ∇_μ are the (horizontal) vector fields on $\mathcal{Q}$ defining covariant differentiation of spinor fields, then the Dirac operator $\gamma^\mu\nabla_\mu$, acting on a Weyl spinor field $\psi = e_A\psi^A : \mathcal{Q} \to S_+$, yields the spinor field $e_{B'}{\nabla^{B'}}_A\psi^A : \mathcal{Q} \to S_-$, where ${\nabla^{B'}}_A = {\sigma^{\mu B'}}_A\nabla_\mu$. Relative to the basis $(e_A, e_{B'})$ of S, the matrices of the linear maps B and C are either block-diagonal or block-antidiagonal, depending on whether they preserve or change the chirality. In particular, if m is even, then B yields an isomorphism of $S_\pm$ onto $S_\pm^*$, $B(e_A) = B_{AB}e^B$ and $B(e_{A'}) = B_{A'B'}e^{B'}$. If m is odd, then $B : S_\pm \to S_\mp^*$, $B(e_A) = B_{AB'}e^{B'}$, etc. Depending on the parity of m, the matrices (B_{AB}) and $(B_{A'B'})$ or $(B_{AB'})$ and $(B_{A'B})$, and their inverses, are used to lower and raise spinorial indices. Similarly, for $l - k \equiv 0$ or $4 \mod 8$ one has $C(e_A) = {C_A}^{\dot{B}}e_{\dot{B}}$, $C(e_{A'}) = {C_{A'}}^{\dot{B}'}e_{\dot{B}'}$ and the invertible matrices $({C_A}^{\dot{B}})$ and $({C_{A'}}^{\dot{B}'})$ are used to convert undotted into dotted indices, etc. Originally, van der Waerden (1929) emphasized the representations γ_+ and $\bar{\gamma}_+$ and introduced dotted indices for the latter. Penrose (1960) pointed out that it suffices to consider the representations γ_+ and γ_- and used primed indices for spinors in S_-. In Minkowski space one has $l - k = 2$ so that $\bar{\gamma}_+ \sim \gamma_-$: primed and dotted indices are equivalent.

6. Pure spinors

Pythagoras and Euclid could have solved the equation $x^2 - y^2 + z^2 = 0$ by noting that it is equivalent to the statement that the symmetric matrix

$$\begin{pmatrix} y+z & x \\ x & y-z \end{pmatrix}$$

is of rank < 2 and, therefore, there is a 'spinor' (p, q) such that $x = 2pq$, $y = p^2 + q^2$ and $z = p^2 - q^2$. Interpreting now the triple (x, y, z) as an element of a real vector space, one can rephrase the solution of the Pythagoras equation to read 'the (tensor) square of a real, two-component spinor equals an isotropic vector in $\mathbb{R}^3$ with a quadratic form of signature $(2,1)$. This observation has interesting generalizations to higher dimensions, giving rise to the notion of pure spinors.

6. 1. The complex case. Continuing to use the notation of Sec. 5. 1, consider a non-zero spinor $\varphi \in S$ associated with $W = \mathbb{C}^{2m+1}$. The vector space

$$N(\varphi) = \{v \in W | \sigma(\eta v)\varphi = 0\}$$

is totally isotropic; its dimension is called the *nullity* of φ. A spinor $\varphi \neq 0$ is said to be *pure* if its nullity is maximal, i. e. equal to m. Let $W = N \oplus P \oplus \mathbb{C}u$ be a Witt decomposition and let $(n_1, \ldots, n_m)$ be a basis of N. Put $a_N = n_1 \cdots n_m$ for m even and $a_N = \eta n_1 \cdots n_m$ for m odd, so that $a_N \in \mathrm{Cl}^+_{2m+1}$. Since σ is faithful, there is a spinor φ_0 such that $\varphi = \sigma(a_N)\varphi_0 \neq 0$ and then $N(\varphi) = N$ so that φ is pure. If ψ is another spinor such that $N(\psi) = N$, then there is $z \in \mathbb{C}^\times$ such that $\psi = z\varphi$: there is a bijective correspondence between the set (in fact, a compact, connected, complex manifold Σ_{2m+1} of complex dimension $\frac{1}{2}m(m+1)$) of directions of pure spinors and the set of maximal isotropic subspaces of W. If $a \in \mathrm{Spin}_{2m+1}(\mathbb{C})$ and φ is pure, then $\sigma(a)\varphi$ is also pure; the induced action of $\mathrm{Spin}_{2m+1}(\mathbb{C})$ on Σ_{2m+1} is transitive.

Let $\varphi \neq 0$ be a spinor of nullity q. Let $(n_1, \ldots, n_q)$ be a basis of $N(\varphi)$. It follows from part (ii) of Prop. 2 and (8) that $n \in N(\varphi)$ implies $e(n)E_{2p}(\varphi, \varphi) = 0$ and $i(n)E_{2p}(\varphi, \varphi) = 0$ for every p such that $p \equiv \nu \mod 2$. Therefore, there is a $(2p-q)$-vector E'_{2p-q} such that $E_{2p}(\varphi, \varphi) = n_1 \wedge \cdots \wedge n_q \wedge E'_{2p-q}$ and $i(n)E'_{2p-q} = 0$ for every $n \in N(\varphi)$. This implies that if $2p < q$ or $2p > 2m+1-q$, then $E_{2p}(\varphi, \varphi) = 0$. In particular, if φ is pure, then $E_{2p}(\varphi, \varphi) \neq 0$ if, and only if, $p = \nu$. A more precise result is contained in

PROPOSITION 3. *A spinor $\varphi \neq 0$ is pure if, and only if, $p \neq \nu$ implies $E_{2p}(\varphi, \varphi) = 0$. If φ is pure, then there is a basis $(n_1, \ldots, n_m)$ of $N(\varphi)$ such that*

$$\text{if } m \text{ is even, then} \quad E_m(\varphi, \varphi) = n_1 \wedge \cdots \wedge n_m;$$

$$\text{if } m \text{ is odd, then} \quad \star E_{m+1}(\varphi, \varphi) = n_1 \wedge \cdots \wedge n_m.$$

The somewhat difficult proof of the 'only if' part of Prop. 3 appears in [**C4, Ch**]; recently, a new proof has been given by Urbantke [**U**] with the help of 'quartic identities' used by physicists in connection with the classification of interaction terms occurring in the Fermi theory of weak nuclear forces. By considering the multivectors $E_{2p}(\varphi, \varphi)$ for low values of m, one obtains, as a corollary of (8) and Prop. 3, that all spinors associated with W of dimension 3 and 5 are pure; in dimensions 7 and 9 pure spinors belong to the cone of equation $E_0(\varphi, \varphi) = 0$.

If φ is pure and $v \in W$ is non-isotropic, $v^2 \neq 0$, then $\sigma(\eta v)\varphi$ is also pure: $N(\sigma(\eta v)\varphi) = vN(\varphi)v^{-1}$. In particular, if $N(\varphi)$ is orthogonal to the unit vector e_{2m+1}, then $N(\sigma(\eta e_{2m+1})\varphi) = N(\varphi)$. Therefore, $\sigma(\eta e_{2m+1})\varphi = \pm\varphi$. Introducing. as in Sec. 5. 1. 2, the $2m$-dimensional space V orthogonal to e_{2m+1}, one obtains, as a corollary of Prop. 3 and (11):

PROPOSITION 4. *Let $W = V \oplus \mathbb{C}e_{2m+1}$ and γ be the Dirac representation of* Cl_{2m} *in S, $\gamma(v) = \sigma(\eta v)$ for $v \in V$. If φ is pure, then $N(\varphi) \subset V$ if, and only if, φ is a Weyl spinor with respect to γ. Assuming that φ is such a spinor, one has $F_p(\varphi, \varphi) = 0$ for every $p \neq m$ and there is a basis $(n_1, \dots, n_m)$ in $N(\varphi)$ so that*

$$F_m(\varphi, \varphi) = n_1 \wedge \cdots \wedge n_m.$$

The m-vector $F_m(\varphi, \varphi)$ is either self-dual ($\Gamma\varphi = \varphi$) or antiself-dual ($\Gamma\varphi = -\varphi$).

In other words, in even-dimensional complex vector spaces, tensor squares of pure spinors define self- or antiself-dual decomposable multivectors of the middle degree. A pure spinor φ associated with the representation $\gamma : \mathrm{Cl}_{2m} \to \mathrm{End}\, S$ can be characterized, without reference to the odd-dimensional space W, by $\dim\{v \in \mathbb{C}^{2m} | \gamma(v)\varphi = 0\} = m$; it follows that it is a Weyl spinor. The set of directions of pure spinors associated with γ is a $\frac{1}{2}m(m-1)$-dimensional complex compact manifold $\Sigma_{2m} = \mathrm{O}_{2m}/\mathrm{U}_m$; it has two connected components, Σ^+_{2m} and Σ^-_{2m}, corresponding to pure spinors of opposite chiralities. An argument similar to the one used in odd dimensions shows that, in dimensions 2, 4 and 6 all Weyl spinors are pure. In dimension 8 pure spinors lie on the cones of equation $F_0(\varphi, \varphi) = 0$ in S_+ and S_-; in dimension 10 pure spinors are characterized by the equation $F_1(\varphi, \varphi) = 0$ in $S_\pm$ and generic Weyl spinors have nullity 1; for $m = 4$ and $m > 5$ generic Weyl spinors have zero nullity. Spinors belonging to one orbit of the group $\mathrm{Spin}_{2m}(\mathbb{C})$ have the same nullity, but the converse is not true: in dimensions $\geqslant 12$ nullity of Weyl spinors provides a rather coarse classification of the orbits. There are no Weyl spinors of nullity q such that $m - 4 < q < m$. All homogeneous polynomial invariants of the spin group vanish on spinors of positive nullity [**TT**].

6. 2. The real case. Changing somewhat the notation, consider now a $2m$-dimensional *real* vector space V, with a quadratic form h of signature $(2\kappa+\varepsilon, 2\lambda+\varepsilon)$, where κ and λ are non-negative integers and $\varepsilon = 0$ or 1. The cases $\varepsilon = 0$ and $\varepsilon = 1$ are referred to as *pseudo-euclidean* and *pseudo-lorentzian*, respectively; the prefix 'pseudo' is dropped when either κ or $\lambda = 0$. The *complexification* of V, $W = \mathbb{C} \otimes_{\mathbb{R}} V = V \oplus \sqrt{-1}V$, is given a quadratic form $\mathbb{C} \otimes h$ which is a natural extension of h so that $\mathrm{Cl}(\mathbb{C} \otimes V, \mathbb{C} \otimes h) = \mathbb{C} \otimes \mathrm{Cl}(V, h)$. If $N \subset W$, then $\bar{N} \subset W$ is defined as the set $\{u + \sqrt{-1}v | u - \sqrt{-1}v \in N\}$. If N is a (complex) vector subspace of W, then there are (real) vector subspaces K and L of V such that

(17) $$N \cap \bar{N} = \mathbb{C} \otimes K \quad \text{and} \quad N + \bar{N} = \mathbb{C} \otimes L.$$

If φ is a pure spinor, then so is φ_c and (14) implies $N(\varphi_c) = \overline{N(\varphi)}$. Put $N = N(\varphi)$; the dimension r of K is called the *real index* of the pure spinor φ. One shows that the real index can assume every of the following values: $\varepsilon, \varepsilon + 2, \dots, \varepsilon + 2\min\{\kappa, \lambda\}$, and only these values. Pure spinors with the least value of the real index are generic in the sense that the manifolds of directions of such spinors, of positive and negative chiralities, are open and dense in Σ^+_{2m} and Σ^-_{2m}, respectively. Let φ be a pure spinor of real index r; the vector space K is totally isotropic and $K^\perp = L \supset K$. The quadratic form h descends to a quadratic form of signature $(2\kappa + \varepsilon - r, 2\lambda + \varepsilon - r)$ on the quotient L/K. Moreover, there is an orthogonal complex structure J on L/K defined as follows: let $w \mod K$ denote the class in L/K containing $w \in L$; every element of L/K is of the form $v + \bar{v} \mod K$, where $v \in N$; put $J(v + \bar{v} \mod K) = \sqrt{-1}(v - \bar{v}) \mod K$.

The most interesting case, from the point of view of applications in geometry and physics, is that of pure spinors in a general position, i. e. of pure spinors of real index ε. If φ is a generic pure spinor associated with a pseudo-euclidean quadratic space (V, h), then the sum $N(\varphi) + \overline{N(\varphi)}$ is direct, $K = \{0\}$, and J is an orthogonal complex structure in $V = L$. The pair (h, J) defines a Hermitean form on V, considered as a complex vector space of dimension $\kappa + \lambda$. The imaginary part of that Hermitean form can be identified with $F_2(\varphi_c, \varphi)/F_0(\varphi_c, \varphi)$. If φ is a generic pure spinor associated with a pseudo-lorentzian quadratic space and $N = N(\varphi)$, then K is a real isotropic line, a 'light ray' for physicists; the quotient L/K is a 'screen space', if one insists on the optical analogy. The line $\mathbb{C} \otimes K$ is generated by $F_1(\varphi_c, \varphi)$ [**KoT, T**].

6. 3. Applications. The preceding algebraic constructions can be applied, 'pointwise', to fibres of appropriate bundles. From now on, all manifolds, bundles and maps are assumed to be smooth. Consider a $2m$-dimensional, paracompact, connected manifold $\mathcal{M}$ with a metric tensor of signature $(2\kappa+\varepsilon, 2\lambda+\varepsilon)$ and a spin structure $\mathcal{Q} \xrightarrow{\chi} \mathcal{P} \to \mathcal{M}$. The last assumption is, in fact, not essential because all that is needed here are bundles of 'projectivized' spinors (i. e. bundles of directions of spinors) which are associated with the $\mathrm{SO}_{2\kappa+\varepsilon,2\lambda+\varepsilon}$-bundle $\mathcal{P} \to \mathcal{M}$. It is often, however, convenient to work with spinor fields and to represent them by the corresponding maps from $\mathcal{Q}$ to the typical fibre S equivariant with respect to the action of the group $G = \mathrm{Spin}_{2\kappa+\varepsilon,2\lambda+\varepsilon}$, as described in Sec. 4. If $\mathcal{E} \to \mathcal{M}$ is a fiber bundle, then $\mathcal{E}_x$ denotes its fibre over $x \in \mathcal{M}$. Recall that vector fields, differential forms, etc. on $\mathcal{M}$ can be equivalently described as sections of appropriate bundles or as equivariant maps from $\mathcal{Q}$ to a suitable vector space. In paricular, a complex vector tangent to $\mathcal{M}$ at x can be identified with a map $v : \mathcal{Q}_x \to W = \mathbb{C} \otimes V$ such that $v(qa) = a^{-1}v(q)a$ for every $q \in \mathcal{Q}_x$ and

$a \in G$. With the pair (φ, ψ) of spinor fields one associates the multivector-valued field $\mathcal{F}(\psi, \varphi) : \mathcal{P} \to \wedge W$, given by $\mathcal{F}(\psi, \varphi)(p) = F(\psi(q), \varphi(q))$, where $q \in \mathcal{Q}$, $p = \chi(q)$ and F is as in Sec. 5. 1. 2. The manifold $\mathcal{M}$ being orientable, it is meaningful to consider spinors of the same chirality over $\mathcal{M}$. Generalizing the constructions due to Atiyah, Hitchin and Singer [**AHS**], Penrose [**P**], and O'Brian and Rawnsley [**OBR**], one defines the total space of the *twistor bundle* $\mathcal{T}_r$ of *real index* r as consisting of directions of all pure spinors on $\mathcal{M}$ of one, say positive, chirality and of real index r. If a section of the twistor bundle exists, then it can be represented (in 'homogeneous coordinates') by a nowhere vanishing field $\varphi : \mathcal{Q} \to S$ of pure spinors; such a section defines a complex vector bundle $\mathcal{N} \to \mathcal{M}$: this is a subbundle of the complexified tangent bundle $\mathcal{W} = \mathbb{C} \otimes T\mathcal{M} \to \mathcal{M}$ such that the fibre of $\mathcal{N} \to \mathcal{M}$ at $x \in \mathcal{M}$ is $\mathcal{N}_x = \{v \in \mathcal{W}_x | \gamma(v(q))\varphi(q) = 0 \quad \text{for} \quad q \in \mathcal{Q}_x\}$, where γ is the Dirac representation of $\mathrm{Cl}_{2\kappa+\varepsilon, 2\lambda+\varepsilon}$ in S.

The considerations of Sec. 6. 3 can be applied, pointwise, to the fibers of the bundle $\mathcal{N} \to \mathcal{M}$, whose fibers are maximal, totally isotropic subspaces of the complexified tangent spaces of $\mathcal{M}$, even if this bundle does not come from a section of the twistor bundle. In any case, given $\mathcal{N} \to \mathcal{M}$, one constructs the real vector bundles $\mathcal{K} \subset \mathcal{L} \subset T\mathcal{M} \to \mathcal{M}$ and a field $\mathcal{J}$ of complex structures in the fibers of the quotient bundle $\mathcal{L}/\mathcal{K} \to \mathcal{M}$; for example, $\mathbb{C} \otimes \mathcal{K}_x = \mathcal{N}_x \cap \overline{\mathcal{N}_x}$, etc. It is convenient to say that the triple $(\mathcal{K}, \mathcal{L}, \mathcal{J})$ is the *flag geometry* associated with $\mathcal{N} \to \mathcal{M}$.

Let $\mathcal{S}(\mathcal{N})$ be the module of sections of the vector bundle $\mathcal{N} \to \mathcal{M}$; a similar notation is used for other vector bundles over $\mathcal{M}$. One says that the complex bundle $\mathcal{N}$ is *integrable* if

$$[\mathcal{S}(\mathcal{N}), \mathcal{S}(\mathcal{N})] \subset \mathcal{S}(\mathcal{N}). \tag{18}$$

In the general case, when $r > 0$, (18) implies

$$[\mathcal{S}(\mathcal{K}), \mathcal{S}(\mathcal{K})] \subset \mathcal{S}(\mathcal{K}), \tag{19}$$

and

$$[\mathcal{S}(\mathcal{K}), \mathcal{S}(\mathcal{L})] \subset \mathcal{S}(\mathcal{L}). \tag{20}$$

Therefore, the bundle $\mathcal{K}$ defines a *foliation* on $\mathcal{M}$. From now on, this foliation is assumed to be *regular* in the sense that the set $\mathcal{M}' = \mathcal{M}/\mathcal{K}$ of all its leaves has the structure of a smooth, $(2m-r)$-dimensional manifold such that the canonical map $\mathcal{M} \to \mathcal{M}'$ is a submersion. The following proposition rephrases well-known facts concerning complex and Cauchy–Riemann (CR) geometries [**J**].

PROPOSITION 5. *Let $\mathcal{N} \to \mathcal{M}$ be a bundle on a $2m$-dimensional (pseudo-) Riemannian manifold $\mathcal{M}$. Assume that the fibers of $\mathcal{N}$ are maximal, totally isotropic subspaces of the fibers of the complexified tangent bundle of $\mathcal{M}$. Let $(\mathcal{K}, \mathcal{L}, \mathcal{J})$ be the flag geometry associated with the bundle $\mathcal{N} \to \mathcal{M}$ and let r be the dimension of the fibers of the bundle $\mathcal{K} \to \mathcal{M}$. Then*

(i) if $r = 0$, *then* $\mathcal{J}$ *defines an almost complex structure on* $\mathcal{M}$ *and* (18) *is its classical integrability condition, equivalent to the vanishing of the Nijenhuis tensor of* $\mathcal{J}$*;*
(ii) if $r > 0$, *condition* (18) *implies the Frobenius integrability condition* (19) *of* $\mathcal{K}$*; the leaves of the foliation defined by* $\mathcal{K}$ *are totally isotropic; they are geodetic by virtue of* (20)*; the quotient manifold* $\mathcal{M}' = \mathcal{M}/\mathcal{K}$ *inherits, by projection, a subbundle* $\mathcal{L}'$ *of its tangent bundle and* $\mathcal{J}$ *descends to* $\mathcal{M}'$, *defining there a CR geometry.*

Assume now that the bundle $\mathcal{N} \to \mathcal{M}$ is defined by a section of the twistor bundle $\mathcal{T}_\varepsilon$, i. e. by a field of generic pure spinors, $r = \varepsilon$. In the pseudo-Euclidean case, a section of the twistor bundle $\mathcal{T}_0$ defines an orthogonal almost complex structure $\mathcal{J}$ on $\mathcal{M}$; therefore, together with g, it defines a pseudo-Hermitean form of signature (κ, λ) at each point of $\mathcal{M}$. The resulting geometry can be called *almost pseudo-Hermitean*; it becomes pseudo-Hermitean if $\mathcal{J}$ is integrable.

In the pseudo-Lorentzian case, a section of the twistor bundle $\mathcal{T}_1$ defines an *almost optical geometry*, i. e. a flag geometry $(\mathcal{K}, \mathcal{L}, \mathcal{J})$ on $\mathcal{M}$ with a metric tensor g of signature $(2\kappa+1, 2\lambda+1)$ such that $\mathcal{K} = \mathcal{L}^\perp$ is a real line bundle and $\mathcal{J}$ is orthogonal with respect to the metric induced by g in the fibres of $\mathcal{L}/\mathcal{K}$. An almost optical geometry satisfying the integrability condition (18) is called *optical*: the trajectories of $\mathcal{K}$ are then isotropic geodesics and the quotient $\mathcal{M}' = \mathcal{M}/\mathcal{K}$ is a $(2m-1)$-dimensional CR manifold. In the proper Lorentzian case $(p = 1, q = 0)$ the integrability condition is equivalent to the statement that the trajectories of $\mathcal{K}$ form a congruence of *isotropic geodesics without shear*. Such congruences play a fundamental role in the study of *algebraically special* gravitational fields; see, e. g., [**PR, RT2**] and the numerous references given there. Cauchy-Riemann structures appeared, in the context of relativity theory, for the first time in relation with Penrose's twistor spaces [**P**]; their connection with curved Lorentzian spaces, admitting non-shearing congruences of isotropic geodesics, has been made explicit in [**RT1**]. The analogies between Hermitean and optical geometries, apparent from the above considerations, and Cartan's results on 3-dimensional CR spaces, have been used in the study of solutions of Einstein's equations; see [**N**] and the references listed there.

Appendix: A derivation of the Vahlen-Ahlfors formula

Let $K = \mathbb{R}$ or $\mathbb{C}$ and let V be an n-dimensional vector space over K with a non-degenerate quadratic form $h : V \to K$. Consider the vector space $V' = V \oplus K^2$ with the quadratic form h' such that, for every $v' = (v, \lambda, \mu) \in V'$, where $v \in V$ and $\lambda, \mu \in K$, one has $h'(v') = h(v) + \lambda\mu$. Let $\operatorname{dir} v'$ be the direction (line) in V' containing $v' \neq 0$ and let

$$Q = \{ \operatorname{dir} v' \in \mathbb{P}(V') | h'(v') = 0 \}$$

be the *projective quadric* associated with V [**CGuT**]. The quadric is a compact submanifold, of dimension n, of the projective space $\mathbb{P}(V')$. In particular, if

$K = \mathbb{R}$ and h is of signature (k, l), where $k + l = n$, then Q is diffeomorphic to $(\mathbb{S}_k \times \mathbb{S}_l)/\mathbb{Z}_2$. If either k or l is 0, then $Q = \mathbb{S}_n$. The geometry of V' induces a *conformal structure* in Q. For every $v \in V$, the vector $i(v) = (v, -h(v), 1) \in V'$ is isotropic. The map

$$j : V \to Q, \quad \text{given by} \quad j(v) = \operatorname{dir} i(v),$$

is a conformal diffeomorphism of V on its image in Q; the image is open and dense in Q. Let G be the connected component of the group $\operatorname{Spin}(V', h')$. The action of G on V' induces a transitive action of G on Q given by

$$\text{(21)} \qquad \operatorname{dir} v' \mapsto f(A) \operatorname{dir} v' = \operatorname{dir} A v' A^{-1}.$$

Denoting by $\operatorname{Aut} Q$ the *Möbius group*, i. e. the connected component of the group of conformal automorphisms of Q, one sees that $f : G \to \operatorname{Aut} Q$ is an epimorphism of groups. The action of G on Q induces *local* conformal transformations of V. Namely, the set

$$V_A = \{\, v \in V | f(A) j(v) \in j(V) \,\}.$$

is open and dense in V and the map

$$\text{(22)} \qquad F(A) : V_A \to V_{A^{-1}}, \quad \text{defined by} \quad j(F(A)v) = f(A) j(v),$$

is a conformal diffeomorphism. The Clifford map $V' \to K(2) \otimes_K \operatorname{Cl}(V, h)$,

$$\text{(23)} \qquad (v, \lambda, \mu) \mapsto \begin{pmatrix} v & \lambda \\ \mu & -v \end{pmatrix},$$

extends to the isomorphism of algebras with units, $\operatorname{Cl}(V', h') \to K(2) \otimes_K \operatorname{Cl}(V, h)$. From now on these algebras are identified. Therefore, a typical element of G is of the form

$$\text{(24)} \qquad A = \begin{pmatrix} a & b \\ c & d \end{pmatrix},$$

where a, d and b, c are suitable, even and odd elements of $\operatorname{Cl}(V, h)$, respectively. The spinor norm of $A \in G$ is 1; therefore,

$$\begin{pmatrix} a & b \\ c & d \end{pmatrix}^{-1} = \begin{pmatrix} d^t & -b^t \\ -c^t & a^t \end{pmatrix},$$

where a^t is the transpose of $a \in \operatorname{Cl}(V, h)$, etc.

PROPOSITION 6. *Let $A \in G$ be as in* (24). *Then*

$$V_A = \{\, v \in V | cv + d \text{ is invertible in } \operatorname{Cl}(V, h) \,\}$$

and the local conformal transformation (22) *is given by the Vahlen-Ahlfors formula,*

$$\text{(25)} \qquad F(A)v = (av + b)(cv + d)^{-1}.$$

To prove the proposition, let $v \in V_A$ and put $u = F(A)v$. According to (23), the isotropic vector $i(v)$ is represented in $K(2) \otimes_K \mathrm{Cl}(V, h)$ by the matrix

$$\begin{pmatrix} v & -h(v) \\ 1 & -v \end{pmatrix} = \begin{pmatrix} v \\ 1 \end{pmatrix} (\, 1 \quad -v \,)\,;$$

there is a similar representation of $i(u)$. The definitions (21) and (22) imply that the vectors $i(u)$ and $Ai(v)A^{-1}$ are parallel: there is $\kappa \in K,\ \kappa \neq 0$, such that

$$\kappa \begin{pmatrix} u & -h(u) \\ 1 & -u \end{pmatrix} = \begin{pmatrix} a & b \\ c & d \end{pmatrix} \begin{pmatrix} v & -h(v) \\ 1 & -v \end{pmatrix} \begin{pmatrix} d^t & -b^t \\ -c^t & a^t \end{pmatrix}.$$

Therefore, $\kappa = (cv + d)(cv + d)^t$, so that $cv + d$ is invertible and (25) holds.

Moreover, if $h(dv) = h_{\mu\nu} dv^\mu dv^\nu$ denotes the quadratic differential form giving V the structure of a (flat) Riemannian space, then $h(du) = \kappa^{-2} h(dv)$.

The Appendix is based on **[RT3]**, where references to the relevant papers by Vahlen and Ahlfors are given.

References

[ABS] M. F. Atiyah, R. Bott and A. Shapiro, *Clifford modules*, Topology **3, Suppl. 1** (1964), 3–38.

[AHS] M. F. Atiyah, N. J. Hitchin and I. M. Singer, *Self-duality in four-dimensional Riemannian geometry*, Proc. Roy. Soc. (London) **A 362** (1978), 425–461.

[AS] M. F. Atiyah and I. M. Singer, *The index of elliptic operators on compact manifolds*, Bull. A. M. S. (1963), 422–433.

[BGV] N. Berline, E. Getzler and M. Vergne, *Heat Kernels and Dirac Operators*, Springer, Berlin, 1992.

[Bo] J.-P. Bourguignon, *L'operateur de Dirac et la géometrie riemannienne*, Rend. Sem. Mat. Univ. Politec. Torino **44** (1986), 317–359.

[BuT] P. Budinich and A. Trautman, *An Introduction to the Spinorial Chessboard*, Trieste Notes in Physics, Springer, Berlin, 1988.

[CGuT] M. Cahen, S.Gutt and A. Trautman, *Spin structures on real projective quadrics*, J. Geom. Phys. **10** (1993), 127–154.

[C1] É. Cartan, *Nombres complexes*, exposé daprès l'article allemand de E. Study, Encycl. Sci. Math., éd. française, 1908 (reprinted in: É. Cartan, *Œuvres Complètes*, Partie II, vol. 1, Gauthier-Villars, Paris, 1953).

[C2] ———, *Les groupes projectifs qui ne laissent invariante aucune multiplicité plane*, Bull. Soc. Math. France **41** (1913), 53–96.

[C3] ———, *Les groupes projectifs continus réels qui ne laissent invariante aucune multiplicité plane*, J. Math. pures et appl. **10** (1914), 149–186.

[C4] ———, *Théorie des spineurs*, Actualités Scientifiques et Industrielles, No. 643 et 701, Hermann, Paris, 1938; English transl. *The Theory of Spinors*, Hermann, Paris, 1966.

[Ch] C. Chevalley, *The algebraic theory of spinors*, Columbia U. P., New York, 1954.

[Cl1] W. K. Clifford, *Applications of Grassmann's extensive algebra*, Amer. J. Math. **1** (1878), 350–358.

[Cl2] ———, *On the classification of geometric algebras*, Mathematical Papers of W. K. Clifford, Chelsea, New York, 1968 (reprint of the first edition published 1882), pp. 397–401.

[Corr] *Correspondence*, Ann. Math. **69** (1959), 247–251, an anonymous note.

[E] L. Euler, *Formulae generales pro translatione quacunque corporum rigidorum*, N. Comm. Acad. Sci. Imp. Petrop. **20** (1776), 189–207.

[F] V. Fock, *L'équation de Dirac et la géometrie de Riemann*, J. de Physique et le Radium **10** (1929), 392–405.

[Gi] P. B. Gilkey, *Invariance Theory, The Heat Equation and The Atiyah-Singer Index Theorem*, Publish or Perish, Wilmington, Delaware, 1984.

[H] A. Haefliger, *Sur l'extension du groupe structural d'un espace fibré*, C. R. Acad. Sci. (Paris) **243** (1958), 558–560.

[Ha] F. R. Harvey, *Spinors and Calibrations*, Academic Press, San Diego, CA, 1990.

[J] H. Jacobowitz, *An Introduction to CR Structures*, American Mathematical Society, Providence, 1990.

[K] E. Kähler, *Innerer und äusserrer Differentialkalkül*, Abh. Deutsch. Akad. Wiss. Berlin (Math.-Phys) **4** (1960).

[Ka] M. Karoubi, *Algèbres de Clifford et K-théorie*, Ann. Sci. Éc. Norm. Sup., 4ème sér. **1** (1968), 161–270.

[KoT] W. Kopczyński and A. Trautman, *Simple spinors and real structures*, J. Math. Phys. **33** (1992), 550–559.

[LM] H. B. Lawson, Jr. and M. L. Michelsohn, *Spin Geometry*, Princeton U. P., Princeton, 1989.

[Li] A. Lichnerowicz, *Spineurs harmoniques*, C. R. Acad. Sci. (Paris) Sér. A-B **257** (1963), 7–9.

[Lp] R. O. Lipschitz, *Untersuchungen ueber die Summen von Quadraten*, Max Cohen und Sohn, Bonn, 1886; French summary of the book in: Bull. Sci. Math. **21** (1886), 163–183.

[N] P. Nurowski, *Optical geometries and related structures*, J. Geom. Phys. (1995), accepted for publication.

[OBR] N. R. O'Brian and J. H. Rawnsley, *Twistor spaces*, Ann. Global Anal. Geom **3** (1985), 29–58.

[P] R. Penrose, *Physical space-time and nonrealizable CR-structures*, Bull. Amer. Math. Soc. (N. S.) **8** (1983), 427–448.

[PR] R. Penrose and W. Rindler, *Spinors and space-time*, vol. 1 and 2, Cambridge U. P., Cambridge, 1984 and 1986.

[Ri] M. Riesz, *L'équation de Dirac en relativité générale*, C. R. XII Congrès des Mathématiciens Scandinaves, Lund, 1953, pp. 241–259.

[RT1] I. Robinson and A. Trautman, *Cauchy-Riemann structures in optical geometry*, Proc. of the Fourth Marcel Grossmann Meeting on General Relativity (R. Ruffini, ed.), Elsevier Science Publ., 1986, pp. 317–324.

[RT2] I. Robinson and A. Trautman, *Optical geometry*, New Theories in Physics; Proc. of the XI Warsaw Symp. on Elementary Particle Physics, Kazimierz, 23–27 May 1988 (Z. Ajduk *et al.*, eds.), World Scientific, Singapore, 1989, pp. 454–497.

[RT3] I. Robinson and A. Trautman, *The conformal geometry of complex quadrics and the fractional-linear form of Möbius transformations*, J. Math. Phys. **34** (1993), 5391–5406.

[Ro] O. Rodrigues, *Des lois géométriques qui régissent les déplacements d'un système solide*, Journal de Math. (Liouville) **5** (1840), 380–384.

[Sa] F. Sauter, *Lösung der Diracschen Gleichungen ohne Spezialisierung der Diracschen Operatoren*, Z. Phys. **63** (1930), 803–814.

[SW] R. F. Streater and A. S. Wightman, *PCT, Spin & Statistics, and All That*, W. A. Benjamin, New York, 1964.

[T] A. Trautman, *Optical structures in relativistic theories*, Astérisque, numéro hors série (1985), 401–420.

[TT] A. Trautman and K. Trautman, *Generalized pure spinors*, J. Geom. Phys. **15** (1994), 1–22.

[U] H. Urbantke, *Pauli-Kofink identities and pure spinors*, Spinors, Twistors, Clifford Algebras and Quantum Deformations (Z. Oziewicz *et al.*, eds.), Kluwer, Dordrecht, 1993, pp. 53–60.

[V1] O. Veblen, *Geometry of two-component spinors*, Proc. Nat. Acad. Sci. **19** (1933), 462–474.

[V2] ———, *Geometry of four-component spinors*, Proc. Nat. Acad. Sci. **19** (1933), 503–517.

[VG] O. Veblen and J. W. Givens, *Geometry of complex domains* (notes from a seminar conducted in 1935-36 at the Institute for Advanced Study in Princeton by O. Veblen and J. von Neumann), Princeton, 1955.

[W] H. Weyl, *Elektron und Gravitation I*, Z. Phys. **56** (1929), 330–352.

INSTYTUT FIZYKI TEORETYCZNEJ, UNIWERSYTET WARSZAWSKI, HOŻA 69, PL-00681 WARSAW, POLAND

E-mail address: amt@fuw.edu.pl

Contemporary Mathematics
Volume **203**, 1997

Dirac's Algebra and Braer-Wall Groups

BIRGER IVERSEN

Contents

It is said that P.A.M. Dirac rediscovered the Clifford algebras. It is certainly true that the construction used by Dirac produces the Clifford algebra of a quadratic form of Sylvester type $(-3, 1)$. In this paper we exploit further the construction of Dirac and show that it induces an involution on the Brauer-Wall group, which interchanges classes of odd and even type in the sense of [Wall]. The main motivation for this paper is the use of Dirac' type constructions in the identification of Spin groups of various types of quadratic forms, see [Deheuvels], [Gilbert and Murray], [Porteus] and [Shafarevich]. Thus, the paper is generated by the examples included in section 1 and 6.

It should be said that the theory work for any field of characteristic different from 2. Investigations of a discriminant type invariant of Dirac's algebra lead us to a theory of the "anti-centre" of a graded algebra. This theory is presented in some details.

In the case where the ground fiels is $\mathbb{R}$ we find that any similarity class of mod-2-graded algebras can be obtained by iteration of Dirac's construction $A \mapsto D(A)$ from one of the four algebras $\mathbb{R}$, $\mathbb{C}$, $\mathbb{H}$, $\mathbb{J}$, where $\mathbb{H}$, resp. $\mathbb{J}$, denotes the quaternion algebra equipped with trivial involution, resp. conjugation with the pure quaternion i. In order to see this we represent the tensor powers of the graded algebra $\mathbb{C}$ in the Brauer-Wall group as follows

$$\mathbb{R}, \quad \mathbb{C}, \quad \mathbb{J}, \quad D(\mathbb{H}), \quad \mathbb{H}, \quad D(\mathbb{J}), \quad D(\mathbb{C}), \quad D(\mathbb{R}), \quad \mathbb{R}$$

1991 *Mathematics Subject Classification.* Primary 11E88 ; Secondary 15A66.

Let me add that this reflects Bott's mod 8 periodicity in the Brauer-Wall group.

1. DIRAC'S ALGEBRA

Let A be a ring equipped with an *involution* i.e a morphism of rings $a \mapsto \tilde{a}$, $\tilde{\tilde{a}} = a$. Let us introduce multiplication on $A \oplus A$ by the formula

$$(1.1) \qquad (a,b)\cdot(c,d) = (ac + b\tilde{d}, ad + b\tilde{c}) \qquad ;(a,b),(c,d) \in A \oplus A$$

This multiplication rule defines a structure of a ring on $A \oplus A$ with $(1,0)$ as the multiplicative neutral element. This ring will be denoted by $D(A)$.

The map $A \to D(A)$, $a \mapsto (a,0)$ is a morphism of rings which we shall use to identify A with a subring of $D(A)$. Put $\Psi = (0,1)$ and observe any element of $D(A)$ can be written in the form $a + \Psi b, a, b \in A$, and that

$$(1.2) \qquad \Psi^2 = 1 \quad , \quad a\Psi = \Psi\tilde{a} \qquad ;a \in A$$

The ring $D(A)$ comes equipped with the *principal involution*

$$(1.3) \qquad a + \Psi b \mapsto a - \Psi b \qquad ;a, b \in A$$

In the following we shall give a number of examples all taken from the theory of Clifford algebras over a field k of characteristic different from 2.

THEOREM 1.4. *For a nonsingular quadratic form Q on the vectorspace E, let ΔQ denote the quadratic form defined on $k \oplus E$ by $\Delta Q(x,y) = x^2 - Q(y)$, $x \in k, y \in E$. The Clifford algebra of ΔQ is given by $C(\Delta Q) = D(C(Q))$.*

PROOF. Let us recall that there is a canonical identification of E with a subspace of the odd part of $C = C(Q)$ such that $x \cdot x = Q(x) \cdot 1$ for all $x \in E$. At the same time we can identify $k \oplus E$ with the subspace of $D(C)$ generated by Ψ and ΨE. Let us calculate the square in $D(C)$ of an element from $\Psi(k \cdot 1 + E)$

$$\Psi(x+y)\Psi(x+y) = x^2 - y^2 \qquad ;x \in k, y \in E$$

This defines a morphism of graded algebras $C(\Delta Q) \mapsto D(C(Q))$. Observe that the kernel of this map is a graded bilateral ideal in $C(\Delta Q)$ and conclude that the map is injective since all bilateral graded ideals in $C(\Delta Q)$ are trivial. The map is surjective for dimension reasons.

EXAMPLE 1.5. The Clifford algebra of type $(-1,0)$ can be identified with $\mathbb{C}$ equipped with ordinary complex conjugation. The assignments

$$\Psi \mapsto \begin{pmatrix} -1 & 0 \\ 0 & 1 \end{pmatrix}, \quad a + ib \mapsto \begin{pmatrix} a & -b \\ b & a \end{pmatrix} \qquad ;a, b \in \mathbb{R}$$

defines an isomorphism of rings $D(\mathbb{C}) \tilde{\to} \mathrm{M}_2(\mathbb{R})$, which identifies the principal involution on $D(\mathbb{C})$ with conjugation by $\begin{pmatrix} 0 & 1 \\ -1 & 0 \end{pmatrix}$ on $\mathrm{M}_2(\mathbb{R})$. Observe that a matrix from the odd part of the algebra has scaler square

$$\begin{pmatrix} x & y \\ y & -x \end{pmatrix}^2 = x^2 + y^2 \qquad ;x, y \in \mathbb{R}$$

In fact, $D(\mathbb{C})$ is the Clifford algebra of a quadratic form of Sylvester type $(0,2)$. If we iterate this construction we find that $D(D(\mathbb{C}))$ may be identified with the algebra $\mathrm{M}_2(\mathbb{C})$ equipped with the involution which is entry by entry conjugation. Let us consider the subspace $i\mathrm{sl}_2(\mathbb{R})$ of $\mathrm{M}_2(\mathbb{C})$ and observe that

$$i\begin{pmatrix} a & b \\ c & -a \end{pmatrix} i\begin{pmatrix} a & b \\ c & -a \end{pmatrix} = -a^2 - bc$$

This identifies our algebra $D(D(\mathbb{C}))$ with the Clifford algebra of the quadratic form $-a^2 - bc$, a form of Sylvester type $(-2,1)$. This algebra can be used to give a convenient model for the hyperbolic plane H^2, compare [Iversen, III].

EXAMPLE 1.6. Let us recall two important anti-involutions on $\mathrm{M}_2(\mathbb{C})$

$$\begin{pmatrix} a & b \\ c & d \end{pmatrix}^* = \begin{pmatrix} \bar{a} & \bar{c} \\ \bar{b} & \bar{d} \end{pmatrix}, \quad \begin{pmatrix} a & b \\ c & d \end{pmatrix}^{\check{}} = \begin{pmatrix} d & -b \\ -c & a \end{pmatrix} \qquad ; a,b,c,d \in \mathbb{C}$$

The composite $R \mapsto R^{*\check{}}$ defines an involution on $\mathrm{M}_2(\mathbb{C})$ from which we construct *Dirac's algebra* $D(\mathrm{M}_2(\mathbb{C}))$. Let us observe that a vector of the form ΨH, where H is a Hermitian matrix, has a square which is a scalar

$$\Psi H \Psi H = H\tilde{H} = \begin{pmatrix} a & b \\ \bar{b} & d \end{pmatrix}\begin{pmatrix} d & -b \\ -\bar{b} & a \end{pmatrix} = ad - b\bar{b} \qquad ; a,d \in \mathbb{R}, b \in \mathbb{C}$$

On this basis we can identify Dirac's algebra with the Clifford algebra of a quadratic form of type $(-3,1)$. This leads to a convenient model for hyperbolic three space H^3, [Iversen, VIII].

EXAMPLE 1.7. Let $\mathrm{M}_{1,1}(\mathbb{H})$ denote the algebra $\mathrm{M}_2(\mathbb{H})$ equipped with inner conjugation by $\begin{pmatrix} 1 & 0 \\ 0 & -1 \end{pmatrix}$ and consider the subspace of vectors of the form $\begin{pmatrix} 0 & h \\ h^* & 0 \end{pmatrix}$ $h \in \mathbb{H}$. Observe that the square of a matrix of this form is $-hh^* = -N(h)$ and use this to identify $\mathrm{M}_{1,1}(\mathbb{H})$ with the Clifford algebra of a form of type $(0,4)$.

In the algebra $D(\mathrm{M}_{1,1}(\mathbb{H}))$ we consider a vector of the form $\Psi\begin{pmatrix} x & h \\ h^* & x \end{pmatrix}$ $x \in \mathbb{R}, h \in \mathbb{H}$, and observe that its square is a scalar matrix

$$\Psi\begin{pmatrix} x & h \\ h^* & x \end{pmatrix}\Psi\begin{pmatrix} x & h \\ h^* & x \end{pmatrix} = x^2 - N(h) \quad ; x \in \mathbb{R}, h \in \mathbb{H}$$

This realises $D(\mathrm{M}_{1,1}(\mathbb{H}))$ as a Clifford algebra of a form of Sylvester type $(-4,1)$. As an application, this may be used to identify $\mathrm{Spin}(-4,1)$ with the group

$$A\begin{pmatrix} 1 & 0 \\ 0 & -1 \end{pmatrix} A^* = I \qquad ; A \in \mathrm{Gl}_2(\mathbb{H})$$

This group is known as the symplectic unitary group $\mathrm{SpU}(2)$, [Shaferevitz, p.145].

2. MOD-2 GRADED ALGEBRAS

By an algebra over k we understand an associative unital algebra of finite dimension over k. A (mod-2) *graded algebra* is an algebra A equipped with a decomposition $A = A_0 \oplus A_1$ into two k – *subspaces* which satisfy

$$1 \in A_0,\ A_i \cdot A_j \subseteq A_{i+j} \qquad ; i,j \in \mathbb{Z}/(2) \tag{2.1}$$

The decomposition of an element $a \in A$ will be written $a = a_0 + a_1$. The *principal involution* $a \mapsto \tilde{a}, a \in A$, is the linear endomorphism of A which multiplies with $+1$ on A_0 and multiplies with -1 on A_1.

DEFINITION 2.2. The *graded centre* of the graded algebra A is the linear subspace $GZ(A) = GZ_0(A) \oplus GZ_1(A)$ where

$$GZ_0(A) = \{a \in A_0 | \forall b \in A\ ab = ba\}, GZ_1(A) = \{a \in A_1 | \forall b \in A\ ab = \tilde{b}a\}$$

We say that a graded algebra A is *central* if $GZ_0(A) = k$ and $GZ_1(A) = 0$. We say that a graded algebra is *simple* if A has no non-trivial graded bilateral ideals.

PROPOSITION 2.3. *The Dirac algebra $D(A)$ of a graded algebra A is graded central simple if and only if A is a graded central simple algebra.*

PROOF. The graded centre of Dirac's algebra is easily calculated

$$GZD(A) = GZ_0(A) + \Psi GZ_1(A) \tag{2.4}$$

To complete the proof, observe that we have a general identity

$$\Psi(a + \Psi b)\Psi = \tilde{a} + \Psi\tilde{b} \qquad ; a, b \in A$$

which shows that a graded bilateral ideal in $D(A)$ must have the form $J \oplus J\Psi$ where J is a graded bilateral ideal in A. □

The graded tensor product of the graded algebras A and B will be denoted $A\hat{\otimes}B$. Let us recall from [Wall] that the graded tensorproduct of two graded central simple algebras is a graded central simple algebra. For integers $p, q \in \mathbb{N}, p \geq 1$, we let $\mathrm{M}_{p,q}(k)$ denote the algebra $\mathrm{M}_{p+q}(k)$ equipped with the involution which is inner conjugation with the p, q-block matrix $\begin{pmatrix} 0 & 1 \\ 1 & 0 \end{pmatrix}$ followed by entry by entry action of the principal involution from A. The transformation $A \mapsto \mathrm{M}_{p,q}(A)$ given by

$$a \mapsto \begin{pmatrix} a & 0 \\ 0 & \tilde{a} \end{pmatrix} \qquad ; a \in A$$

This together with the natural inclusion $\mathrm{M}_{p,q}(k) \to \mathrm{M}_{p,q}(A)$ defines an isomorphism

$$\mathrm{M}_{p,q}(k)\hat{\otimes}_k A \tilde{\to} \mathrm{M}_{p,q}(A) \tag{2.5}$$

In particular, put $A = \mathrm{M}_{r,s}(k)$ in 2.5 and perform a simultaneous permutation of columns and rows to get an isomorphism

$$\mathrm{M}_{p,q}(k)\hat{\otimes}\mathrm{M}_{r,s}(k) \tilde{\to} \mathrm{M}_{m,n}(k) \qquad ; m = pr + qs, n = ps + qr \tag{2.6}$$

Let us say that graded central simple algebras A and B are *graded similar* if we can find pairs of integers (p,q) and (r,s) as above such that

$$\mathrm{M}_{p,q}(k)\hat{\otimes}A\tilde{\rightarrow}\mathrm{M}_{r,s}(k)\hat{\otimes}B$$

The graded tensor product $\hat{\otimes}$ induces a group structure on the set of similarity classes of graded algebras over k. This group is called the *Brauer-Wall group* of k.

3. DUALITY IN THE BRAUER-WALL GROUP

In this section we shall see that the Dirac construction induces an operator on the Brauer-Wall group. Quite precisely, for integers $p,q\in\mathbb{N}, p\geq 1$, we have a canonical isomorphism

$$D(A\hat{\otimes}\mathrm{M}_{p,q}(k))\tilde{\rightarrow}D(A)\hat{\otimes}\mathrm{M}_{p,q}(k) \tag{3.1}$$

PROOF. Let us introduce $\Phi\in\mathrm{M}_{p,q}(D(A))$ and $\phi:\mathrm{M}_{p,q}(A)\rightarrow\mathrm{M}_{p,q}(D(A))$

$$\Phi=\begin{pmatrix}\Psi & 0\\ 0 & -\Psi\end{pmatrix},\ \phi\begin{pmatrix}a & b\\ c & d\end{pmatrix}=\begin{pmatrix}a & \Psi\tilde{b}\\ \Psi c & \tilde{d}\end{pmatrix}$$

Observe that $\tilde{\Phi}=-\Phi, \Phi^2=1$ and that $\phi(\tilde{R})=\Phi\phi(R)\Phi^{-1}$ for all $A\in\mathrm{M}_{p,q}(A)$. It follows that $R+\Psi S\mapsto\phi(R)+\Phi\phi(S)$ defines an isomorphism of graded rings $D(\mathrm{M}_{p,q}(A))\tilde{\rightarrow}\mathrm{M}_{p,q}(D(A))$. Conclusion by 2.5. □

Let us establish the key formula

$$D(D(A))\tilde{\rightarrow}A\hat{\otimes}\mathrm{M}_{1,1}(k)) \tag{3.2}$$

PROOF. Let us first introduce the graded algebra $M(A)$ which is the algebra $\mathrm{M}_2(A)$ equipped with the involution which is conjugation by $\begin{pmatrix}0 & 1\\ 1 & 0\end{pmatrix}$ followed by entry by entry action of the involution from A. Conjugation by the matrix $\begin{pmatrix}1 & -1\\ 1 & 1\end{pmatrix}$ on $\mathrm{M}_2(A)$ is an isomorphism of graded rings

$$M(A)\tilde{\rightarrow}\mathrm{M}_{1,1}(A)\tilde{\rightarrow}A\hat{\otimes}\mathrm{M}_{1,1}(k))$$

Let us introduce the representation $\gamma:D(A)\rightarrow\mathrm{M}_{1,1}(A)$ given by

$$\gamma(a+\Psi b)=\begin{pmatrix}a & b\\ \tilde{b} & \tilde{a}\end{pmatrix}\qquad ;a,b\in A$$

Observe that $\Gamma=\begin{pmatrix}1 & 0\\ 0 & 1\end{pmatrix}$ satisfies $\Gamma^2=1,\tilde{\Gamma}=-\Gamma$ and $\Gamma\gamma(R)\Gamma^{-1}=\gamma(R)$ for all $R\in D(A)$. It follows that $R+\Psi S\mapsto\gamma(R)+\Gamma\gamma(S), R,S\in D(A)$, defines an isomorphism of graded rings $D(D(A))\tilde{\rightarrow}M(A)$. □

For $a \in k^*$, let $Sqr(a)$ denote $k[T]/(T^2 - a)$ equipped with the k-involution which transforms the residue class $\sqrt{a}$ of T mod $(T^2 - a)$ into $-\sqrt{a}$. Let us show that

$$(3.3) \qquad D(Sqr(a)) \tilde{\to} (\mathrm{M}_2(k), \begin{pmatrix} 0 & 1 \\ a & 0 \end{pmatrix}) \qquad ; a \in k^*$$

PROOF. To this end we we introduce $\Sigma \in \mathrm{M}_2(k)$ and a morphism $\sigma : Sqr(a) \to \mathrm{M}_2(A)$ given by

$$\Sigma = \begin{pmatrix} 1 & 0 \\ 0 & -1 \end{pmatrix}, \ \sigma(x + \sqrt{a}y) = \begin{pmatrix} x & y \\ ya & x \end{pmatrix} \qquad ; x, y \in k$$

Observe that $\Sigma^2 = 1, \tilde{\Sigma} = -\Sigma$ and $\Sigma\sigma(R)\Sigma^{-1} = \sigma(\tilde{R})$ for all $R \in Sqr(a)$. It follows that the map $R + \Psi S \mapsto \sigma(R) + \Sigma\sigma(S)$, $R, S \in Sqr(a)$, defines an isomorphism of graded rings as required. □

$$(3.4) \qquad D(Sqr(a)) = Sqr(1)\hat{\otimes}Sqr(-a) \qquad ; a \in k$$

PROOF. Identify $Sqr(-a)$ and $Sqr(1)$ with graded subalgebras of the graded algebra (3.3) through

$$\sqrt{a} \mapsto \begin{pmatrix} 0 & 1 \\ a & 0 \end{pmatrix}, \ \sqrt{1} \mapsto \begin{pmatrix} 1 & 0 \\ 0 & -1 \end{pmatrix}$$

Observe that the matrices $\begin{pmatrix} 0 & 1 \\ a & 0 \end{pmatrix}$ and $\begin{pmatrix} 1 & 0 \\ 0 & -1 \end{pmatrix}$ anti-commutes and conclude on this basis that we have an isomorphism of the required sort. □

$$(3.5) \qquad Sqr(a)\hat{\otimes}Sqr(-a) = \mathrm{M}_{1,1}(k) \qquad ; a \in k^*$$

PROOF. Let us identify $Sqr(a)$ and $Sqr(-a)$ with subalgebras of $\mathrm{M}_{1,1}(k)$ by means of

$$\sqrt{a} \mapsto \begin{pmatrix} 0 & 1 \\ a & 0 \end{pmatrix}, \ \sqrt{-a} \mapsto \begin{pmatrix} 0 & 1 \\ -a & 0 \end{pmatrix}$$

Observe that the two matrices $\begin{pmatrix} 0 & 1 \\ a & 0 \end{pmatrix}$ and $\begin{pmatrix} 0 & 1 \\ -a & 0 \end{pmatrix}$ anti-commutes and conclude on this basis that we have an isomorphism of the required sort. □

As a consequence of (3.4) and (3.5) we see that $D(Sqr(1))$ is the neutral element in $\mathrm{BW}(k)$. Let us also show that

$$(3.6) \qquad \mathrm{M}_{p,q}(k)\hat{\otimes}Sqr(a) \tilde{\to} \mathrm{M}_{p+q}(k)\hat{\otimes}Sqr(a) \qquad ; a \in k^*$$

PROOF. To this end we let $x \mapsto \tilde{x}$ denote conjugation in $Sqr(a)$ as well as entry by entry conjugation on $\mathrm{M}_{p+q}(Sqr(a))$. We ask the reader to verify that

$$\tilde{S}\begin{pmatrix} 1 & 0 \\ 0 & -1 \end{pmatrix} S^{-1} = \begin{pmatrix} 1 & 0 \\ 0 & 1 \end{pmatrix} \qquad ; S = \begin{pmatrix} 1 & 0 \\ 0 & \sqrt{a} \end{pmatrix}$$

And conclude that $X \mapsto SXS^{-1}$ induces an isomorphism as required. □

LEMMA 3.7. *Let A and H be a graded algebras. If $[H]_1 = 0$, then*

$$D(A\hat{\otimes}H) = D(A)\hat{\otimes}H$$

PROOF. Let us write down an explicit map

$$a \otimes h + \Psi(b \otimes k) \mapsto a \otimes h + \Psi b \otimes k$$

The remaining details are left to the reader. □

4. THE ANTI-CENTRE

In the first part of this section we develop the general notion of the anti-centre of a graded central algebra. The anti-centre of a graded central simple algebra will be enriched with a structure of graded quadratic form, which will be identified with the "graded Arf-invariant" [Knus p.156].

DEFINITION 4.1. Let R be a graded algebra. The *anti-centre* of R is the graded subspace $AZ(R) = AZ_0(R) \oplus AZ_1(R)$ given by

$$AZ_0(R) = \{r \in R_0 | \forall x \in R, xr = r\tilde{x}\} \; AZ_1(R) = \{r \in R_1 | \forall x \in R, xr = rx\}$$

Direct calculation shows that the anti-centre of Dirac's algebra is given by

$$AZ(D(R)) = AZ_1(R) + \Psi AZ_0(R) \tag{4.2}$$

By standard calculations we find for graded algebras R and S that

$$AZ(R\hat{\otimes}S) = AZ(R)\hat{\otimes}AZ(S) \tag{4.3}$$

PROPOSITION 4.4. *Let R be a graded central algebra. Then, for all $a, b \in AZ(R)$ we have that $ab = ba \in k$. Moreover, the function $a \mapsto a^2$ defines a quadratic form on $AZ(R)$ for which $AZ_0(R)$ is orthogonal to $AZ_1(R)$.*

PROOF. Let us observe that $r_0r_1 = r_1r_0 = 0$ for $r_0 \in AZ_0(R)$ and $r_1 \in AZ_1(R)$ as follows by observing that r_0r_1 and r_1r_0 belong to $GZ_1(R) = 0$. Thus

$$ab = (a_0 + a_1)(b_0 + b_1) = a_0b_0 + a_1b_1 \qquad ; a, b \in AZ(R)$$

The result follows by observing that a_0b_0 and a_1b_1 belong to $GZ_0(R) = k$. □

REMARK 4.5. The result of 4.4 can be interpreted in terms of the notation of a *graded quadratic form*, by which we understand a quadratic space E equipped with a direct sum decomposition $E = E_0 \oplus E_1$ into two orthogonal subspaces. Given two graded quadratic forms we can form $E\hat{\otimes}F$ through the convention

$$< e_i \otimes f_j, e_r \otimes f_s >= (-1)^{jr} < e_i, e_r >< f_j, f_s >$$

THEOREM 4.6. *The anti-centre of a graded central simple algebra R has dimension 1 and any non-zero $r \in AZ(R)$ has non-zero square: $r^2 \neq 0$.*

PROOF. Let us make the general remark that the ordinary centre $Z(R)$ of R is a graded subalgebra of R, in fact $Z(R) = GZ_0(R) \oplus AZ_1(R)$. Our investigation will follow the natural division of graded simple central algebras into two types.

Even type: $AZ_1(R) = 0$. It follows from the remark above that the underlying ungraded algebra $[R]$ is central. The Radical of $[R]$ is a bilateral ideal invariant under the principal involution, and we conclude that $\mathrm{Rad}([R]) = 0$. It follows that $[R]$ is semi-simple, indeed central simple. The principal involution of R is conjugation by a unit $r \in R$ by the Skolem-Noether theorem. This gives $\tilde{x} = rxr^{-1}$ or $\tilde{x}r = rx$ for all $x \in R$; in particular for $x = r$ we find that $\tilde{r} = r$. This shows that $r \in AZ_0(R)$. Since conjugation with r is an involution, we can write $r^2 = a \in k^*$. Let us show that $AZ(R)$ is generated by r. For $s \in AZ_0(R)$ we have $s(rxr^{-1}) = xs$ for all $x \in R$, which show that sr belongs to $Z(R) = k$. Write $sr = b$ and conclude that $s = a^{-1}br$.

Odd type: $AZ_1(R) \neq 0$. Pick a non-zero element $r \in AZ_1(R)$ and let us show that $r^2 \neq 0$: Note that Rr is a graded bilateral ideal and that $r^2 = 0$ implies that $Rr \neq R$. It follows that r^2 belongs to $GZ_0(R) = k$ and that we can write $r^2 = a$ where $a \in k^*$. Observe that $R_1 = rR_0$: $R_1 = aR_1 = r^2R_1 \subseteq rR_0$. We can now use $r \in AZ_1(R), r^2 = a$, to define an isomorphism $R_0 \hat{\otimes} Sqr(a) \tilde{\rightarrow} R$. It follows from the presence of such an isomorphism that R_0 is a central simple algebra. From the same source it follows that $AZ(R)$ is generated by r, see (4.3). □

By a *graded discriminant* we understand a non-singular graded quadratic form (4.6) of dimension 1. Tensor product $\hat{\otimes}$ organises the set of isomorphism classes of graded discriminants into a group $\mathrm{Disc}^g(k)$. This group can be described as the set $\mathbb{F}_2 \times k^*/k^{*2}$ equipped with the composition rule given by

$$(4.7) \qquad (m, a) \cdot (n, b) = (m + n, (-1)^{mn}ab) \qquad ; m, n \in \mathbb{F}_2, a, b \in k^*/k^{*2}$$

As an example, consider an n-dimensional quadratic form Q over k with discriminant $d \in k^*/k^{*2}$. The anti-centre of the Clifford algebra $C(Q)$ is

$$(4.8) \qquad AZ(C(Q)) = (n, (-1)^{n(n-1)/2}d)$$

To see this, pick an orthogonal basis $e_1, \dots, e_n$ for Q with $< e_i, e_i >= a_i, i = 1, \dots, n$. Conclude from (4.3) and (4.7) that the anti-centre is given by

$$(1, a_1)(1, a_2) \dots (1, a_n) = (n, (-1)^{n(n-1)/2}a_1a_2 \dots a_n)$$

For a graded algebra R we see by inspection that $AZ(R)$ is contained in the centraliser $Z_R(R_0)$. For a graded central simple algebra R with $R_1 \neq 0$ it follows from the proof of Theorem 4.6 that we can reconstruct the graded algebra $Z_R(R_0)$ from $AZ(R)$ as the graded sum

$$(4.9) \qquad Z_R(R_0) = k \oplus AZ(R) \qquad ; R_1 \neq 0$$

With this terminology we can state a theorem of Wall in the form of an exact sequence induced by the anti-centre, compare [Hahn p.203].

$$0 \to \mathrm{Br}(k) \to \mathrm{BW}(k) \to \mathrm{Disc}^g(k) \to 0 \tag{4.10}$$

5. STRUCTURE THEOREM FOR BW(k)

A graded central simple algebra R is of *even type* if $AZ_0(R) \neq 0$ and of *odd type* if $AZ_1(R) \neq 0$. From the discussion in the previous section it follows that the type of a graded algebra is invariant under graded similarity.

STRUCTURE THEOREM 5.1. *A graded central simple algebra R is graded similar to an algebra of the following form*

$$\begin{aligned} \textit{odd type:} &\quad Sqr(a)\hat{\otimes}H \\ \textit{even type:} &\quad D(Sqr(a))\hat{\otimes}H \end{aligned}$$

where $a \in k^*$ and H is a central simple skew field over k (with trivial involution). The isomorphism class of H and the residue class $a \in k^*/k^{*2}$ are uniquely determined by the class of R in $BW(k)$.

PROOF. Odd type: We have already seen during the proof of (4.6) that R has the form $Sqr(a)\hat{\otimes}[R]_0, a \in k^*$, where $[R]_0$ is central simple. Let us write $[R]_0 = \mathrm{M}_n(H)$, where H is a central skew field over k to get that

$$R \tilde{\to} Sqr(a)\hat{\otimes}\mathrm{M}_n(H) \tilde{\to} Sqr(a)\hat{\otimes}H\hat{\otimes}\mathrm{M}_n(k)$$

This shows that R is graded similar to an algebra of the form $Sqr(a)\hat{\otimes}H$. From the isomorphism $R \tilde{\to} Sqr(a)\hat{\otimes}[R]_0$ it follows easily that

$$[\mathrm{M}_{p,q}(k)\hat{\otimes}R]_0 = \mathrm{M}_{p+q}([R]_0) \qquad ; p,q \in \mathbb{N}, p \geq 1 \tag{5.2}$$

This show that the class $[R]_0 \in \mathrm{Br}(k)$ is a graded similarity invariant of R. Uniqueness of the class $a \in k^*/k^{*2}$ follows from the theory of the anti-centre.

Even type: The result follows from the structure of algebras of odd type using duality in the Brauer group, in particular (3.2) and (4.2).

PRODUCT FORMULAS 5.3. *For $a, b \in k^*$ we let $(\frac{a,b}{k}) \in Br(k)$ denote the quaternion symbol, i.e. the class of the ungraded algebra underlying $Sqr(a)\hat{\otimes}Sqr(b)$. These symbols satisfies the following identities*

$$\begin{aligned} Sqr(a)\hat{\otimes}Sqr(b) &= D(Sqr(-ab))\hat{\otimes}(\frac{a,b}{k}) \\ Sqr(a)\hat{\otimes}D(Sqr(b)) &= Sqr(ab)\hat{\otimes}(\frac{-a,b}{k}) \\ D(Sqr(a))\hat{\otimes}D(Sqr(b)) &= D(Sqr(ab))\hat{\otimes}(\frac{a,b}{k}) \end{aligned}$$

PROOF. Consider $a, b, c \in k^*$ and let us show that

$$Sqr(a)\hat{\otimes}Sqr(b)\hat{\otimes}Sqr(c) \tilde{\to} Sqr(-abc)\hat{\otimes}(\frac{-ac,-bc}{k}) \tag{5.4}$$

To this end we let α, β, γ denote the canonical square roots for a, b, c. The even part of the triple tensor product is generated by $\alpha \otimes \gamma$ and $\beta \otimes \gamma$ which satisfy

$$(\alpha \otimes \gamma)^2 = -ac, \; (\beta \otimes \gamma)^2 = -bc, \; (\alpha \otimes \gamma)(\beta \otimes \gamma) = -(\beta \otimes \gamma)(\alpha \otimes \gamma)$$

Besides, $\alpha \otimes \beta \otimes \gamma$ belongs to the anti-centre and has square $-abc$. The first of the three formulas follows from (5.4) with $c = -1$ and (3.4) and (3.5). The second formula follows from (5.4) by substituting $-b$ for b and 1 for c using (3.4) and (3.5) again. The third formula follows from the second upon substituting $-a$ for a and multiplying the formula with $Sqr(1)$ using (3.4) and (3.5). □

EXAMPLE 5.5. The group BW($\mathbb{R}$) has order 8 and is generated by $Sqr(-1)$. The powers $Sqr(-1)^p$ can be represented by the graded algebras $\mathbb{R}, \mathbb{C}, \mathbb{H}, \mathbb{J}$, where $\mathbb{H}$, resp. $\mathbb{J}$, denotes the quaternion algebra equipped with trivial involution, resp. conjugation with the pure quaternion i.

p	0	1	2	3	4	5	6	7	8
$Sqr(-1)^p$	$\mathbb{R}$	$\mathbb{C}$	$\mathbb{J}$	$D(\mathbb{H})$	$\mathbb{H}$	$D(\mathbb{J})$	$D(\mathbb{C})$	$D(\mathbb{R})$	$\mathbb{R}$

6. THE QUATERNION MODEL FOR H^5

In this section we shall use the ideas developed in the previous sections to write down the real Clifford algebra $C(-5, 1)$. Let us start with the algebra $M_2(\mathbb{H})$ and its $*$–operator given by

$$\begin{pmatrix} a & b \\ c & d \end{pmatrix}^* = \begin{pmatrix} a^* & c^* \\ b^* & d^* \end{pmatrix} \qquad ; a, b, c, d \in \mathbb{H}$$

A matrix $X \in M_2(\mathbb{H})$ with $X^* = X$ is called *Hermitian*. The Hermitian matrices form a six dimensional vectorspace W over $\mathbb{R}$, which comes equipped with the quadratic form Q of type $(-5, 1)$ given by

$$(6.1) \qquad Q(\begin{pmatrix} a & b \\ b^* & d \end{pmatrix}) = ad - bb^* \qquad a, d \in \mathbb{R}, b \in \mathbb{H}$$

The group $Gl_2(\mathbb{H})$ operates on the space W by the formula

$$(6.2) \qquad G.X = GXG^* \qquad ; G \in Gl_2(\mathbb{H}), X \in W$$

We ask the reader to verify that this action relates to the form Q as follows

$$(6.3) \qquad Q(GXG^*) = Q(GG^*)Q(X) \qquad ; G \in Gl_2(\mathbb{H}), X \in W$$

It follows that the function $Q(GG^*)$ is multiplicative in G. Let us recall that the group $Gl_2(\mathbb{H})$ modulo its commutator group is generated by real scalar matrices and let us observe that $Q(\lambda I) = \lambda^4, \lambda \in \mathbb{R}$. On this basis we can identify the character $Q(GG^*)$ and the determinant of Dieudonné

$$(6.4) \qquad Q(GG^*) = det(G) \quad ; G \in Gl_2(\mathbb{H})$$

We shall make use of the involution $X \mapsto \breve{X}$ of the subspace W of $\mathrm{M}_2(\mathbb{H})$ given by

$$\begin{pmatrix} a & b \\ c & d \end{pmatrix}^{\breve{}} = \begin{pmatrix} d & -b \\ -c & a \end{pmatrix} \qquad ; a, b, c, d \in \mathbb{H}$$

This involution is related to the action of $\mathrm{Gl}_2(\mathbb{H})$ on W through the formula

$$(6.5) \qquad (GXG^*)\breve{} = G^{*-1}\breve{X}G^{-1}det(G) \qquad ; G \in \mathrm{Gl}_2(\mathbb{H}), X \in W$$

PROOF. Verify that $X\breve{X} = Q(X)$ for all $X \in W$ and observe that the formula (6.5) is linear in $X \in W$. It follows that this suffices to prove (6.5) in the case where $Q(X) \neq 0$. But this case is easily handled using (6.3) and (6.4). □

It is now time to turn the graded algebra $\mathrm{M}_{2,2}(\mathbb{H})$ into a Clifford algebra by introducing the linear subspace of matrices of the form

$$(6.6) \qquad \begin{pmatrix} 0 & X \\ \breve{X} & 0 \end{pmatrix} \qquad ; X \in \mathrm{M}_2(\mathbb{H}), X^* = X$$

Let us observe that the square of this matrix is $Q(X)$ times the identity matrix and conclude that the algebra at hand is the real Clifford algebra $C(-5, 1)$. The *principal anti-involution* τ of the Clifford algebra is given by

$$(6.7) \qquad \tau\begin{pmatrix} A & B \\ C & D \end{pmatrix} = \begin{pmatrix} D^* & B^* \\ C^* & A^* \end{pmatrix} \qquad ; A, B, C, D \in \mathrm{M}_2(\mathbb{H})$$

To see this, observe that τ is an anti-involution of $\mathrm{M}_4(\mathbb{H})$ which fixes all matrices of the form (6.6). We are going to determine the group Cl_+ consisting of units in the even part of the Clifford algebra which conjugates the space matrices of the form (6.6) into itself. We claim that Cl_+ is the group of matrices of the form

$$(6.8) \qquad \begin{pmatrix} G & 0 \\ 0 & gG^{*-1} \end{pmatrix} \qquad ; G \in \mathrm{Gl}_2(\mathbb{H}), g \in \mathbb{R}^*, g^2 = det(G)$$

Indeed, for a matrix of this form we find using 6.5 that

$$\begin{pmatrix} G & 0 \\ 0 & gG^{*-1} \end{pmatrix}\begin{pmatrix} 0 & X \\ \breve{X} & 0 \end{pmatrix}\begin{pmatrix} G^{-1} & 0 \\ 0 & g^{-1}G^* \end{pmatrix} = g^{-1}\begin{pmatrix} 0 & GXG^* \\ (GXG^*)\breve{} & 0 \end{pmatrix}$$

Conversely, for a matrix $U \in Cl_+$ we have $\tau(U)U = g, g \in \mathbb{R}^*$. In other words

$$\begin{pmatrix} G & 0 \\ 0 & K \end{pmatrix}\begin{pmatrix} K^* & 0 \\ 0 & G^* \end{pmatrix} = g$$

This gives at once that U has the form (6.8). The matrix (6.8) has *spinor norm* g, and its action on W is given by $GXG^*g^{-1}, X \in W$. It follows that we can identify the Spin and the group $\mathrm{Sl}_2(\mathbb{H})$ of matrices X with $det(X) = 1$

$$(6.9) \qquad \mathrm{Spin}(-5, 1) = \mathrm{Sl}_2(\mathbb{H})$$

To a non-zero vector $\begin{pmatrix} z \\ w \end{pmatrix}$ in $\mathbb{H}^2$ we assign a Hermitian matrix

$$(6.10) \qquad L\begin{pmatrix} z \\ w \end{pmatrix} = \begin{pmatrix} zz^* & zw^* \\ wz^* & ww^* \end{pmatrix} = \begin{pmatrix} z \\ w \end{pmatrix}\begin{pmatrix} z^* \\ w^* \end{pmatrix}^T \qquad ; z, w \in \mathbb{H}$$

Observe that Hermitian matrices of this form are isotropic vectors for the quadratic form Q. Besides we ask the reader to verify the transformation rule

$$(6.11) \qquad L\begin{pmatrix} zs \\ ws \end{pmatrix} = ss^* \begin{pmatrix} z \\ w \end{pmatrix} \qquad ; s \in \mathbb{H}^*$$

Observe that L induces an isomorphism between $P^2(\mathbb{H})$ and the isotropic cone $PC(W)$, where W is the space of Hermitian matrices in $\mathrm{M}_2(\mathbb{H})$. Note also that

$$(6.12) \qquad L(\sigma(R)) = \sigma L(R)\sigma^* \qquad ; \sigma \in \mathrm{Sl}_2(\mathbb{H}), R \in \mathbb{H}^2$$

In conclusion, we have identified the natural action of $\mathrm{Sl}_2(\mathbb{H})$ on $P^2(\mathbb{H})$ with the action of $\mathrm{Sl}_2(\mathbb{H})$ on the space ∂H^5 of isotropic lines in W.

BIBLIOGRAPHY

R. Deheuvels, *Formes quadratiques et groupes classiques*, Presses Universitaire de France, Paris 1981.

J.E. Gilbert, M.A.Murray, *Clifford algebras and Dirac operators in harmonic analysis*, Cambridge University Press, Cambridge 1989.

A.J.Hahn, *Quadratic Algebras, Clifford algebras and Arithmetic Witt groups*, Springer-Verlag, New York 1991.

B. Iversen, *Hyperbolic geometry*, London Math. Student text 25, Cambridge University Press, Cambridge 1992.

M.A. Knus, *Quadratic and Hermitian forms over rings*, Grundlehren der Math. Wissenshaften 294, Springer-Verlag, Berlin 1991.

T.Y. Lam, *The algebraic theory of quadratic forms*, W.A. Benjamin, Reading Massachusetts 1973.

I.R. Porteous, *Topological geometry*, Cambridge University Press, 2nd ed. Cambridge 1981.

W. Scharlau, *Quadratic and Hermitian forms*, Springer Verlag, Berlin 1985.

I.R. Shafarevich, *Algebra I. Encyclopaedia of Mathematical Sciences vol. 11*, Springer Verlag, Berlin 1985.

C.T.C. Wall, *Graded Brauer groups*, J. reine angew. Math. **213** (1964), 187-199.

DEPARTMENT OF MATHEMATICS, UNIVERSITY OF AARHUS, NY MUNKEGADE, DK-8000 AARHUS C

E-mail address: birger@mi.aau.dk

Contemporary Mathematics
Volume **203**, 1997

Linear Endomorphisms of Clifford Algebras

D. Kastler and M. Mebkhout

ABSTRACT. Linear endomorphisms of Clifford algebras are described by bilateral translations.

In the early stage of the Connes-Lott model it was believed that tensoring over the quaternions occured in the theory [**1**]. Later this (a priori fascinating) eventuality proved to be elusive. But it induced us for a time to perform an extensive search for quaternionic structures, a pursuit for which it was a relief to escape the awkward use of 16×16 matrices, using instead the following handy form of linear endomorphisms of $\mathbb{H}$:

PROPOSITION 1. Denoting by $\mathbb{H}$ the real vector space of quaternions, and with the operator $M_{q,q'}\colon \mathbb{H} \to \mathbb{H}$ defined as follows:

$$M_{q,q'}q'' = qq''q' \,, \qquad q, q', q'' \in \mathbb{H}, \tag{1}$$

one gets a basis of the real vector space of $\mathbb{R}$-linear operators of $\mathbb{H}$ by taking the set

$$\begin{aligned} \{M_{\mathbb{1},\mathbb{1}},\ M_{I,I},\ M_{J,J},\ M_{K,K},\ M_{\mathbb{1},I},\ M_{I,\mathbb{1}},\ M_{J,K},\ M_{K,J},& \\ M_{\mathbb{1},J},\ M_{J,\mathbb{1}},\ M_{K,I},\ M_{I,K},\ M_{\mathbb{1},K},\ M_{K,\mathbb{1}},\ M_{I,J},\ M_{J,I}\}.& \end{aligned} \tag{2}$$

Consequently the linear endomorphisms of $\mathbb{H}$ are the $M_{q,q'}$, $q, q' \in \mathbb{H}$.

Proof: The $M_{q,q'}$ are clearly $\mathbb{R}$-linear. The fact that the 16 operators (2) are a linearly independent set results from the following table giving the values of these

1991 *Mathematics Subject Classification.* 15A66, 11E88.

operators on the basis elements $\mathbb{1}$, I, J, K, of $\mathbb{H}$:

$$
(3)\qquad
\begin{array}{lcccc}
 & \mathbb{1} & I & J & K \\
 & & & & \\
M_{\mathbb{1},\mathbb{1}} & \mathbb{1} & I & J & K \\
M_{I,I} & -\mathbb{1} & -I & J & K \\
M_{J,J} & -\mathbb{1} & I & -J & K \\
M_{K,K} & -\mathbb{1} & I & J & -K \\
 & & & & \\
M_{\mathbb{1},I} & I & -\mathbb{1} & K & -J \\
M_{I,\mathbb{1}} & I & -\mathbb{1} & -K & J \\
M_{J,K} & -I & -\mathbb{1} & -K & -J \\
M_{K,J} & I & \mathbb{1} & -K & -J \\
 & & & & \\
M_{\mathbb{1},J} & J & -K & -\mathbb{1} & I \\
M_{J,\mathbb{1}} & J & K & -\mathbb{1} & -I \\
M_{K,I} & -J & -K & -\mathbb{1} & -I \\
M_{I,K} & J & -K & \mathbb{1} & -I \\
 & & & & \\
M_{\mathbb{1},K} & K & J & -I & -\mathbb{1} \\
M_{K,\mathbb{1}} & K & -J & I & -\mathbb{1} \\
M_{I,J} & -K & -J & -I & -\mathbb{1} \\
M_{J,I} & K & -J & -I & \mathbb{1}
\end{array}
$$

The claimed linear independence then follows from invertibility of the matrix

$$
\begin{pmatrix}
-1 & 1 & 1 & 1 \\
1 & -1 & 1 & 1 \\
1 & 1 & -1 & 1 \\
1 & 1 & 1 & -1
\end{pmatrix}
$$

Conversations with colleagues seeming to indicate that this result is not notorious, we proposed to M. Laoues to look at bilateral translations in Clifford algebras $Cl(V)$, V a n-dimensional pseudo-euclidean vector space. For n even these maps were found to generalize the property above, whilst a slightly more involved result holds in the odd case [**2**]. We thought it worthwhile to mention this elementary tool to the engeneers of the Clifford industry in a book devoted to the memory of W.K. Clifford.

References

[1] A. Connes, J. Lott, Particle models and Non-commutative Geometry, *Nucl. Phys. B (Proc. Suppl.)*, **18 B** (1990) 29-47.

[2] M. Laoues, Linear operators in Clifford algebras, *Letters in Mathematical Physics*, **23** (1991) 325-331.

Centre de Physique Théorique, Unité Propre de Recherche 7061, CNRS Luminy, Case 907, F-13288 Marseille – Cedex 9

Part II

RIEMANNIAN SURFACES

Contemporary Mathematics
Volume **203**, 1997

Uniformizations of Riemann Surfaces: Poincaré Theta Series, Riemann's Theta Function and Theta Constants

I. KRA

1. Introduction

Although his major interests were in other directions, William Kingdon Clifford made a number of advances in the theory of Riemann surfaces. My biggest contribution to a conference in his honor whose participants are mostly mathematical physicists might be the exposition of a small part of the theory of moduli of Riemann surfaces – not only because of its inherent beauty, but also because it might have some applications to string theory. It is only slightly surprising that some of the open problems in the field were already identified when Clifford made some of his most important contributions to mathematics and other sciences.

We will be discussing the following problem. Given a Riemann surface X of finite conformal type (p, n) with singularities (perhaps), find complete invariants for it (this is usually a point in $\mathbb{C}^{3p-3+n}$). Ideally, we should have a recipe for computing this invariant from the presentation of the surface, and conversely from this invariant we should be able to reconstruct the surface X and any interesting property of X (for example, its canonical metric of constant curvature and its period matrix). Needless to say, this problem is far from solution. From the conformal invariant of type of a surface, one constructs a rougher topological invariant; the negative of the Euler characteristic

$$-\chi = 2p - 2 + n > 0.$$

There are no (orientable) surfaces with $\chi > 2$. There is only one Riemann surface with $\chi = 2$; the Riemann sphere

$$S^2 = \{(x, y, t) \in \mathbb{R}^3;\ x^2 + y^2 + t^2 = 1\} \cong \hat{\mathbb{C}} = \mathbb{C} \cup \{\infty\}$$

– the unique compact surface with a metric of positive curvature. The zero curvature situation corresponds to the case of the complex plane $\mathbb{C}$ or punctured Riemann sphere ($\chi = 1$), the cylinder or punctured plane or twice punctured sphere ($(p, n) = (0, 2)$), and the one (complex) parameter family of tori ($(p, n) = (1, 0)$).

1991 *Mathematics Subject Classification.* Primary 30F10, 30F35, 20H10, 11F27; Secondary 32G15, 14H05, 30F40.

Research supported in part by NSF Grant DMS 9500557.

So far we have discussed only the classical case; studied and "solved" well before this century.

The interesting open case concerns negative curvature surfaces. For most problems it suffices to consider surfaces whose only singularities are nodes. Thus for surfaces X that we are discussing, a nonsingular point on X has a neighborhood conformally equivalent to

$$\{z \in \mathbb{C};\ |z| < 1\};$$

while a singular point on X lies in the intersection of a neighborhood of $0 \in \mathbb{C}^2$ with the coordinate axes

$$\{(z, w) \in \mathbb{C}^2;\ zw = 0\}.$$

An equivalent problem is that of determining whether two given surfaces X and Y are or are not conformally equivalent. This is a particularly difficult question since, for example, X might be given as D/G, the quotient of a plane domain D by a group of Möbius transformations G, and Y might be expressed as a projective algebraic curve. It is only slightly less ambitious to construct the compactified moduli space[1] $\overline{\mathbf{R}(p,n)}$[2] of surfaces of conformal type (p, n) and a fibration

$$\pi : \overline{\mathbf{W}(p,n)} \rightarrow \overline{\mathbf{R}(p,n)}$$

with $\pi^{-1}(t)$ the Riemann surface corresponding to $t \in \overline{\mathbf{R}(p,n)}$ (factored by its automorphism group), and then to describe the equation of the algebraic curve $\pi^{-1}(t)$. It is traditional and much easier to work with the *Teichmüller curve*

$$\pi : \mathbf{V}(p,n) \rightarrow \mathbf{T}(p,n)$$

(here a point t in the base stands for a "marked" surface and the fiber over t consists of the surface), and various intermediate moduli spaces lying below $\mathbf{T}(p,n)$ and above $\mathbf{R}(p,n)$. The points of the *Teichmüller space* $\mathbf{T}(p,n)$ represent surfaces of conformal type (p,n)[3] marked by a choice of generators for the fundamental group of the surface. There is a group $\text{Mod}(p,n)$, the *Teichmüller modular group*, that acts on $\mathbf{T}(p,n)$ to produce the *Riemann space* $\mathbf{R}(p,n)$.

This is a good point at which we should define what we mean by a uniformization of a Riemann surface X of conformal type (p, n). If X is given as the quotient of a domain $D \subset \hat{\mathbb{C}}$ by a discontinuous group G, then we *uniformize* X by describing the algebraic curve corresponding to its compactification $\overline{X}$ and locating the punctures $\overline{X} - X$ on this curve. Conversely, we *uniformize* an algebraic curve (or more generally, the complement of a finite set on an algebraic curve) by describing it as the quotient of a plane domain by a group of Möbius transformations. That such dual descriptions of Riemann surfaces exist is usually called the uniformization theorem and forms a central chapter in any study of Riemann surfaces (see, for example, [**23**]). In particular, we can always choose D to be either $\hat{\mathbb{C}}$, $\mathbb{C}$ or the upper half plane

$$\mathbb{H}^2 = \{z \in \mathbb{C};\ \Im z > 0\}.$$

In the latter ("generic") case, G is a discrete subgroup of $\text{PSL}(2, \mathbb{R})$.

---

[1]The points of this space are conformal equivalence classes of Riemann surfaces.

[2]The bar in the symbol $\overline{\{\cdot\}}$ is meant to denote that the object under consideration $\{\cdot\}$ is being compactified; usually by the addition of lower dimensional subspaces at infinity.

[3]That is, n-punctured closed surfaces of genus p. Punctures are distinguished from holes analytically and quasiconformally; not topologically.

We are assuming that the reader is familiar with the basic facts concerning algebra, topology and analysis (including functional analysis). In addition some familiarity with Riemann surfaces and actions of groups of manifolds is needed. I am trying to address someone whose mathematical background is roughly at the level of a graduate student who has completed one full year of study and possesses considerably more maturity and sophistication[4]. This paper is organized as follows. In §2 we discuss what is meant by good coordinates for moduli spaces and describe briefly the important results at the foundation of the subject; §3 sets up the notation and background for a discussion of the most basic ideas in the theory of Kleinian groups. The next section, §4, discusses examples of the the key constructions of families of Kleinian groups. We devote §5 to a study of the genus one case. This classical case, presented in a modern setting, illustrates the direction for the general theory. We also define a set of transcendental functions (θ-functions and θ-constants) that will be used to uniformize surfaces represented by subgroups of $\mathrm{PSL}(2,\mathbb{Z})$. In §6, we describe the elementary plumbing constructions that can be used to produce all tori. These constructions can be generalized to arbitrary finite type surfaces. The topological aspects of this construction are controlled by a (generalization of a) trivalent graph described in §7. The analytic aspects of the construction are controlled by a set of plumbing parameters. These and the deformation spaces that they lead to are described in §8. Various Teichmüller spaces are discussed in §9. A detailed description of good real analytic coordinates for moduli spaces of punctured tori are treated in §10, and §11 is devoted to the uniformization of surfaces represented by certain subgroups of the (classical) modular group $\mathrm{PSL}(2,\mathbb{Z})$ using theta constants.

2. Intrinsic coordinates for moduli spaces

We will describe the Earle-Marden coordinates [**18**] on moduli spaces. We are limiting our discussion to complex analytic (except in §10) coordinates[5]. The earliest complex analytic coordinates for Teichmüller spaces for compact surfaces[6] were constructed from period matrices by Ahlfors [**3**]. The most general construction of complex moduli [**10**] is based on the idea of Bers [**8**] of uniformizing surfaces by quasiFuchsian groups[7]. This *Bers embedding*[8] as well as all other "good" coordinates on the Teichmüller space $\mathbf{T}(p,n)$ are constructed by choosing a reasonable family of Kleinian groups that represents all surfaces of type (p,n)[9]. The Bers embedding leads to nonintrinsic or variational coordinates on $\mathbf{T}(p,n)$ since they depend on the choice of a base point. The search for intrinsic coordinates produced results in the seventies. Among the successful attempts were those due to Maskit [**35**] (see also [**28**]), Bers [**15**], Earle [**17**] and Earle-Marden [**18**]. We describe in some detail only the last set, because it deals most naturally with noded surfaces

[4]Not always requiring, for example, precise definitions of all concepts.

[5]Based on the assumption that the space of complex analytic structures on a topological surface, should itself have a complex structure. For a discussion of real analytic coordinates on $\mathbf{T}(p,n)$ we refer the reader to the monograph [**1**].

[6]The case $n=0$.

[7]This method leads to complex structures on all Teichmüller spaces – including the infinite dimensional ones.

[8]Every such embedding endows the Teichmüller space with an interesting relative boundary.

[9]And perhaps, other surfaces as well.

(Riemann surfaces with singularities[10]). This and indeed most, if not all, constructions of moduli spaces are based on fundamental work of Ahlfors, Bers and Maskit [**2**], [**7**], [**5**] and [**32**]; among these:

(a) (The generalized Riemann Mapping Theorem (RMT)) Existence and uniqueness of homeomorphic solutions of the Beltrami equation: For each $\mu \in L^\infty(\mathbb{C})$ with $||\mu||_\infty < 1$, there exists a unique normalized (fixing three points on $\hat{\mathbb{C}}$; for example: 0, 1 and ∞) μ-conformal (a homeomorphic solution of the Beltrami equation $w_{\bar{z}} = \mu w_z$) automorphism of $\hat{\mathbb{C}}$, w^μ, [**36**] (see [**5**] for a "modern" proof).[11]

(b) Holomorphic dependence of solution on parameters, [**5**]: If μ varies holomorphically in the Banach space $L^\infty(\mathbb{C})$, then w^μ varies holomorphically in an appropriate Frèchet space; in particular for fixed $\mu \in L^\infty(\mathbb{C})$ and fixed $z \in \mathbb{C}$,

$$t \mapsto w^{t\mu}(z)$$

is a holomorphic function (in the ordinary classical sense on the disc $\left\{t \in \mathbb{C};\ t < \frac{1}{||\mu||_\infty}\right\}$).

(c) Klein-Maskit combination theorems; that is, construction of "complicated" Kleinian groups from "simpler" groups by three operations: free products[12], amalgamated free products (AFP's) and Higman-Neumann-Neumann (HNN) extensions [**33**].

(d) Bers (μ) trick [**8**]: The uniformization of a fixed finite set of surfaces S_o (of finite conformal type) by a Kleinian group G_o leads by variation of complex structure to a uniformization of an arbitrary set of surfaces S_t that are quasiconformally equivalent to S_o by a group G_t quasiconformally conjugate to G_o.

3. Kleinian groups

We will be working with finitely generated *Kleinian* groups; that is, subgroups G of PSL$(2,\mathbb{C})$ acting discontinuously on a part Ω[13] of the Riemann sphere $\hat{\mathbb{C}} = \mathbb{C} \cup \{\infty\}$. For convenience, Kleinian groups are represented pictorially by their fundamental domains. We define a *fundamental domain* for the action of G on Ω as an open set $\omega \subset \Omega$ bounded by finitely many arcs with the property that no two points of ω are G-equivalent and every point of Ω is G-equivalent to at least one point in the closure of ω in Ω. Thus the orbit space Ω/G is a finite union of Riemann surfaces obtained from ω by some boundary identifications. By a major result due to Ahlfors [**4**], Ω/G can be compactified by the addition of finitely many points. A special roles in the theory is occupied by the special classes of *Fuchsian* (PSL$(2,\mathbb{C})$-conjugate to a discrete subgroup of PSL$(2,\mathbb{R})$) and *quasiFuchsian* (quasiconformally conjugate to a Fuchsian group) groups. In general, the region of discontinuity Ω of

[10] Points corresponding to theses surfaces suffice to compactify the moduli space of nonsingular surfaces.

[11] Quasiconformality can be defined without the use of the Beltrami equation. The uniqueness statement is easy to prove. Existence requires hard work. See [**29**] and [**30**, Ch. I] for details on proofs and historical remarks.

[12] The classical case treated by Klein [**25**].

[13] The biggest open set on which the Kleinian group G acts discontinuously is usually denoted by $\Omega = \Omega(G)$.

a Kleinian group G is not connected, and the components of Ω are permuted by G. The group G is a *function* group if there is a component Δ of Ω that is invariant under G; the function group G is *b*-group if its invariant component Δ is simply connected. Finitely generated function groups are much better understood than arbitrary Kleinian groups [**34**]. We will entirely ignore the connection of Kleinian groups to three dimensional hyperbolic geometry. For this direction, see [**31**] and [**38**].

For simplicity, all Kleinian groups are assumed to be torsion free (that is, we are assuming that our groups do not contain any elements of finite order; other than the identity, of course). By a celebrated theorem of Selberg [**37**] every finitely generated matrix (hence in particular, Kleinian) group contains a normal torsion free subgroup of finite index. This result allows us to reduce many questions concerning Kleinian groups to the special case of torsion free groups.

4. Some examples of Kleinian groups; Use of the Klein-Maskit combination theorems

We illustrate with the simplest cases of Kleinian groups, and use the combination theorems – also in the simplest cases – to uniformize various Riemann surfaces.

4.1. Rank two parabolic groups. For $\tau \in \mathbb{H}^2 = \{z \in \mathbb{C};\ \Im z > 0\} \cong \mathbf{T}(1,0)$, let G_τ be the group generated by the translations

$$A : z \mapsto z + 1 \text{ and } B : z \mapsto z + \tau.$$

Möbius transformations are also represented by elements of $\mathrm{SL}(2,\mathbb{C})$[14]. Thus[15]

$$A = \begin{bmatrix} 1 & 1 \\ 0 & 1 \end{bmatrix} \text{ and } B = \begin{bmatrix} 1 & \tau \\ 0 & 1 \end{bmatrix}.$$

The group G_τ acts properly discontinuously on $\mathbb{C}$ (that is, $\Omega(G_\tau) = \mathbb{C}$); and the *period* parallelogram $\mathcal{P}$ with vertices $0,\ 1,\ 1+\tau,\ \tau$ (illustrated in Figure 1a) forms a fundamental domain for the rank two group of translations G_τ. Thus $T_\tau = \mathbb{C}/G_\tau$ is a torus (a surface of type $(1,0)$) with *modulus* τ. Given any $x \in \mathbb{C}$, we can construct a *symmetric* parallelogram $\mathcal{P}_x$ centered at x with vertices $x - \frac{1+\tau}{2}$, $x + \frac{1-\tau}{2}$, $x - \frac{1-\tau}{2}$ and $x + \frac{1+\tau}{2}$, that also serves as a fundamental domain for G_τ.

The most general marked[16] rank two parabolic group G is conjugate to a unique G_τ. If $G = C\ G_\tau\ C^{-1}$, where the Möbius transformation C sends infinity to the finite point $x \in \mathbb{C}$, then $\Omega(G) = \hat{\mathbb{C}} - \{x\}$. It is easy to write down generators for the group G; we can choose

$$C = \begin{bmatrix} x & -1 + xd \\ 1 & d \end{bmatrix},$$

and conclude that

$$\begin{bmatrix} 1 - x\alpha_1 & x^2\alpha_1 \\ -\alpha_1 & 1 + x\alpha_1 \end{bmatrix} \text{ and } \begin{bmatrix} 1 - x\alpha_2 & x^2\alpha_2 \\ -\alpha_2 & 1 + x\alpha_2 \end{bmatrix}$$

---

[14]The group of Möbius transformations is isomorphic to $\mathrm{PSL}(2,\mathbb{C}) \cong \mathrm{SL}(2,\mathbb{C})/\{\pm I\}$.

[15]By abuse of language, we use the same symbol to represent a Möbius transformation and its associated 2×2 matrix.

[16]The marking consists of a choice of generators for G that are mapped to A and $B \in G_\tau$ respectively. The marking is needed for the uniqueness statement.

generate the group G. Further, this group represents a torus with modulus $\frac{\alpha_2}{\alpha_1}$. A fundamental domain ω for G consists of the exterior of 4 circles passing through x as shown in Figure 1b. Function theory on the torus is well understood. We describe one approach in §5. Given a torus and a point on it, then there exists a unique conformal involution of the torus fixing the given point. This allows us to identify the function theory on the zero curvature surfaces of type $(1,0)$ with that on the negative curvature surfaces $(1,1)$. This simple observation serves as a rough guide for determining which results in genus one translate to higher genus.

4.2. Free products: I. It is a result of Klein **[25]** that under appropriate hypothesis the free product Γ of two Kleinian groups Γ_1 and Γ_2 is again Kleinian. Further, the surfaces represented by Γ can be read off from those represented by Γ_1 and Γ_2. Thus, for example (Figure 2a), the free product Γ of G_τ and G will represent a surface Y of genus two. Necessary and sufficient geometric conditions for a Kleinian group Γ to be the free product of two other Kleinian groups are not known. However, sufficient conditions are easy to obtain. In Klein's original formulation, he assumes that a fundamental domain of one group is contained in the exterior of the fundamental domain of the other group. It is easy to describe this construction topologically in the example we have illustrated. We have removed a disc from each of the tori represented by the groups G_τ and G and identified (sewn together) the corresponding boundary circles. It is much harder to obtain the complex analytic description of the construction; for example, the period matrix[17] of Y. It must involve transcendental functions. From our construction it would appear that coordinates for some moduli[18] space for genus two surfaces would be: the modulus of G_τ (that is, τ), the modulus of G ($\frac{\alpha_2}{\alpha_1}$) and the common fixed point (x) of this group. It would take us too far afield to explain why this is not so.

4.3. Free products: II. As a second example, we look at the free product Γ (Figure 2b) of a Fuchsian group G_1 of type $(1,1)$ and a rank two parabolic group G_2. The function group Γ will represent two Riemann surfaces: one of type $(2,1)$; the second of type $(1,1)$. When we vary the group Γ, the resulting group is still a free product; but the Fuchsian subgroup corresponding to G_1 has in general been replaced by a quasiFuchsian one. While we can control the construction so that the subgroup corresponding to G_1 stays Fuchsian (see §10), the resulting groups do not lead to complex analytic moduli, but only to real analytic ones. The topological interpretation of this construction is identical to the one for the previous case. The Fuchsian group G_1 represents two punctured tori T_1 and T_2. The group G_2 represents a torus T. We have, once again, removed a disc from one of the punctured tori (say T_1) and from the torus and identified the resulting boundaries to produce a punctured genus two surface and a punctured torus.

4.4. AFP: Maskit I. The general amalgamated free product (AFP) construction produces a group Γ as the product of two groups Γ_1 and Γ_2 across a common (often, cyclic) subgroup H. In Figure 3, we illustrate an AFP of two Fuchsian groups of type $(0,3)$ across a common cyclic parabolic subgroup $< A >$. We label the surfaces represented by Γ_1 (Γ_2) as S_1 and S_3 (S_2 and S_4). The motion

[17]The *period matrix* of a surface of genus $p \geq 1$ is a $p \times p$ symmetric matrix with positive definite imaginary part that describes, using suitable normalizations, all integrals of holomorphic 1-forms along closed paths.

[18]Technically, for a quotient of the Teichmüller space $\mathbf{T}(2,0)$.

A represents a puncture on one of the surfaces represented by each of Γ_1 and Γ_2; say, S_1 and S_2. The resulting group represents three surfaces: one of type $(0,4)$ (obtained topologically by removing punctured discs from S_1 and S_2 and identifying their boundary curves) and two of type $(0,3)$ (the surfaces S_3 and S_4). This construction has two novel features. Variation of the complex structure of Γ leads to another group that is constucted from the same type of ingredients (we do NOT leave the Fuchsian portion of the quasiFuchsian universe), and the group represents only one surface with moduli (the four punctured sphere)[19]. The group Γ can hence be used to produce intrinsic coordinates for $\mathbf{T}(0,4)$. See [**35**], [**27**] and [**28**].

4.5. HNN: Maskit II. The general Higman-Neumann-Neumann (HNN) construction deals with a Kleinian group G that contains two subgroups H_1 and H_2 (usually cyclic) which are not conjugate in G, but are $\mathrm{PSL}(2,\mathbb{C})$-conjugate. One chooses an element $C \in \mathrm{PSL}(2,\mathbb{C})$ that conjugates one of these to the other (say $H_2 = CH_1C^{-1}$) and constructs a new group Γ by adjoining the motion C to the group G. We have illustrated in Figure 4, an HNN extension of a Fuchsian triangle group G (that is, of a group of type $(0,3)$) by a motion that conjugates one parabolic cyclic subgroup of G onto another. The resulting group Γ represents a once punctured torus and a thrice punctured sphere. This construction enjoys all the nice features listed an the end of the previous section and provides the basis for a good model of $\mathbf{T}(1,1)$. Every once punctured torus admits a unique conformal involution. This allows us to pick out three distinguished points on the punctured torus; namely, the fixed points of the involution. The quotient of the punctured torus by the involution is a punctured sphere with three distinguished points; allowing us to identify function theory on surfaces of type $(1,1)$ and $(0,4)$.

An arbitrary surface of hyperbolic type (p,n) can be built up from $2p-2+n$ triangle groups by $2p-3+n$ AFP's followed by p HNN's. The topology/combinatorics of this construction is controlled by a graph (see §7); the analytic aspects by plumbing parameters (see §6). The full details may be found in [**28**]. The construction also yields an injective holomorphic map of $\mathbf{T}(p,n)$ into the cartesian product of $3p-3+n$ one dimensional Teichmüller spaces; each of which is either $\mathbf{T}(0,4)$ or $\mathbf{T}(1,1)$. This map is NOT surjective if $3p+n > 4$; but DOES define a so far unstudied (probably interesting) relative boundary of $\mathbf{T}(p,n)$.

5. Elliptic function theory

For the purposes of orienting the reader, we discuss briefly the situation for surfaces of type $(1,0)$ – the well traveled elliptic paradise. We have already observed that the Teichmüller space $\mathbf{T}(1,0)$ can be naturally identified with the upper half plane $\mathbb{H}^2$. It is well known (see also below) that the Riemann space of tori, $\mathbf{R}(1,0) = \mathbb{H}^2/\,\mathrm{SL}(2,\mathbb{Z})$, is conformally equivalent to the plane $\mathbb{C}$ (with 2 distinguished points). The semi-direct product $\mathbb{Z}^2 \propto \mathrm{SL}(2,\mathbb{Z})$ acts on $\mathbb{C}\times\mathbb{H}^2$:

$$(z,\tau)\in\mathbb{C}\times\mathbb{H}^2,\ (n,m)\in\mathbb{Z}^2,\ \gamma=\begin{bmatrix} a & b\\ c & d\end{bmatrix}\in\ \mathrm{SL}(2,\mathbb{Z}),$$

$$(n,m)(z,\tau)=(z+n+m\tau,\tau),$$

[19]Thrice punctured spheres are rigid; a reflection of the fact that the group of Möbius transformations is triply transitive.

$$\gamma(z,\tau)=\left(\frac{z}{c\tau+d},\frac{a\tau+b}{c\tau+d}\right).$$

This is a good place to introduce some more terminology and to define some important functions. These functions will, among other things, make it useful to study the above group actions. A *characteristic*[20] $\chi=\begin{bmatrix}\epsilon\\ \epsilon'\end{bmatrix}$ is merely a vector in $\mathbb{R}^2$. There is an action of $\mathrm{SL}(2,\mathbb{Z})$ on characteristics defined by

$$\chi\gamma=\begin{bmatrix}a\epsilon+c\epsilon'-ac\\ b\epsilon+d\epsilon'+bd\end{bmatrix}.$$

This is NOT a group action of $\mathrm{SL}(2,\mathbb{Z})$ on $\mathbb{R}^2$. But we can introduce (see [**24**]) an equivalence relation on $\mathbb{R}^2$ so that the above action becomes a group action of $\mathrm{PSL}(2,\mathbb{Z})$ on classes of characteristics ($\mathbb{R}^2$ modulo the equivalence relation). Riemann's *theta* (θ-) function with characteristic χ is defined by the convergent series on $\mathbb{C}\times\mathbb{H}^2$:

$$\theta[\chi](z,\tau)=\sum_{n\in\mathbb{Z}}\exp 2\pi\imath\left[\frac{1}{2}\left(n+\frac{\epsilon}{2}\right)^2\tau+\left(n+\frac{\epsilon}{2}\right)\left(z+\frac{\epsilon'}{2}\right)\right].$$

The semi-direct product of $\mathbb{Z}\oplus\mathbb{Z}$ with $\mathrm{SL}(2,\mathbb{Z})$ acts on the set of θ-functions. For fixed values of the second variable $\tau\in\mathbb{H}^2$, $\theta[\chi](\cdot,\tau)$ is almost invariant under the group of translations:

$$\theta\begin{bmatrix}\epsilon\\ \epsilon'\end{bmatrix}(z+n+m\tau,\tau)=\exp\left\{\frac{n\epsilon-m\epsilon'}{2}-mz-\frac{m^2}{2}\tau\right\}\theta\begin{bmatrix}\epsilon\\ \epsilon'\end{bmatrix}(z,\tau).$$

The value of the theta function at $z=0$ is called a *theta constant.* It is a holomorphic function on $\mathbb{H}^2$. We abbreviate

$$\theta[\chi](0,\cdot)=\theta[\chi].$$

The transformation theory for theta functions under the modular group is more interesting and more mysterious. With the usual symbols for the variables,

$$\frac{\exp\left\{\pi\imath\left(\frac{-cz^2}{c\tau+d}\right)\right\}\theta\begin{bmatrix}\epsilon\\ \epsilon'\end{bmatrix}\left(\frac{z}{c\tau+d},\frac{a\tau+b}{c\tau+d}\right)}{\theta\begin{bmatrix}a\epsilon+c\epsilon'-ac\\ b\epsilon+d\epsilon'+bd\end{bmatrix}(z,\tau)}=\kappa(\chi,\gamma)\,(c\tau+d)^{\frac{1}{2}},$$

where $\kappa(\chi,\gamma)$ is a constant of absolute value one that can be computed. It is a root of unity if χ is rational. The action of $\mathrm{SL}(2,\mathbb{Z})$ on theta constants is easier to understand:

$$\theta[\chi](0,\gamma(\tau))\gamma'(\tau)^{\frac{1}{4}}=\kappa(\chi,\gamma)\theta[\chi\gamma](0,\tau).$$

We can use θ-functions to uniformize all tori simultaneously by producing a map from the factor space

$$\mathbf{V}(1,0)=(\mathbb{C}\times\mathbb{H}^2)/(\mathbb{Z}\oplus\mathbb{Z})$$

to $\mathbf{P}\mathbb{C}^2$ by sending $(z,\tau)\in\mathbb{C}\times\mathbb{H}^2$ to the projective equivalence class of the point

$$\left(\theta^3\begin{bmatrix}1\\1\end{bmatrix},\theta^2\begin{bmatrix}0\\0\end{bmatrix}\theta\begin{bmatrix}1\\1\end{bmatrix},\theta\begin{bmatrix}0\\0\end{bmatrix}\theta\begin{bmatrix}0\\1\end{bmatrix}\theta\begin{bmatrix}1\\0\end{bmatrix}\right)(z,\tau)$$

[20] "Mark" in the terminology of Clifford [**16**].

in $\mathbb{C}^3 - \{0\}$. This map is injective on each fiber T_τ over $\tau \in \mathbb{H}^2$, and leads to the theta identity

$$\theta^4\begin{bmatrix}0\\0\end{bmatrix}\theta^2\begin{bmatrix}0\\1\end{bmatrix}(z,\tau)\,\theta^2\begin{bmatrix}1\\0\end{bmatrix}(z,\tau)=$$
$$\left(\theta^2\begin{bmatrix}1\\0\end{bmatrix}\theta^2\begin{bmatrix}0\\0\end{bmatrix}(z,\tau)-\theta^2\begin{bmatrix}0\\1\end{bmatrix}\theta^2\begin{bmatrix}1\\1\end{bmatrix}(z,\tau)\right)\times$$
$$\left(\theta^2\begin{bmatrix}0\\1\end{bmatrix}\theta^2\begin{bmatrix}0\\0\end{bmatrix}(z,\tau)+\theta^2\begin{bmatrix}1\\0\end{bmatrix}\theta^2\begin{bmatrix}1\\1\end{bmatrix}(z,\tau)\right).$$

The theta identity identifies the torus T_τ with the projective algebraic curve

$$\theta^4\begin{bmatrix}0\\0\end{bmatrix}Y^2Z=X\left(\theta^2\begin{bmatrix}1\\0\end{bmatrix}X-\theta^2\begin{bmatrix}0\\1\end{bmatrix}Z\right)\left(\theta^2\begin{bmatrix}0\\1\end{bmatrix}X+\theta^2\begin{bmatrix}1\\0\end{bmatrix}Z\right).$$

We note that the limiting case of this equation (as $\tau \to \imath\infty$) is

$$Y^2 + X^2 = 0.$$

6. Elementary plumbing

The construction of T_τ can be interpreted as a "plumbing." Our previous (classical) construction started with the simply connected (holomorphic universal) cover of the torus. Now we start with the infinite cylinder $\mathcal{C} = \mathbb{C}/ < A >$. This is topologically the most complicated surface (its fundamental group is isomorphic to $\mathbb{Z}$) of zero curvature without moduli (it is complex analytically equivalent to the punctured plane, also the twice punctured sphere, and is distinct from the punctured disc). The cylinder carries a metric of zero curvature. Perfectly good local coordinates each vanishing at one puncture of $\mathcal{C}$ are

$$z = e^{2\pi\imath\zeta} \text{ and } w = e^{-2\pi\imath\zeta},\ \zeta \in \mathbb{C}.$$

Fix a $t \in \mathbb{C},\ 0 < |t| < 1$. Form the truncated cylinder

$$\mathcal{C}' = \mathcal{C} - \{0 < |z| \leq |t|\} - \{0 < |w| \leq |t|\}.$$

Introduce an equivalence relation $\mathcal{R}$ on $\mathcal{C}'$

$$P\,\mathcal{R}\,Q \quad \text{if and only if} \quad z(P)w(Q) = t.$$

The *plumbed* surface is

$$S_t = \mathcal{C}'/\mathcal{R}.$$

It is an easy exercise to show that

$$T_\tau = S_{e^{2\pi\imath\tau}}.$$

We conclude from Abel's theorem or from the uniformization theorem that all tori are constructed by this "elementary plumbing." Our construction identifies a *central* curve on S_t; namely

$$\{|z| = 1\} = \{|w| = 1\}.$$

The length of central curve in the ("correct") metric (on S_t)

$$ds = \frac{-2\pi|dz|}{\log|t|}$$ [21]

[21] We normalized the metric so that the torus S_t has area 1.

is

$$\sqrt{\frac{-2\pi}{\log|t|}}.$$

These lengths approach zero as $t \to 0$. Hence it makes sense to extend our construction to include the case $t = 0$ by defining S_0 as the singular surface obtained from $\mathcal{C}$ by identifying the two punctures to form a node. The length of the curves "transverse" to the central curve are

$$\frac{|\log t|}{2\pi}\sqrt{\frac{-2\pi}{\log|t|}}^{22}.$$

It is easily checked that these curves become infinitely long in the limit ($t = 0$). Thus creation of a node exhibits some "catastrophic behaviour." We have already observed that

$$\mathbf{R}(1,0) = \mathbf{T}(1,0)/\mathrm{Mod}(1,0) = \mathbb{H}^2/\mathrm{PSL}(2,\mathbb{Z});$$

because one easily checks that

$$T_{\tau_1} \cong T_{\tau_2} \quad \text{if and only if} \quad \tau_2 = \frac{a\tau_1 + b}{c\tau_1 + d}$$

for some integers

$$a,\ b,\ c,\ d \text{ with } ad - bc = 1.$$

It follows that $\mathbf{R}(1,0)$ is a Riemann surface (really an orbifold with an underlying Riemann surface structure) with signature $(0,3;2,3,\infty)$. It is compactified by the addition of the puncture[23] – the modulus of the noded torus S_0. In the process of compactifying $\mathbf{R}(1,0)$, we constructed an intermediate moduli space: the punctured unit disc

$$\mathbf{T}(1,0)/ < \text{Dehn twists about central curve} > \cong \mathbb{H}^2/ < A >^{24}.$$

In this intermediate moduli space, we can see (some) degenerations ($t \to 0$). The limiting case corresponds to the puncture on the punctured unit disc.

7. Trivalent graphs

The next goal is clearly to carry out as much of above program (the one in §6) in the hyperbolic (rather than the elliptic) universe. The generalization from parabolic groups (that represent only the twice punctured sphere and tori) to more general Kleinian groups (representing arbitrary, with few well understood exceptions, surfaces) constitutes a major chapter of 20th century mathematics. See the expository articles by Bers [**11**] and [**15**]. The line of research (for the construction of compactified moduli spaces $\overline{\mathbf{R}(p,n)}$) we are following has two distinct parts: a topological/combinatorial one and an analytic one. We start with the much better understood topological aspects. We fix a hyperbolic

$$\text{type } = (p,n);$$

[22]The multivaluedness of the answer reflects the fact that there are countably many (homotopically) distinct curves transverse to the central curve.

[23]The other distinguished points on $\mathbf{R}(1,0)$ also have special meaning.

[24]A *Dehn twist* about a simple closed curve c on a surface S is a smooth version of the following operation. Cut the surface S along the curve c. In a neighborhood of one side of c, twist through 2π radians and glue back. In this operation, the complex structure has remained fixed, but the generators of the fundamental group of S changed.

here

$$p, \text{ the genus and } n, \text{ the number of punctures,}$$

are nonnegative integers with $2p-2+n>0$. The topology of our construction can be described either in terms of partitions of a surface or (alternatively) in terms of graphs. A (maximal) *partition* $\mathcal{P}$ of a surface S of type (p,n) is a collection of $3p-3+n$ disjoint simple closed curves on S with the property that no such curve bounds a disc or punctured disc on S and no pair of curves bound a cylinder. The complement in S of the curves in $\mathcal{P}$ is a disjoint union of $2p-2+n$ *pairs of pants* (thrice punctured spheres). From the partition $\mathcal{P}$, we construct an *admissible* graph $\mathcal{G}$ of type (p,n) consisting of $2p-2+n$ vertices, one for each pair of pants, and $d=3p-3+n$ edges; where an edge joins two vertices if and only if the edge corresponding to the partition curve is the common boundary of the two pairs of pants corresponding to the two vertices. There are $0\le e\le 3$ edges emanating from each vertex V. We now adjoin $3-e$ *phantom* edges to the vertex V (these edges start at the vertex and are open at the other end – they march off to infinity), and obtain our graph. It is easily checked that an admissible graph of type (p,n) has n phantom edges[25], and that a tubular neighborhood of $\mathcal{G}$ embedded in $\mathbb{R}^3$ is a topological model of the surface S with the partition $\mathcal{P}$. Figure 5 shows a partition of a surface of genus two and the corresponding graph. Figure 6 illustrates the simplest graphs – the unique graphs of type $(0,3)$, $(0,4)$, $(0,5)$ and $(1,1)$ as well as the two graphs of type $(1,2)$. The thrice punctured sphere (alternately the graph of type $(0,3)$), corresponding to the empty set in the construction of the integers, can be used as the basic building block in a construction of all surfaces of type (p,n). As usual, the topological aspects are rather straight forward. Given two graphs of types (p_1,n_1), $n_1>0$ and (p_2,n_2), $n_2>0$, we join two phantom edges to form an edge and thereby obtain the *AFP* of the two graphs, a graph of type (p_1+p_2,n_1+n_2-2). By joining two phantom edges on the same graph of type (p,n), $n>1$, we obtain the *HNN* of the graph, a graph of type $(p,n-2)$.

In our situation, the basic building block, the thrice punctured sphere corresponds topologically to graph of type $(0,3)$ and analytically to the $\mathrm{PSL}(2,\mathbb{C})$-conjugacy class of (torsion free) triangle groups. A "marked" triangle group is determined by a triple of distinct points in $\hat{\mathbb{C}}$ and can hence be denoted by the symbol $F(a,b,c)$[26]. From the parameters a, b and c, we can write down formulae for parabolic elements A, B and C that generate $F(a,b,c)$ subject to the defining relation $ABC=I$. Groups representing an arbitrary surface are built (inductively in a finite number of steps) from this family of triangle groups. We will describe three such constructions leading to three distinct families of Kleinian groups, but up to some elementary "translations" just one set of coordinates on Teichmüller space, [**26**]. We need, of course, to bring a parameter space into the picture.

8. Hyperbolic plumbing

At each puncture of a thrice punctured sphere, we introduce a distinguished local coordinate. This coordinate is uniquely determined up to a $\mathbb{Z}_2$ ambiguity. It comes from horocycles – not from the embedding of the thrice punctured sphere

[25] The graph $\mathcal{G}$ is trivalent in the sense described above.

[26] The generalization that includes torsion leads to a definition of groups depending on three symbols, integers ≥ 2 or ∞, and three distinct points in $\hat{\mathbb{C}}$. See [**6**].

into $\hat{\mathbb{C}}$. The group $F(\infty, 0, 1)$ acts on the upper half plane (as well as the lower half plane)[27] and it is generated by the parabolic motions[28]

$$A = \begin{bmatrix} 1 & 2 \\ 0 & 1 \end{bmatrix}, \; B = \begin{bmatrix} 1 & 0 \\ -2 & 1 \end{bmatrix}, \; C = \begin{bmatrix} 1 & -2 \\ 2 & -3 \end{bmatrix}$$

with fixed points ∞, 0 and 1, respectively, subject to the defining relation described above. For the puncture determined by infinity on the thrice punctured sphere $\mathbb{H}^2/F(\infty, 0, 1)$, the distinguished local coordinate is

$$z = \exp \pi \imath \zeta, \; \zeta \in \mathbb{C}, \; \Im\zeta > 1.$$

The reason for starting with the thrice punctured sphere is rather simple. It is the most complicated (topologically; its fundamental group is isomorphic to the free (noncommutative) group on two generators: $\mathbb{Z} \times \mathbb{Z}$) Riemann surface with a complete negative curvature metric and without moduli (it has a unique complex structure). A graph $\mathcal{G}$, which we fix once and for all, controls the topological (combinatorial) aspects of the construction of a surface of type (p, n) from $2p - 2 + n$ thrice punctured spheres. Plumbing parameters t from a simply connected domain $\mathbb{D}(\mathcal{G}) \subset \mathbb{C}^{3p-3+n}$ control the analytic (geometric) aspects. In contrast to the situation for the flat case $(p, n) = (1, 0)$, the exact[29] shape of this domain is not known. The points in $\mathbb{D}(\mathcal{G})$ represent nonsingular Riemann surfaces with a partition determined by the graph $\mathcal{G}$ along with the singular surfaces obtained by squeezing the partition curves to points. The nonsingular surfaces correspond to the subspace $\mathbb{D}_o(\mathcal{G}) \subset \mathbb{D}(\mathcal{G})$ of vectors without any zero components. Thus $\mathbb{D}_o(\mathcal{G})$ is the intersection of $\mathbb{D}(\mathcal{G})$ with the complement of the $3p - 3 + n$ coordinate hyperplanes:

$$\mathbb{D}_o(\mathcal{G}) = \{(z_1, \ldots, z_{3p-3+n}) \in \mathbb{D}(\mathcal{G});\; z_j \neq 0, \; j = 1, \ldots, 3p - 3 + n\}.$$

In the case of tori, a classical theorem (either the uniformization theorem or Abel's theorem) tells us that the plumbing construction gives us all tori. For hyperbolic type we need to use the full Ahlfors-Bers machinary involving quasiconformal maps – in particular, the results described in (a), (b) and (d) of §2.

It is convenient to introduce two topological invariants for the surface S_t. We let k be the number of nodes in S_t (= the number of zero components of the vector $t \in \mathbb{C}^{3p-3+n}$) and N be the number of connected components of the nonsingular points in S_t, which we will call the *parts* of the surface S_t.

We can obtain certain estimates for the size of $\mathbb{D}(\mathcal{G})$. Under appropriate normalizations,

$$\text{ball of radius } e^{-2\pi} \subset \mathbb{D}(\mathcal{G}) \subset \text{ball of radius } e^{-\frac{\pi}{2}}.$$

There are no known numerical conditions for a point $t \in \mathbb{C}^{3p-3+n}$ to belong to $\mathbb{D}(\mathcal{G})$. The plumbing constructions, however, have direct Kleinian group analogues. They can be translated to a set of Klein-Maskit combination constructions (involving only AFP's and HNN's). One starts with a collection of $2p - 2 + n$ triangle groups, a graph $\mathcal{G}$ of type (p, n), and a point $t \in \mathbb{D}_o(\mathcal{G})$. One constructs, as described in the next section, a group Γ_t inductively in $3p - 3 + n$ steps.

---

[27]We have in terminology to be introduced below

$$F(\infty, 0, 1) = \Gamma(2).$$

[28]Möbius transformations with one fixed point on $\hat{\mathbb{C}}$.

[29]For example, we know that this domain is Kobayashi complete; but we cannot evaluate its hyperbolic metric. Nor can we describe its group of complex analytic automorphisms.

These lead to "horocyclic" (= logarithms of plumbing parameters) coordinates for Teichmüller spaces and "plumbing" coordinates for (compactified) moduli spaces. The horocyclic coordinates are intrinsic (nonvariational, not base point dependent) coordinates on the Teichmüller spaces. The compactified moduli space $\overline{\mathbf{R}(p,n)}$ is covered by finitely many $\mathbb{D}(\mathcal{G})$. The points in $\mathbb{D}(\mathcal{G})$ representing nonsingular surfaces form a quotient of a Teichmüller space:

$$\mathbb{D}(\mathcal{G}) \supset \mathbb{D}_o(\mathcal{G}) = \mathbf{T}(p,n)/\mathbb{Z}^d,$$

where $\mathbb{Z}^d$ is the lattice generated by "Dehn twists" about partition curves acting by translation on horocyclic coordinates.

9. Teichmüller spaces and families of Kleinian groups

Most of the constructions of Teichmüller and moduli spaces employ various families of Kleinian groups; among these are Fuchsian groups, quasiFuchsian groups, terminal regular *b*-groups, and families of nonfunction groups introduced by Bers [**15**] and Earle and Marden [**18**]. We will after some description of other families of groups, concentrate on this last one. The first two families are in a sense the most natural, they are independent of the graph $\mathcal{G}$ that we fixed, but it is more difficult to extract the information we want from these groups. They are also, in some sense, more difficult to construct.

9.1. Spaces of Fuchsian groups. By the general uniformization theorem (see, for example [**23**, Ch. IV]), a Riemann surface X of finite hyperbolic type (this means that X can be compactified by the addition of finitely many points and its Euler characteristic is negative) can always be uniformized by a Fuchsian group (finitely generated, torsion free and of the first kind) in the sense that we can find such a subgroup $G \subset \mathrm{PSL}(2,\mathbb{R})$ so that $X \cong \mathbb{H}^2/G$. The group G is unique up to conjugation (in $\mathrm{PSL}(2,\mathbb{R})$). Thus one can define a space of conjugacy classes of marked Fuchsian groups of a fixed type (p,n) (of fixed signature, if one were to allow torsion) and obtain a model of $\mathbf{T}(p,n)$. The group $\mathrm{PSL}(2,\mathbb{R})$ has a real, but not an obvious complex, analytic structure. Thus we are only lead to a real analytic structure on Teichmüller space[30].

9.2. Spaces of quasiFuchsian groups and the Bers slices. A Fuchsian group represents a surface and its mirror image. It acts discontinuosly on and fixes each component of the complement of a circle in $\hat{\mathbb{C}}$. Bers made a simple but important observation: change the circle to its image under a quasiconformal map. The complement of this *quasi-circle* can now be made to represent two arbitrary surfaces of the same type. That such groups (called *quasiFuchsian*) exist is a simple consequence of the generalized RMT (see [**9**]). We can thus naturally define a space of $\mathrm{PSL}(2,\mathbb{C})$-conjugacy classes of appropriately marked quasiFuchsian groups of type (p,n). This yields a complex analytic model of $\mathbf{T}(p,n)\times\mathbf{T}(p,n)$. Riemann's moduli problem is essentially to produce a complex analytic model of $\mathbf{T}(p,n)$ or a *slice* of $\mathbf{T}(p,n) \times \mathbf{T}(p,n)$. Bers observed that one can vary the complex structure on one side of the quasi-circle while keeping the complex structure on the other side fixed; thus producing the desired slice. This ingenious device is only an existence

[30]The *Fricke spaces* of [**1**, Ch. II] yield a model for this real analytic structure on Teichmüller space.

proof, at best a construction involving transcendental methods. The resulting coordinates for Teichmüller space are variational[31]; they depend on a choice of a base point; such choice depending on $3p - 3 + n$ complex parameters. The search for nonvariational coordinates, based on only finitely many choices, leads us to

9.3. Terminal (regular) *b*-groups – the direct Kleinian group translation of nonsingular plumbing. We are working with a fixed hyperbolic type (p, n) and a fixed admissible graph $\mathcal{G}$ of this type. We assume that $3p - 3 + n > 0$ to avoid modifications required for the zero dimensional Teichmüller space. A group Γ_t is constructed algebraically from plumbing parameters $t \in \mathbb{D}_o(\mathcal{G})$ using $v = 2p - 2 + n$ AFP's and p HNN extensions. The group represents the surface S_t on its invariant component Δ and v thrice punctured spheres on the noninvariant components. The invariant component is simply connected. Hence we have constructed a *b*-group; because it also represents the "correct" number of thrice punctured spheres, this group is called *terminal*. *Accidental* parabolic motions are used to represent curves in partition. The idea of using these groups to construct intrinsic coordinates for $\mathbf{T}(p, n)$ is due to Maskit [**35**]; the connection to plumbing parameters and the development of this approach to the theory of moduli can be found in [**28**]. It is quite easy to describe a set of parameters t that yield groups of Möbius transformations. As remarked earlier, in this family of groups as well as the next two, the difficult point is to determine for what t the construction has geometric significance and produces a group with "the right" properties. Degeneration of surfaces is not seen in this picture, since only nonsingular surfaces are represented by terminal *b*-groups[32]. We have illustrated in Figures 7 and 8, the fundamental domains of *b*-groups of type $(0, 4)$ and $(2, 0)$, respectively.

9.4. A family due to Bers. In order to have the groups reflect (at least some) degenerations, Bers [**15**][33] develops an alternate set of nonvariational coordinates for Teichmuller space. He starts with a fixed function group Γ_0[34] that represents a fixed surface of type $(0, 2p+n)$ on its (non simply connected) invariant component Δ and $(2p-2+n)$ thrice punctured spheres on the rest of its region of discontinuity $\Omega - \Delta$[35]. We assign a vertex of $\mathcal{G}$ to each of these thrice punctured spheres, and use the edges of $\mathcal{G}$ to "artificially" pair, subject to some mild directions provided by $\mathcal{G}$, $6p - 6 + 2n$ of the $6p - 6 + 3n$ punctures of $(\Omega - \Delta)/\Gamma_0$ to form a singular surface of genus p with $3p - 3 + n$ nodes and n punctures. The group Γ_0 will now play two roles. It is a "holding group" for our construction and the group corresponding to the parameter $0 \in \mathbb{D}(\mathcal{G})$. Let $t \in \mathbb{D}(\mathcal{G})$ be arbitrary. We construct a group Γ_t that will represent the surface S_t (which may have nodes). Each nonzero component of t is used to make an HNN-construction which replaces a node by a tube. This operation does not affect the genus nor the number of punctures, but does decrease the number of nodes by one. The group Γ_t will, in general (for $t \neq 0$), not be a function group; it will have $N + 1$ inequivalent components. Precisely one of these will not be simply connected and its stabilizer Γ_0 will represent the fixed

[31] This approach also works for surfaces that are not of finite type.

[32] These groups can be made to represent noded surfaces; but it would be "the wrong" noded surfaces for our purposes.

[33] Slightly different approaches [**13**], [**12**], [**14**] were announced earlier.

[34] This group can be constructed as a free product of $2p - 2 + n$ triangle groups $F(a, b, c)$.

[35] Each component of this set is simply connected, in fact, a disc that is stabilized by a $F(a, b, c)$.

sphere with $2p-2+n$ punctures that started the Bers construction. Each of the other components of $\Omega(\Gamma_t)$ will be simply connected, will be stablized by a terminal *b*-group, and will represent a part of S_t.

9.5. Another family representing two surfaces, Earle-Marden slices (coordinates). As in the last construction, we will describe a surface S_t for each point $t \in \mathbb{D}(\mathcal{G})$, not just $\mathbb{D}_o(\mathcal{G})$. The advantage over the previous family of groups is that a node will be represented by the same parabolic fixed point rather than by two artificially paired fixed points. The group Γ_t will represent a fixed (nonsingular) surface of type (p, n) and a variable surface of this type (that can be singular). We start with a fixed terminal *b*-group of given graph type[36]. This group represents a fixed surface S_o of type (p, n) and $2p-2+n$ thrice punctured spheres S. We pair two punctures on S to form a node if and only if they are determined by the same parabolic fixed point. The resulting singular surface of type (p, n) with $3p-3+n$ nodes and $2p-2+n$ parts is S_0. Let us produce a list of $3p-3+n$ inequivalent parabolic fixed points

$$x_1,\ \ldots,\ x_{3p-3+n}$$

that represent punctures on S (nodes on S_0). Now, for a given $t \in \mathbb{D}(\mathcal{G})$, we perform $3p-3+n-k$ HNN constructions to produce the group Γ_t that represents the surface S_t. Each HNN construction has produced a rank two parabolic subgroup. The group Γ_t is not a function group. It has $N+1$ component subgroups; each a terminal *b*-group. It represents the nonsingular surface S_o on one of these components and the surface S_t on the rest of the components (if t has no nonzero components, then S_t has precisely two component subgroups); an illustration for a genus two surface with one node is provided by Figure 9.

Let P_{x_i} be the stabilizer of x_i in Γ_t; this is a rank two parabolic group if the edge represents a curve on S_t, and a rank one group if edge represents a node on S_t (if and only if the corresponding plumbing parameter is zero). We form the *relative Poincaré series*

$$\Phi_i(z) = \sum_{\gamma \in P_{x_i}\backslash\Gamma_t} \frac{\gamma'(z)^2}{(\gamma(z)-x_i)^4},\ z \in \Omega(\Gamma_t).$$

The resulting sums are quadratic differentials; that is, they satisfy

$$\Phi_i(g(z))(g'(z))^2 = \Phi_i(z),\ i = 1,\ \ldots,\ 3p-3+n,\ z \in \Omega(\Gamma_t),\ g \in \Gamma_t.$$

The automorphic forms

$$\Phi_1,\ \ldots,\ \Phi_{3p-3+n}$$

can be used to embed S_t into $\mathbf{P}\mathbb{C}^{3p-2+n}$ provided $3p+n > 6$. More functions are needed in the exceptional cases (one can use higher order (for example, cubic) differentials). By properly normalizing the automorphic forms, the embedding can be made to depend holomorphically on $t \in \mathbb{D}(\mathcal{G})$. It should be emphasized that this statement also holds for t representing singular surfaces.

[36]For example the group corresponding to $t = (\exp 2\pi\imath,\ \ldots,\ \exp 2\pi\imath)$, illustrated in Figure 8 for type $(2, 0)$.

10. Punctured tori

We describe a construction of a Fuchsian group representing an arbitrary once punctured torus (hence the group also represents its mirror image). This construction will depend real analytically on parameters. It parallels the construction in the elliptic case, and leads to a complete (computable) invariant for the family of punctured tori. Our construction will depend on two real positive parameters a and b (points of $\mathbb{R}^+$). The role of the first parameter is rather easy to see. Fix an $a \in \mathbb{R}$, $a > 0$. The ordered quadruple $\infty, 0, 1$ and $1 + a$ determines a non-Euclidean quadrilateral $\mathcal{P}$ bounded by the simple closed curve consisting of the infinite length geodesic segments c_1 from 0 to ∞ (a vertical ray in $\mathbb{H}^2$ starting at 0), c_2 from ∞ to $1 + a$ (a vertical ray in $\mathbb{H}^2$ ending at $1 + a$), c_3 from $1 + a$ to 1 (a semicircle in $\mathbb{H}^2$ orthogonal to $\mathbb{R}$ joining $1 + a$ to 1) and c_4 from 1 to 0 (a semicircle in $\mathbb{H}^2$ orthogonal to $\mathbb{R}$ joining 1 to 0). We now pick real Möbius transformations A and B with the properties

$$A(\infty) = 1 + a, \ A(0) = 1, \ B(0) = \infty, \text{ and } B(1) = 1 + a.$$

It is convenient to represent these motions, which automatically satisfy

$$A(c_1) = c_3 \text{ and } B(c_2) = c_4,$$

by matrices in $\mathrm{PGL}(2, \mathbb{R})$. Thus (see Figure 10)

$$A = \begin{bmatrix} (1+a)b & 1 \\ b & 1 \end{bmatrix} \text{ and } B = \begin{bmatrix} 1+(1+a)c & -1 \\ c & 0 \end{bmatrix},$$

for some b and $c \in \mathbb{R}^+$. We are interested in obtaining a representation in $\mathrm{PSL}(2, \mathbb{R})$ of the fundamental group of a punctured torus. One can choose free generators A and B for the fundamental group so that $P = B^{-1}A^{-1}BA$ represents a simple closed curve around the puncture. As an element of $\mathrm{PSL}(2, \mathbb{R})$, P must hence have trace ± 2. It requires a little work to prove that, in fact, trace $(P) = -2$ is the only possibility. This implies that $c = b$. One can now show that, with this last assumption, the group $G(a, b)$ generated by A and B is a discrete subgroup of $\mathrm{PSL}(2, \mathbb{R})$ and hence acts discontinously on $\mathbb{H}^2$. The geodesic quadrilateral $\mathcal{P}$ is a fundamental domain for $G(a, b)$ which hence is easily seen to represent by its action on $\mathbb{H}^2$ a marked punctured torus S. The Bers (μ) trick shows that every punctured torus is so represented; this last step can be carried in out in much more general set ups. The same conclusion may be reached by other methods using known properties of elliptic curves.

We now fill in the puncture on S and obtain a marked torus T. This torus T has a natural origin – the puncture on S. We can hence conclude that $T = T_\tau$ for a unique $\tau \in \mathbb{H}^2$. To determine τ, we form the Poincaré series

$$\Phi(z) = \sum_{\gamma \in G(a,b)} \varphi(\gamma(z))\gamma'(z)^2,$$

where

$$\varphi(z) = \frac{a(1+a)}{z(z-1)(z-1-a)}, \ z \in \mathbb{H}^2.$$

The automorphic form Φ is not identically 0[37] and projects to a quadratic differential on the compact surface $\overline{\mathbb{H}^2/G(a,b)}$ without zeros or poles.

[37]This seemingly simple fact is not immediately obvious, and requires considerable machinary to establish.

To obtain a "good" meromorphic function on $\overline{\mathbb{H}^2/G(a,b)}$, we need to produce a holomorphic quadratic differential for $G(a,b)$ whose projection to $\mathbb{H}^2/G$ has a double pole at the puncture. The relative Poincaré series

$$\Phi_1(z) = \sum_{\gamma \in <P> \backslash G(a,b)} \frac{\gamma'(z)^2}{\gamma(z)^4},\ z \in \mathbb{H}^2,$$

will do. Now the projection F of $\frac{\Phi_1}{\Phi}$ to $\overline{\mathbb{H}^2/G(a,b)}$ is a function of degree two. By the Riemann-Roch theorem (or more elementary arguments), there is a linear relation between the constant function, this function F, and the Weierstrass $\wp$-function. It is

$$F = a\wp + b,\ a \in \mathbb{C} - \{0\},\ b \in \mathbb{C}.$$

Once we have selected an origin on a torus, we have uniquely determined the three half periods. In our case, the three points

$$x_1 = \frac{1 + \imath\sqrt{ab}}{1 + ab},\ x_2 = \frac{\imath}{\sqrt{b}} \text{ and } x_3 = 1 + \imath\sqrt{a}$$

in $\mathbb{H}^2$ project to the half periods on $\overline{\mathbb{H}^2/G(a,b)}$. It follows that

$$\lambda = \frac{F(x_3) - F(x_2)}{F(x_1) - F(x_2)}.$$

and

$$\jmath = \frac{4}{27}\frac{(1-\lambda+\lambda^2)^3}{\lambda^2(1-\lambda)^2}.$$

are the classical invariants for the torus $\overline{\mathbb{H}^2/G(a,b)}$. As functions of $\tau \in \mathbb{H}^2$, λ is a holomorphic universal covering map of $\mathbb{C} - \{0,1\}$, and $\jmath$ is the holomorphic universal covering map (in the sense of orbifolds) of the sphere $\hat{\mathbb{C}}$ with three distinguished points 1, 0 and ∞ with ramification numbers 2, 3 and ∞, respectively. The invariants $\jmath$ and λ have interesting analytic interpretations. Two tori T_{τ_1} and T_{τ_2} are conformally equivalent if and only if $\jmath(\tau_1) = \jmath(\tau_2)$; they are equivalent by a conformal map that conjugates the involution $\tau \mapsto \tau + \frac{1}{2}$ on T_{τ_1} to the corresponding involution on T_{τ_2} (the conformal map is said to preserve the *level two* structure).

The divisor[38] of a Poincaré series is difficult (perhaps, impossible) to compute. Thus the method illustrated here has limited applicability – to surfaces of low genus with few punctures and to surfaces with many symmetries. In special cases, for subgroups of $\mathrm{PSL}(2,\mathbb{Z})$, we have an alternate approach (using different functions) that resolves this particular problem. It is described in

11. Uniformizations of $\overline{\mathbb{H}^2/\Gamma(k)}$

We will discuss only some "sample" results, based on the recent work in [**24**], [**22**], [**19**], [**20**] and [**21**]. Let k be an odd prime, and $\Gamma(k)$, the *principal congruence level* k subgroup of Γ=$\mathrm{PSL}(2,\mathbb{Z})$ defined by matrices $A = \begin{bmatrix} a & b \\ c & d \end{bmatrix} \in \mathrm{SL}(2,\mathbb{Z})$ satisfying

$$a \equiv 1 \bmod k \equiv d,\ b \equiv 0 \bmod k \equiv c.$$

[38]Describing the location of the zeros (in general, also the poles) of the function.

We study the action of $\Gamma(k)$ on $\mathbb{H}^2$. The quotient $\mathbb{H}^2/\Gamma(k)$ is a surface of genus $1+\frac{(k-6)(k^2-1)}{24}$ with $\frac{k^2-1}{2}$ punctures (its Euler characteristic is $-\frac{k(k^2-1)}{12}$). Set for $l=0,\ 1,\ \ldots,\ \frac{k-3}{2}$,

$$\varphi_l(\tau)=\theta\begin{bmatrix}\frac{2l+1}{k}\\ 1\end{bmatrix}(0,k\tau),\ \tau\in\mathbb{H}^2.$$

These functions extend by continuity to $\mathbb{Q}\cup\{\infty\}$.

(a) Their divisors are well defined on the quotient $\overline{\mathbb{H}^2/\Gamma(k)}$ and are in a sense "supported" at $\frac{k-1}{2}$ "distinguished" punctures on $\mathbb{H}^2/\Gamma(k)$.
(b) They span a vector space $V(k)$ of $\frac{1}{4}$-forms.
(c) Actually, a Hilbert space of dimension $\frac{k-1}{2}$ with *Petersson* inner product:

$$<\varphi,\psi>=\int\int_{\mathbb{H}^2/\Gamma(k)}(\Im z)^{-\frac{3}{2}}\varphi(z)\overline{\psi(z)}\frac{|dz\wedge\overline{dz}|}{2},\ \varphi,\psi\in V(k).$$

(d) The functions φ_l form an orthogonal basis for $V(k)$.
(e) The group $\Gamma/\Gamma(k)$ acts by isometries on $\mathbf{P}V(k)$.
The function field[39] of $\overline{\mathbb{H}^2/\Gamma(k)}$, $\mathcal{K}(\overline{\mathbb{H}^2/\Gamma(k)})$, is generated by the projections to $\overline{\mathbb{H}^2/\Gamma(k)}$ of

$$\frac{\varphi_1}{\varphi_0},\ \frac{\varphi_2}{\varphi_1}.$$

More information is available for small k. After some work, one can identify $\overline{\mathbb{H}^2/\Gamma(7)}$ as the Klein surface, defined by the projective algebraic equation in three variables

$$X^3Z+Y^3X+Z^3Y=0.$$

This last equation is equivalent to the θ-identity

$$\left(\omega^{-3}\theta\begin{bmatrix}\frac{1}{7}\\ 1\end{bmatrix}\right)^3\left(\omega^{9}\theta\begin{bmatrix}\frac{5}{7}\\ 1\end{bmatrix}\right)+\left(\omega^{-3}\theta\begin{bmatrix}\frac{1}{7}\\ 1\end{bmatrix}\right)\left(\omega^{9}\theta\begin{bmatrix}\frac{3}{7}\\ 1\end{bmatrix}\right)^3$$
$$+\left(\omega^{9}\theta\begin{bmatrix}\frac{3}{7}\\ 1\end{bmatrix}\right)\left(\omega^{9}\theta\begin{bmatrix}\frac{5}{7}\\ 1\end{bmatrix}\right)^3=0,\ \omega=\exp\left(\frac{\pi\imath}{28}\right).$$

The functions in $V(k)$ are useful in studying projective embeddings of the surface $\overline{\mathbb{H}^2/\Gamma(k)}$. The map

$$\tau\mapsto(\varphi_0(\tau),\ \varphi_1(\tau),\ \ldots,\ \varphi_{\frac{k-5}{2}}(\tau),\ \varphi_{\frac{k-3}{2}}(\tau)),\ \tau\in\mathbb{H}^2$$

defines a holomorphic map of maximal rank

$$\Phi:\overline{\mathbb{H}^2/\Gamma(k)}\to\mathbf{P}\mathbb{C}^{\frac{k-3}{2}}\cong\mathbf{P}V(k)$$

(the same map can be defined by a subset of holomorphic 1-forms on the surface); the resulting homomorphism

$$\Theta:\Gamma/\Gamma(k)\to\ \text{Aut}\ \mathbf{P}V(k)$$

is injective. We conjecture that so is Φ.

The above is part of an ongoing investigation. Most of the results of this section extend to composite k. Also to other (related) families of groups. The theories for even and odd k have different flavors. However, they apply only to surfaces uniformized by some subgroups of $\mathrm{PSL}(2,\mathbb{Z})$. The extension of our methods to include $\mathrm{PSL}(2,\mathbb{R})$ subgroups seems currently out of reach.

[39]Field of meromorphic functions on the compact surface.

References

[1] W. Abikoff, *The Real Analytic Theory of Teichmüller Space*, Lecture Notes in Math., vol. 820, Springer-Verlag, 1980.

[2] L.V. Ahlfors, *On quasiconformal mappings*, J. Analyse Math. **3** (1953-54), 1–58.

[3] ———, *The complex analytic stucture of the space of closed Riemann surfaces*, Analytic Functions, Princeton University Press, 1960, pp. 45–66.

[4] ———, *Finitely generated Kleinian groups*, Amer. J. Math. **86** (1964), 413–429 and **87** (1965), 759.

[5] L.V. Ahlfors and L. Bers, *Riemann's mapping theorem for variable metrics*, Ann. of Math. **72** (1960), 345–404.

[6] P. Arés Gastesi, *Coordinates for Teichmüller spaces of b-groups with torsion*, Ann. Acad. Sci. Fenn. **20** (1995), 279–300.

[7] L. Bers, *Quasiconformal mappings and Teichmüller's theorem*, Analytic Functions, Princeton University Press, 1960, pp. 89–119.

[8] ———, *Simultaneous uniformization*, Bull. Amer. Math. Soc. **66** (1960), 94–97.

[9] ———, *Uniformization by Beltrami equations*, Comm. Pure Appl. Math. **14** (1961), 215–228.

[10] ———, *A non-standard integral equation with applications to quasiconformal mappings*, Acta Math. **116** (1966), 113–134.

[11] ———, *Uniformization, moduli and Kleinian groups*, Bull. London Math. Soc. **4** (1972), 257–300.

[12] ———, *On spaces of Riemann surfaces with nodes*, Bull. Amer. Math. Soc. **80** (1974), 1219–1222.

[13] ———, *Spaces of degenerating Riemann surfaces*, Discontinuous Groups and Riemann Surfaces, Ann. of Math. Studies 79, Princeton Univ. Press, 1974, pp. 43–59.

[14] ———, *Deformations and moduli of Riemann surfaces with nodes and signatures*, Math. Scand. **36** (1975), 12–16.

[15] ———, *Finite dimensional Teichmüller spaces and generalizations*, Bull. Amer. Math. Soc. **(N.S.)5** (1981), 131–172.

[16] W.K. Clifford, *Theory of marks of multiple theta functions*, Mathematical Papers by William Kingdon Clifford, Chelsea Publishing Company, 1968, pp. 356–367.

[17] C.J. Earle, *Some intrinsic coordinates on Teichmüller space*, Proc. Amer. Math. Soc. **83** (1981), 527–531.

[18] C.J. Earle and A. Marden, *Geometric complex coordinates for Teichmüller space*, to appear.

[19] H.M. Farkas and Y. Kopeliovich, *New theta constant identities*, Israel J. Math. **82** (1993), 133–140.

[20] ———, *New theta constant identities II*, Proc. Amer. Math. Soc. **123** (1995), 1009–1020.

[21] H.M. Farkas, Y. Kopeliovich, and I. Kra, *Uniformization of modular curves*, Comm. Anal. Geom. **4** (1996), 207–259.

[22] H.M. Farkas and I. Kra, *Automorphic forms for subgroups of the modular group. II: Composite level*, to appear in J. Analyse Math.

[23] ———, *Riemann Surfaces (second edition)*, Graduate Texts in Mathematics, vol. 71, Springer-Verlag, 1992.

[24] ———, *Automorphic forms for subgroups of the modular group*, Israel J. Math. **82** (1993), 87–131.

[25] F. Klein, *Neue Beträge zur Riemann'schen Functiontheorie*, Math. Ann. **21** (1883), 141–218.

[26] I. Kra, *Horocyclic coordinates for Riemann surfaces and moduli spaces. II: Fiber spaces over deformation spaces and the strong deformation spaces of Bers*, in preparation.

[27] ———, *Non-variational global coordinates for Teichmüller space*, Holomorphic Functions and Moduli II, Springer-Verlag, 1988, pp. 221–249.

[28] ———, *Horocyclic coordinates for Riemann surfaces and moduli spaces. I: Teichmüller and Riemann spaces of Kleinian groups*, J. Amer. Math. Soc. **3** (1990), 499–578.

[29] O. Lehto, *Quasiconformal homeomorphisms and Beltrami equations*, Discrete Groups and Automorphic Functions, Academic Press, 1977, pp. 121–142.

[30] ———, *Univalent Functions and Teichmüller Spaces*, Graduate Texts in Mathematics, vol. 109, Springer-Verlag, 1987.

[31] A. Marden, *The geometry of finitely generated Kleinian groups*, Ann. of Math. **99** (1974), 383–462.

[32] B. Maskit, *On a class of Kleinian groups*, Ann. Acad. Sci. Fenn. **442** (1969), 8 pp.

[33] ———, *On Klein's combination theorem, III*, Advances in the Theory of Riemann Surfaces, Ann. of Math. Studies 66, Princeton Univ. Press, 1971, pp. 297–316.

[34] ———, *Decomposition of certain Kleinian groups*, Acta Math. **130** (1973), 243–263.

[35] ———, *Moduli of marked Riemann surfaces*, Bull. Amer. Math. Soc. **80** (1974), 773–777.

[36] C.B. Morrey, *On the solution of quasilinear elliptic partial differential equations,* Trans. Amer. Math. Soc. **43** (1938), 126–166.

[37] A. Selberg, *On discontinuous groups in higher dimensional symmetric spaces*, Contributions to Function Theory, Bombay, 1960, pp. 147–164.

[38] W.P. Thurston, *Three dimensional manifolds, Kleinian groups and hyperbolic geometry*, Bull. Amer. Math. Soc. **6** (1982), 357–381.

State University of New York at Stony Brook Stony Brook, NY 11794

Fundamental domain for G_τ ($\cong \mathbb{Z}^2$)

and for a group conjugate to G_τ
(with ∞ replaced by common f. p. $x \in \mathbb{C}$).

FIGURE 1a

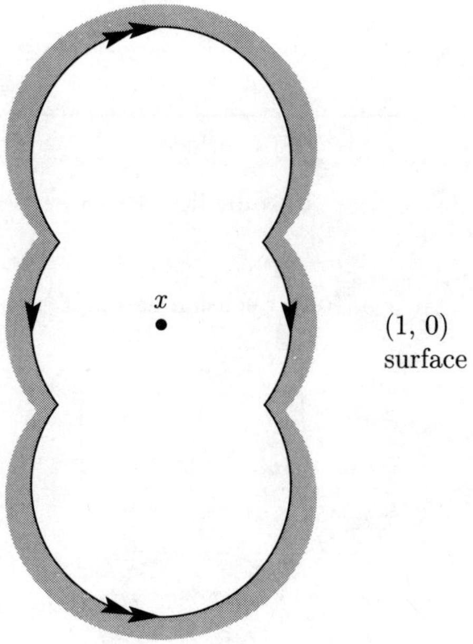

FIGURE 1b

Free product I
(of 2 $\mathbb{Z}^2$'s)

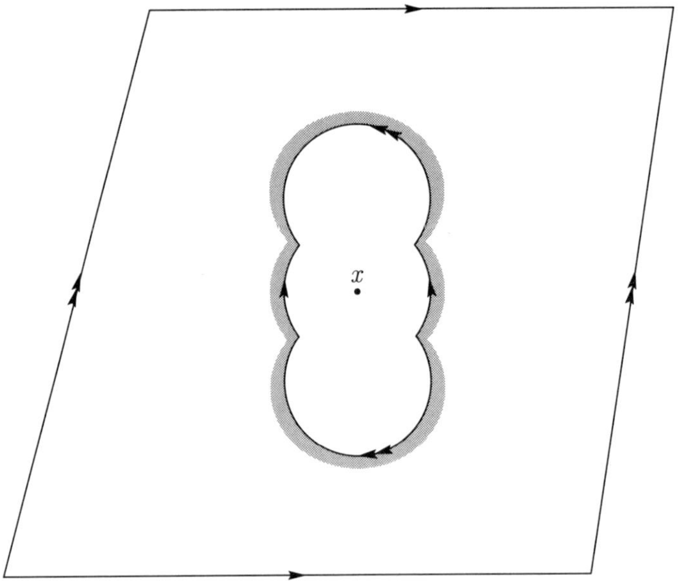

(2, 0) surface

FIGURE 2a

Free product II
(of a Fuchsian and a $\mathbb{Z}^2$)

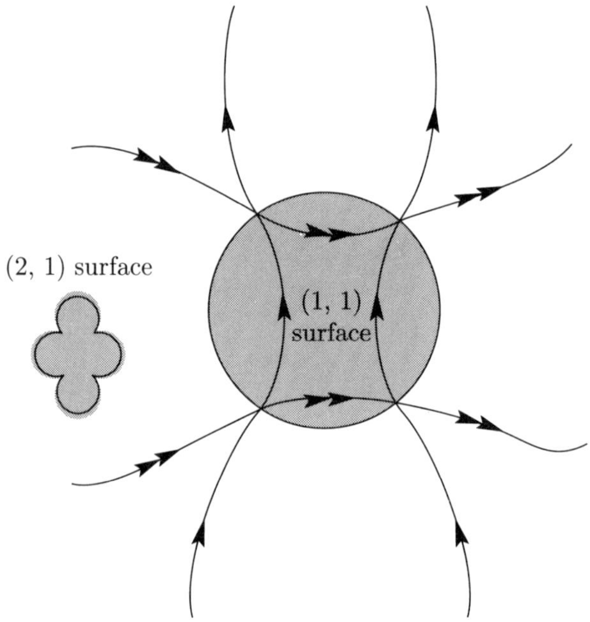

FIGURE 2b

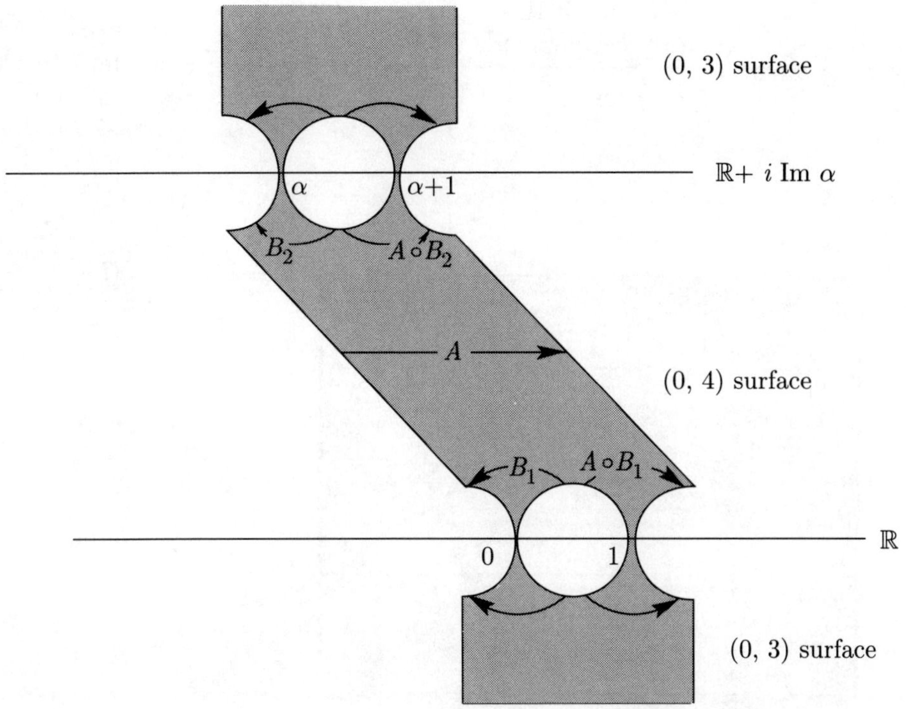

AFP (Maskit I)
(of 2 Fuchsian groups)

FIGURE 3

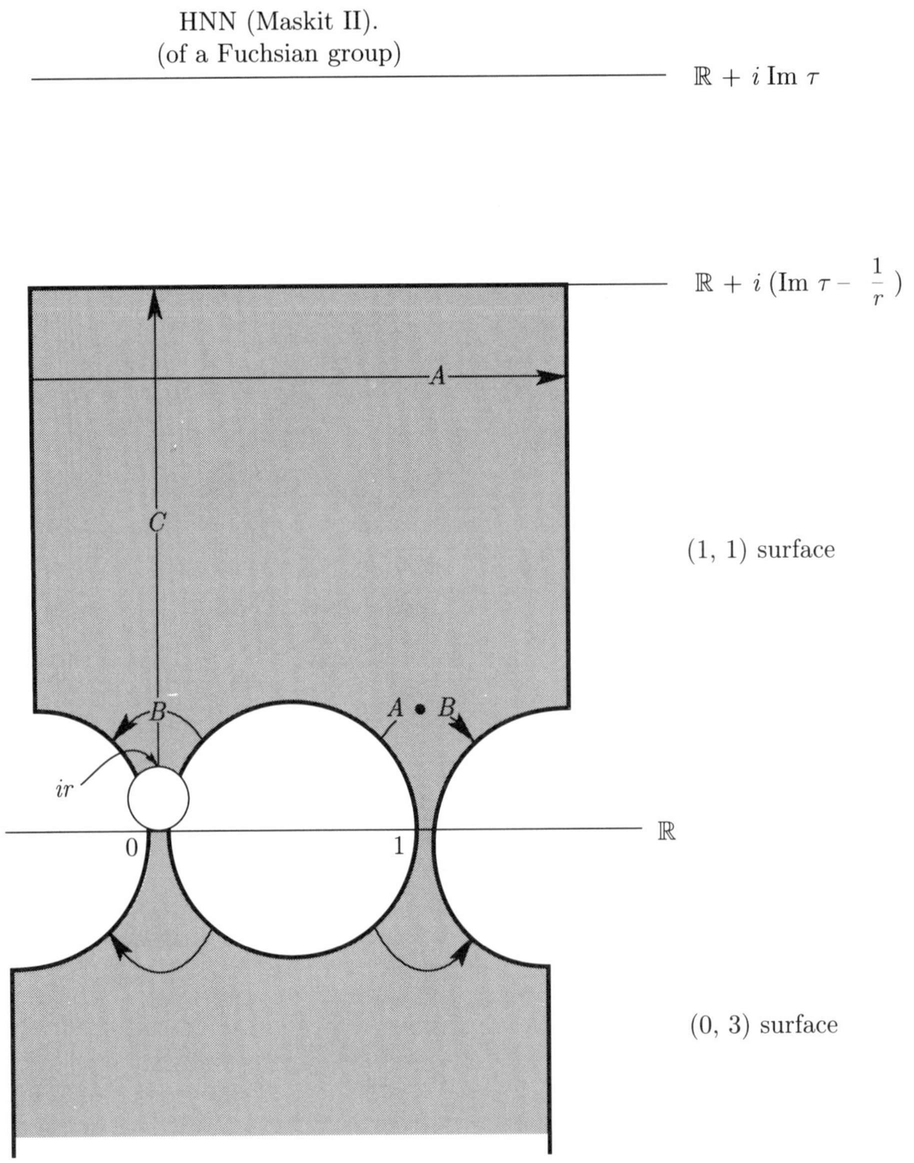
HNN (Maskit II).
(of a Fuchsian group)
$\mathbb{R} + i \operatorname{Im} \tau$
$\mathbb{R} + i (\operatorname{Im} \tau - \frac{1}{r})$
A
C
(1, 1) surface
B
A • B
ir
0
1
$\mathbb{R}$
(0, 3) surface

FIGURE 4

Graphs and partitions.

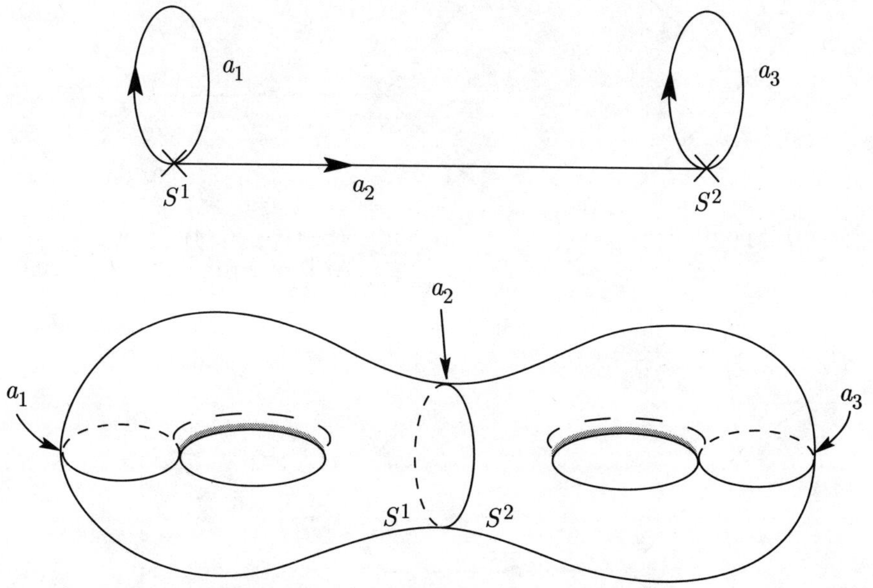

FIGURE 5

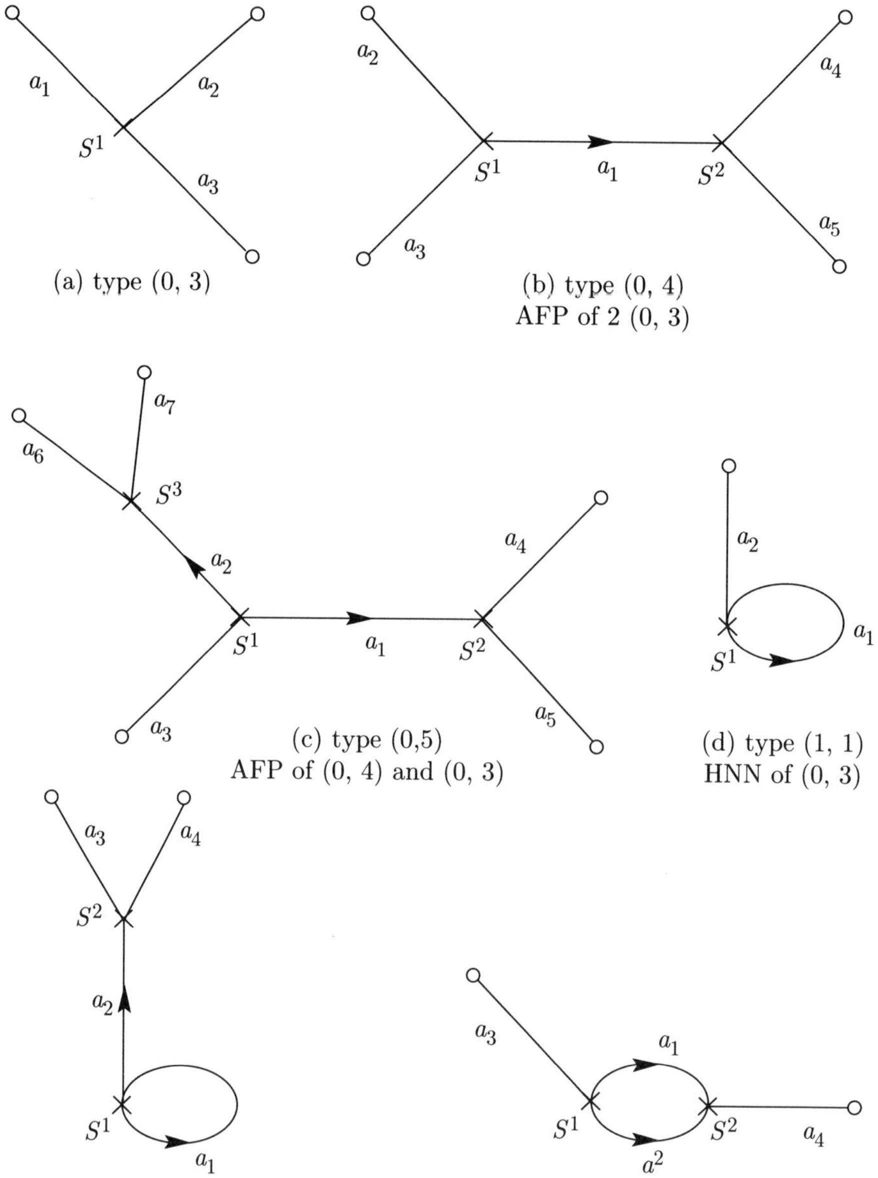

(a) type (0, 3)

(b) type (0, 4)
AFP of 2 (0, 3)

(c) type (0,5)
AFP of (0, 4) and (0, 3)

(d) type (1, 1)
HNN of (0, 3)

(e) type (1, 2)

(f) type (1, 2)

FIGURE 6

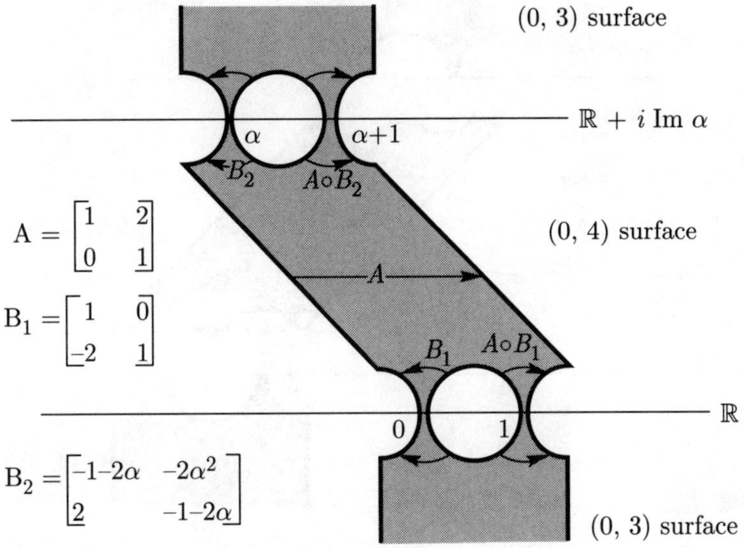

$$\langle A, B_1\rangle = F(\infty, 0, 1)$$
$$\langle A, B_2\rangle = F(\infty, \alpha, \alpha+1)$$

FIGURE 7

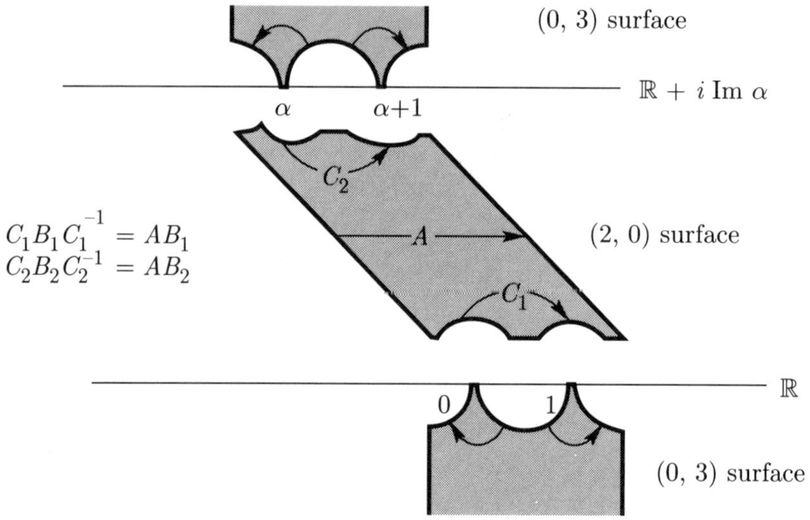

Group that yields horocyclic (Maskit) coordinates
= "holding group" for Earle-Marden construction

FIGURE 8

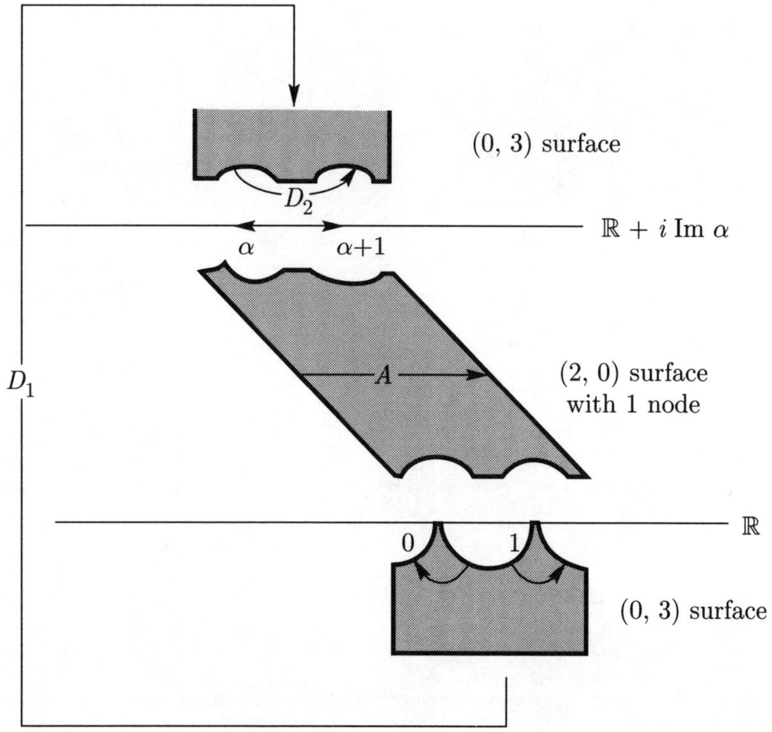

$$D_1 = \begin{bmatrix} 1 & 2\tau \\ 0 & 1 \end{bmatrix} \qquad s = e^{2\pi i\tau} \text{ is good coordinate}$$

FIGURE 9

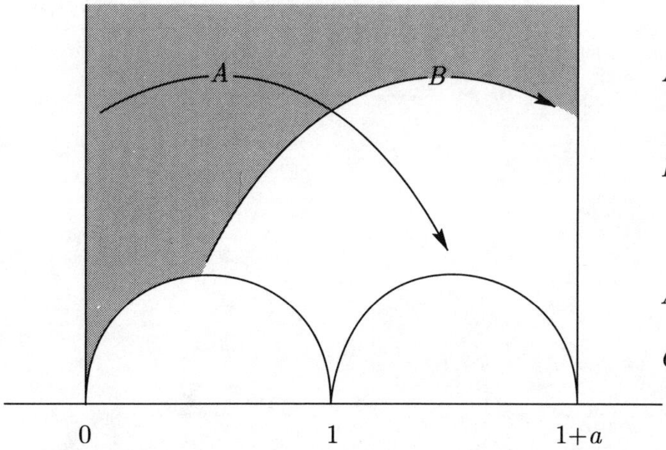

A Fundamental domain for $G(a,\ b)$

$$A = \begin{bmatrix} (1+a)b & 1 \\ b & 1 \end{bmatrix}$$

$$B = \begin{bmatrix} 1+(1+a)b & -1 \\ b & 0 \end{bmatrix}$$

$$A^{-1}(\infty) = -\frac{1}{b}$$

$$G(a,\ b) = \langle A,\ B\rangle$$

$$\cong \mathbb{Z} \times \mathbb{Z}$$

$P = B^{-1} \circ A^{-1} \circ B \circ A$
represents puncture

FIGURE 10

Contemporary Mathematics
Volume **203**, 1997

Adapted metrics and Möbius transformations defined over Clifford Algebras

William Abikoff

ABSTRACT. Adapted metrics for a discrete group G of conformal automorphisms of a finite volume n-dimensional Riemannian orbifold X are classes of G invariant metrics of X whose local behavior reflects the global conformal geometry of X/G. Here we show that the Vahlen-Ahlfors representation of the conformal automorphisms of hyperbolic space lead to a simple estimate of the metric in terms of the hyperbolic metric. In addition, we extend a related estimate of Ahlfors from the case $n = q = 2$ to $q \geq n = 2$.

1. Introduction

In the 1870's Willam Kingdon Clifford conducted profound studies in many fields including complex analysis, elliptic geometry and algebra. His untimely death in 1879 immediately preceded the work and competition of Klein and Poincaré in the early 1880s. Klein's development of the uniformization theorem — the gap in his proof was filled by Brouwer's proof of invariance of domain in 1912 — showed that each Riemann surface is, in modern terminology, a Clifford-Klein space form. However he, along with many other people, was trying to prove that each (projective) plane algebraic curve can be parametrized by a pair of meromorphic functions defined in a plane domain. The latter result was proved by Poincaré as part of his beautiful geometrization of function theory. In the course of that work, he introduced Poincaré series, and it is the quotients of those series which are the parametrizing functions. Here we utilize the series, which Poincaré used to majorize his series, to describe growth properties of a class of metrics, called *adapted metrics*, on certain n-dimensional orbifolds.[1] We will focus our attention to orbifolds admitting a hyperbolic metric; the bounds we obtain are given in terms of that metric.

The adapted metrics undo the basic philosophy of the Clifford-Klein space forms. A given manifold or orbifold X, possessing the structure of a space form, has almost trivial local geometry. On a local level, except at orbifold singularities, space looks the same from every point and, indeed, in every direction from that point. It follows that the local solutions of differential equations and other local

1991 *Mathematics Subject Classification.* 30F45,51M10,11E88.

[1]As we shall later see, the $q = n = 2$ had already been discussed, albeit briefly, by Ahlfors.

structure gives almost no information about global structure. One of the most intriguing features of adapted metrics is that they reintroduce the "average" global behavior into the local structure of the metric without changing the conformal class of the metric. The metrics were introduced in [**1**] and a broader study of them is given in [**2**]. Several others are in progress.

In the spirit of Clifford's work, here we shall restrict our attention to applications of the Clifford algebras to a growth estimate for adapted metrics. We shall derive the estimate as an elementary consequence of the Vahlen-Ahlfors representation of groups of hyperbolic isometries using matrices defined over a real Clifford algebra (see e.g. Ahlfors [**4**]). After substantially completing this work, I learned of the beautiful extension and refinement of Ahlfors' treatment given by Robinson and Trautman [**7**]. We have attempted to merge some of the concepts of the latter work into the summary of Ahlfors' description that we give in Section 2.[2]

I started to reexamine the relationship of Clifford algebras and adapted metrics while I read was reading Clifford's collected papers. His life and work were celebrated at the Conference on New Trends in Topological and Geometric Methods. I would like to thank Irwin Kra for pointing out Ahlfors' estimate to me and also to thank Andrzej Trautman for fascinating discussions at the Conference.

2. Statement of results

Suppose (X, g_0) is an n-dimensional Riemannian manifold having finite volume and G is a group of conformal transformations of X which acts (properly) discontinuously. If J_γ is the determinant of the jacobian of $\gamma \in G$ and $q \in \mathbb{R}^+$ has the property that $\sum_{\gamma \in G} |J_\gamma|^{q/n} < \infty$ for all $x \in X$, then the G-**adapted** or (G, q)-**adapted metric of weight** q is

$$g_q := ds_q^2 = \left[\sum_{\gamma \in G} |J_\gamma|^{q/n} \right]^{2/q} \cdot \quad g_0.$$

g_q is a G-invariant metric on X. Thus G lies in the conformal group of (X, g_q) and the metric projects to X/G.

J_γ is the n-th power of the linear stretching by γ, which we denote γ'. We may therefore write

$$g_q = \left[\sum_{\gamma \in G} |\gamma'|^q \right]^{2/q} \cdot g_0 = \sigma_q^{2/q} \cdot g_0. \tag{2.1}$$

The conformal weight which multiplies g_0 is precisely the square of the ℓ^q-norm of the sequence of stretchings (γ').

Here, we will concern ourselves with the Riemannian manifold (Δ^n, g_E) where Δ^n is the open unit disk in $\mathbb{R}^n$ and g_E is the Euclidean metric restricted to Δ^n. Δ^n also admits a hyperbolic metric g_h which is conformal to g_E. If we denote by $|w|$ the Euclidean norm of $w \in \Delta^n$, we may write $g_h = \rho^2 g_E$, where $\rho(w) = 2(1 - |w|^2)^{-1}$.

It is well-known, see e.g. Maskit [**5**], that the *(relative) Ford region* for G

$$F(G) = \bigcap_{\gamma \in G \setminus \{\mathrm{id}\}} \{w \in \Delta^n \mid \ |\gamma'(w)| < 1\}$$

[2]Trautman's article [**8**] in these Proceedings contains a new derivation of the Vahlen-Ahlfors representation.

is a fundamental region (tile), for the action of G on Δ^n, whenever $\gamma(0) = 0$, for some $\gamma \in G$, implies that $\gamma = \text{id}$.

In 1882, Poincaré [**6**] constructed automorphic functions for Fuchsian groups as quotients of automorphic q-forms. An automorphic q-form for a Fuchsian group G acting in the unit disk $\Delta \subset \mathbb{C}$ is a complex valued function f on Δ which satisfies the functional equation $f \circ \gamma(\gamma')^q = f(z)$. For $\gamma \in G$, γ is a möbius transformation of the form $\gamma : z \mapsto (az + b)(\bar{b}z + \bar{a})^{-1}$ where we may take $|a|^2 - |b|^2 = 1$. As usual, the notation ignores the fact that a should be written a_γ and similarly for b. By using a clever area argument, Poincaré showed that

$$\sum_{\gamma \in G} |\gamma'(z)|^2 < \infty$$

whenever G acts properly discontinuously in Δ. It then follows immediately that the exponent 2 may be replaced by any $q \geq 2$. The argument generalizes with no significant change to groups of isometries of the unit ball model of n-hyperbolic space Δ^n. The corresponding result in dimension n is

$$\sum_{\gamma \in G} |\gamma'(w)|^n < \infty \tag{2.2}$$

for any $z \in \Delta^n$.

Here are the precise statements of our results and one proof.

First observe that, by Equation (2.1), the q-adapted metric, $g_q = \sigma_q^{2/q} g_E$ is the Euclidean metric scaled by the square of the ℓ^q norm of the sequence of linear stretchings by the elements of G. The ℓ^q–norm of a sequence is a decreasing function of q since it has negative derivative with respect to q. In particular, if the q_0-adapted metric exists, it dominates the q-metric for all $q > q_0$.

THEOREM 2.1. *In the hyperbolic plane, modelled on the unit disk Δ, we have the following estimate. For any group G acting discontinuously on Δ, there exists a constant $M = M(G) > 0$ so that*

$$\sigma_q^{1/q} := \left\{ \sum_{\gamma \in G} |\gamma'(w)|^q \right\}^{1/q} \leq M\rho. \tag{2.3}$$

The Theorem follows immediately from Ahlfors' proof (see [**3**]) that Equation 2.3 holds for $q = 2$.

There are other known results about adapted metrics on surfaces (see [**2**]). For example, they have negative curvature. Finitely punctured compact surfaces have finite diameter. Punctures lie at finite distance from any fixed compact subset of a surface and there are curves retracting into punctures having lengths tending to zero. It then follows that the metric completion of a punctured disk is a disk.

Work due to Beardon, Nicholls, Patterson, Sullivan and, later, Bishop and Jones and others, has made explicit the relationship between the exponent of convergence of Poincaré series and the geometry of the quotient hyperbolic manifolds and orbifolds.

First we set the stage. The exponent of convergence $q_0 = q_0(G)$ of a Poincaré series is the infimum of the values of q for which σ_q converges. (Note that convergence at one point in Δ implies convergence at any other point.) Convergence at the exponent of convergence is strongly related to the geometry of the quotient manifold. Under various hypotheses, divergence at the exponent of convergence is equivalent to ergodicity of the geodesic flow while convergence implies that the group is geometrically infinite.

For Fuchsian groups of the second kind acting on $\Delta = \Delta^2$, neither Ahlfors' paper nor its consequence above give any information about the growth of the metric when q is sufficiently near q_0. They give no information when $n > 2$ regardless of the value of q.

As an elementary consequence of the Vahlen-Ahlfors representation of möbius transformations in dimension n, we shall easily show

THEOREM 2.2. *Suppose G is a group of hyperbolic isometries acting discontinuously on hyperbolic n-spacc modcllcd on thc unit ball $\Delta^n \in \mathbb{R}^n$. If the q-metric for G is finite, then, there exists a constant $C'_q = C'_q(G)$ so that*

$$\sigma_q^{1/q} < C'_q \rho^2 \tag{2.4}$$

where $\rho|dz|$ is the hyperbolic line element in Δ^n and $|dz|^2$ is the Euclidean metric. Further, $C'_q(G)$ is a strictly decreasing function of q.

An explicit value for C'_q is given in Section 4.

In general we should not expect to have an Ahlfors-style estimate for g_q when q is near or at the exponent of convergence and there is any degeneracy to Δ^n/G. From the known results on the relationship between degeneracy and the behavior of the exponent of convergence, we would expect that adapted metrics provide some magnification of distances in degenerate ends of the quotients of finitely generated groups (when q is near or equal to q_0). This would be a much desired modification of the hyperbolic metric since current techniques lose control of some aspects of both the geometry and topology in an end when there are closed hyperbolic geodesics whose lengths tend to zero.

3. Möbius transformations defined over a real Clifford algebra

Here we summarize the representation of möbius transformations in space as linear fractional transformations with coefficients in a real Clifford algebra. We also give some geometric consequences of the form of the transformations. The representation was first given by Vahlen [**9**] in 1902; in several stages, it was refined by Ahlfors (see [**4**]). Some of our notation and side remarks follow Robinson and Trautman [**7**]; nonstandard notation should be blamed on the author. Their ideas are oriented toward physics but create a general setting which displays the naturality of the representation.

Let $\mathcal{C}_n$ denote the Clifford algebra over $\mathbb{R}$ generated by $e_1, e_2, \dots, e_{n-1}$. (More precisely, we form the real Clifford algebra using the usual scalar product in $\mathbb{R}^n$.) Vectors in $\mathcal{C}_n$ form the set V of linear combinations of the e_i. Since a Clifford algebra has an identity, we identify $a_0 \in \mathbb{R}$ with the real multiples of the identity. We call $z \in \mathcal{C}_n$ a *number* or *Clifford number of size n* if $z \in \mathbb{R} \otimes V$, i.e. when $z = a_0 + \sum_1^{n-1} a_i e_i$. The Clifford numbers are denoted $\mathcal{N}^n$ and are identified with $\mathbb{R}^n$ via $z \longleftrightarrow (a_0, a_1, \dots, a_n)$. The Clifford numbers inherit the Euclidean norm $|\ \ |$ from this identification. We can think of a Clifford number as a generalized complex number by writing $z = a_0 + v$ and denoting $\Re z := a_0$ and $\Im z := v$. The one point compactification of $\mathcal{N}^n$ is $\widehat{\mathcal{N}^n} := \mathcal{N}^n \cup \{\infty\}$. There are three major involutions on $\mathcal{C}_n$. They are defined first on the finite (or even pairwise) products of the generating elements e_i and then extended by linearity. Suppose $E := e_{i_1} e_{i_2} \cdots e_{i_k}$ is a product of the e_i with the indices i_j increasing with j. Then we define the degree $\deg E := k$. The degree of the empty product is taken as 0. The basic involutions are then given by:

(α) $E \mapsto E^* := (-1)^{\deg E} E$
(β) $E \mapsto \bar{E} := e_{i_k} e_{i_{k-1}} \cdots e_{i_2} e_{i_1}$
(γ) $E \mapsto \tilde{E} := \overline{E^*} = (\bar{E})^*$

The set of units in $\mathcal{C}_n$ form a multiplicative group called the *Clifford group*; it is denoted $\mathcal{C}_n^\times$ or just $\mathcal{C}^\times$. If $z \in \mathcal{C}_n$ is a non-zero Clifford number, then $z\bar{z} = |z|^2$ from which it follows that $z \in \mathcal{C}_n^\times$.

We form the set of 2 by 2 matrices with entries in $\mathcal{C}^\times$ and denote it by $M_2 = M_2(\mathcal{C}^\times)$. For $\gamma \in M_2$, write

$$\gamma := \begin{pmatrix} a & b \\ c & d \end{pmatrix} \quad \text{and} \quad \gamma_1 := \begin{pmatrix} d^* & -b^* \\ -c^* & a^* \end{pmatrix}. \tag{3.1}$$

We associate to the matrix γ and the (extended) number $z \in \widehat{\mathcal{N}^n}$, the map $z \mapsto (az+b)(cz+d)^{-1}$ when the image point makes sense. We also denote the map as γ. With a little care, γ will be bijective and $\gamma^{-1} = \gamma_1$. The *pseudo-determinant* of γ is

$$\Delta(\gamma) = ad^* - bc^*.$$

We may then define

$$\mathrm{GL}_2(\mathcal{C}^\times) = \mathrm{GL}_2(\mathcal{C}_n^\times) := \{\gamma \in M_2 \,|\, \Delta(\gamma) \in \mathbb{R} \setminus \{0\}\} \qquad \text{and}$$

$$\mathrm{PSL}_2^+(\mathcal{C}^\times) = \mathrm{PSL}_2^+(\mathcal{C}^\times) := \{\gamma \in \mathrm{GL}_2(\mathcal{C}^\times) \,|\, \Delta(\gamma) = 1\}/\{\pm \mathrm{Id}\}.$$

The linear stretching of $\gamma \in \mathrm{GL}_2(\mathcal{C}^\times)$, in the notation of Equation (3.1), is

$$|\gamma'(z)| = \frac{|\Delta(\gamma)|}{|cz+d|^2}.$$

We remark, following Ahlfors, that $\gamma \in \mathrm{GL}_2(\mathcal{C}^\times)$ precisely when γ acts bijectively on both $\widehat{\mathcal{N}^n}$ and on $\widehat{\mathcal{N}^{n+1}}$. (Because we are interested in the action of möbius transformations, which are mappings whose actions extend upward in dimension, this will not prove to be a restriction.)

The importance of $\mathrm{PSL}_2^+(\mathcal{C}_n^\times)$ lies in the fact that its elements act as möbius transformations on $\widehat{\mathcal{N}^n} = \widehat{\mathbb{R}^n}$, the one-point compactification of $\mathcal{N}^n$. Using the Poincaré extension to the upper half of $(n+1)$-space, $\mathcal{N}^{n+1}$, we may identify $\mathrm{PSL}_2^+(\mathcal{C}_n^\times)$ with the orientation preserving isometries of hyperbolic $(n+1)$-space $\mathcal{H}^{n+1}$ in the upper half-space model.

For purposes of studying adapted metrics, the upper half space model is not generally useful. Adapted metrics are generally infinite on infinite volume manifolds and the upper half space, with the Euclidean metric suffers from that problem. So we push the action into the unit ball Δ^{n+1}. To do so, we simply conjugate $\mathrm{PSL}_2^+(\mathcal{C}_n^\times)$ by the Cayley map

$$\kappa : z \mapsto (z - e_n)(z + e_n)^{-1} e_n.$$

4. The estimates

The proof of the main estimate is a trivial consequence of the Vahlen-Ahlfors representation and the geometric meaning of the terms in the formula for $|\gamma'|$. Suppose $\gamma : z \mapsto (az+b)(cz+d)^{-1}$. Again we suppress the dependence of a and b on γ. As we have already noted,

$$|\gamma'(z)| = \frac{1}{|cz+d|^2} = \frac{|c|^{-2}}{|z+c^{-1}d|^2}.$$

Here, as in single variable complex analysis, $|z+c^{-1}d|$ is the distance from z to the pole $\gamma^{-1}(\infty)$ of γ. $|c|^{-1}$ is the radius of the sphere, the **isometric sphere**, on which γ acts as an isometry.

Now, the pole of γ lies outside the unit sphere, hence

$$|z+c^{-1}d| \geq 1-|z| \geq \frac{1}{2}(1-|z|^2) > \lambda/4 \tag{4.1}$$

where, as noted before, λ is the ratio of the hyperbolic line element and the Euclidean one. It follows immediately that

$$\sigma_q = \sum_{\gamma\in G} |\gamma'|^q \leq \sum_{\gamma\in G} |c_\gamma|^{-2q} 16^q \cdot \lambda^2 q$$

and

$$ds_q^2 = \left[\sum_{\gamma\in G} |\gamma'|^q\right]^{2/q} \cdot |dz|^2 = C_q' \lambda^2(z)|dz|^2.$$

Here

$$C_q' = \frac{1}{16}\|(c_\gamma^{-2})\|_\infty^2$$

which is the value of C_q' promised in the statement of results.

5. Concluding remarks

The most important values of q are the minimal possible one for which the series converges and, it seems now, also $q=\infty$. As a metric space in the ∞-metric, $\mathcal{H}/G$ is a bounded Euclidean orbifold with codimension one singular set. It has diameter no greater than 2. Estimates of the metric there are not particularly interesting. However at the other end of the possible values of q, we would like to get lower bounds for the metric in Δ^n in terms of the hyperbolic metric as we approach limit points of the group G. (At points of discontinuity of the action of G on $\partial\Delta$ the metric is bounded.) In this way, we can hope to get a finer handle on the geometry of the ends of geometrically infinite hyperbolic orbifolds.

It should also be noted that these same techniques remain valid, when suitably modified, for conformal groups of bounded domains. When the limit set of the group is the full boundary, we again get estimates of the metric in terms of distance to the boundary. However the domain may admit no hyperbolic metric.

References

[1] W[illiam] Abikoff, *Adapted metrics for Kleinian groups*, Abstracts of the Amer. Math. Soc. (1995).
[2] W[illiam] Abikoff and X[ingwang] Xu. To appear.
[3] L[ars] V. Ahlfors, *Eine Bemerkung über fuchssche gruppen*, Math. Z. **84** (1964), 244–245.

[4] L[ars] V. Ahlfors, Möbius transformations and Clifford analysis, 65–73, Springer-Verlag, Berlin–New York, 1985, pp. 65–73.
[5] B[ernard] Maskit, *Kleinian groups*, Springer-Verlag, 1988.
[6] H. Poincaré, *Memoire sur les fonctions Fuchsiennes*, Acta Math. **1** (1882), 193–294.
[7] I[vor] Robinson and A[ndrzej] Trautman, *The conformal geometry of complex quadrics and the fractional-linear form of Möbius transformations*, J. Math. Phys. **34** (1993), 5391–5406.
[8] A[ndrzej] Trautman, *Clifford and the 'square root idea'*, In these proceedings.
[9] K. Vahlen, *Ueber Bewegungen und complexe Zahlen*, Math. Ann. **55** (1902), 585–593.

University of Connecticut at Storrs
E-mail address: abikoff@math.uconn.edu

Part III

INFORMATION GEOMETRY

Contemporary Mathematics
Volume **203**, 1997

Information Geometry

S.-I. AMARI

ABSTRACT. Information geometry has emerged from the study of invariant properties of the manifold of probability distributions. It proposes a Riemannian manifold with two torsion-free dually coupled affine connections. This is called the dual structure. Affine differential geometry and Kählerian manifolds give special cases of the dual structure and statistical models give more general cases. A dually flat manifold has special properties such as dual orthogonal foliations, generalized Pythagoras theorem, the invariant divergence, and the Legendre structure. The present paper reviews the mathematical structure of information geometry and some unsolved mathematical problems are shown. Information geometry is regarded as mathematical sciences having vast developing areas of applications as well as givining new ideas to pure mathematics. The paper touches upon such applications, too.

CONTENTS

1. Introduction

Information geometry originated from the study of the manifold of probability distributions. Consider the set of all the Gaussian distributions as a simple example. Any Gaussian distribution is specified by the mean value μ and variance σ^2, so that it is regarded as a two-dimensional manifold where (μ, σ^2) is a coordinate system. What is the *natural* geometric structure to be introduced in such a manifold? Information geometry gives a definite answer to this problem, and an invariant geometrical structure can be introduced into the manifold consisting

1991 *Mathematics Subject Classification.* Primary 53A15, 62B10; Secondary 95A17.

of a smooth family of general probability distributions. Information geometry is useful not only for applications to statistics, information theory, systems theory where probability distributions play a major role, but also gives a new concept to differential geometry.

Let us consider a manifold M with a Riemannian metric g, $\{M, g\}$. Let $\nabla^{(0)}$ be the covariant derivative due to the Levi-Civita connection. For vector fields X, Y, Z, it satisfies the metric condition

$$X\langle Y, Z\rangle = \langle \nabla^{(0)}_X Y, Z\rangle + \langle Y, \nabla^{(0)}_X Z\rangle, \tag{1.1}$$

where $\langle\cdot,\cdot\rangle$ is the inner product duc to g. This is the Riemannian structure. It was proved that the manifold of probability distributions has a unique invariant Riemannian metric. Moreover, it has a pair of torsion-free affine connections or covariant derivatives ∇ and ∇^* which satisfy the duality condition,

$$X\langle Y, Z\rangle = \langle \nabla_X Y, Z\rangle + \langle Y, \nabla^*_X Z\rangle. \tag{1.2}$$

When the Riemann-Christoffel curvatures R and R^* due to ∇ and ∇^* vanish (but $R^{(0)}$ does not necessarily vanish), the manifold is said to be dually flat. The dually flat manifold has rich structures related to the Legendre transformation and a generalized Pythagoras theorem holds with respect to the invariant divergence. Information geometry gives a powerful method to information sciences and physics based on the dual structure of affine connections.

It is rather surprising that the dual structure has not studied sufficiently from the point of view of pure mathematics. It originated from statistics and information sciences (see Rao [1945], Chentsov [1972], Amari [1985], Murray and Rice [1993], Amari and Nagaoka [1994]). However, a special case of such a structure originated from the theory of hypersurfaces in affine differential geometry (Blaschke [1923], Nomizu and Sasaki [1994]). Another special case was proposed by Shima [1980; 1986; 1995] as the Hessian manifolds which are real manifolds keeping some properties of Kählerian structures.

The present paper reviews the geometry of the dual structure and presents mathematical problems to be solved further. The paper also touches upon wide ranges of applications of information geometry.

2. Riemannian Manifold with Dual Affine Connections

Let $\{S, g, \nabla\}$ be a structure on an n-dimensional manifold S, with Riemannian metric g and affine connection ∇. Let T_p be the tangent space at point $p \in S$. For tangent vectors $X_p, Y_p \in T_p$, the Riemannian metric defines their inner product $g : T_p \times T_p \to \mathrm{R}$,

$$g(X_p, Y_p) = \langle X_p, Y_p\rangle. \tag{2.1}$$

The affine connection is represented by the covariant derivative $\nabla : \mathcal{T} \times \mathcal{T} \to \mathcal{T}$, where $\mathcal{T}$ is the set of all tangent vector fields. The covariant derivative of vector field Y in the direction X is denoted by $\nabla_X Y$. An affine connection is said to be metric when it satisfies

$$X\langle Y, Z\rangle = \langle \nabla_X Y, Z\rangle + \langle Y, \nabla_X Z\rangle. \tag{2.2}$$

The Levi-Civita connection is the only torsion-free metric connection. In this case, $\{S, g, \nabla\}$ is called the Riemannian structure.

Let ξ^i, $i = 1, \cdots, n$, be a local coordinate system of an n-dimensional manifold S. It defines the natural basis vector fields denoted by

$$\mathrm{e}_i = \frac{\partial}{\partial \xi^i} = \partial_i.$$

The metric tensor is given in this coordinate system ∇ by

$$g_{ij} = \langle \mathrm{e}_i, \mathrm{e}_j \rangle \tag{2.3}$$

and the parameters of the affine connection by

$$\Gamma_{ijk} = \langle \nabla_{\mathrm{e}_i} \mathrm{e}_j, \mathrm{e}_k \rangle \tag{2.4}$$

in this coordinate system.

Let us consider a manifold S which has a Riemannian metric g and two affine connections ∇ and ∇^*. This gives a structure $\{S, g, \nabla, \nabla^*\}$. The two affine connections are said to be dual or conjugate with respect to the metric, when

$$X\langle Y, Z\rangle = \langle \nabla_X Y, Z\rangle + \langle Y, \nabla^*_X Z\rangle \tag{2.5}$$

holds for any vector fields X, Y and Z. Refer to Amari [1985], Murray and Rice [1993], Amari and Nagaoka [1994] for the details of the following discussions. By substituting the natural vector fields e_i, e_j, e_k for X, Y and Z, we have

$$\partial_i g_{jk}(\xi) = \Gamma_{ijk}(\xi) + \Gamma^*_{ikj}(\xi) \tag{2.6}$$

for the duality of ∇ and ∇^* in the component form. When $\nabla = \nabla^*$, they are self-dual and (2.5) reduces to the metric condition (2.2). Therefore, the Riemannian geometry $\{S, g, \nabla\}$ with the Levi-Civita connection is the self-dual case of the general dual Riemannian structure.

In general,

$$T_{ijk} = \Gamma^*_{ijk} - \Gamma_{ijk} \tag{2.7}$$

is a tensor. It is easy to prove

$$T_{ijk} = \nabla_{\mathrm{e}_i} \langle \mathrm{e}_j, \mathrm{e}_k \rangle = \nabla_i g_{jk}. \tag{2.8}$$

In particular, when both ∇ and ∇^* are torsion-free, T_{ijk} is a symmetric tensor. Hence, the torsion-free dual structure is characterized by $\{S, g, T\}$, where g and T are symmetric tensors. On the other hand, given symmetric g and T, we can define, for a scalar parameter α, a family of torsion-free connections by

$$\Gamma^{(\alpha)}_{ijk} = \{ij; k\} - \frac{\alpha}{2} T_{ijk}, \tag{2.9}$$

where

$$\Gamma^{(0)}_{ijk} = \{ij; k\} = \frac{1}{2}(\partial_i g_{jk} + \partial_j g_{ik} - \partial_k g_{ij}) \tag{2.10}$$

is the Levi-Civita connection. In this case, for any α, $\{S, g, \nabla^\alpha, \nabla^{-\alpha}\}$ defines the dual structure called the α-structure. The Riemannian structure is the self-dual case with $\alpha = 0$.

The duality can be expressed in the following integral form. Let c be a curve connecting two points p and q. Let Π_c and Π^*_c be the parallel transport operators of vectors from T_p to T_q along c by ∇ and ∇^*, respectively. Then,

$$\langle X, Y\rangle_p = \langle \Pi_c X,\ \Pi^*_c Y\rangle_q \tag{2.11}$$

is locally equivalent to (2.5), where $\langle\ ,\ \rangle_p$ is the inner product at p.

When ∇ and ∇^* are dual, their Riemann-Christoffel curvatures R and R^* are related to each other by the duality. When R vanishes, R^* automatically vanishes and vice versa.

THEOREM 2.1. *For the α-structure, its $\pm\alpha$-curvatures satisfy*

$$R^{(\alpha)}_{ijkl} = -R^{(-\alpha)}_{ijlk}, \tag{2.12}$$

where

$$R^{(\alpha)}_{ijkl} = R(\mathbf{e}_i, \mathbf{e}_j, \mathbf{e}_k, \mathbf{e}_l)$$

is the Riemann-Christoffel curvature tensor.

The dual structure can be extended to any linear fibre bundles with inner product. Moreover, it can be incorporated with the conformal structure. The structure $\{S, \tilde{g}, \tilde{T}\}$ is said to be conformally equivalent to $\{S, g, T\}$ (Okamoto, Amari and Takeuchi [1991]), when there exists a positive scalar field σ such that

$$\tilde{g}_{ij} = \sigma g_{ij} \tag{2.13}$$

$$\tilde{T}_{ijk} = \sigma(T_{ijk} + g_{ij}s_k + g_{ik}s_j + g_{jk}s_i), \tag{2.14}$$

where

$$s_i = \frac{\partial}{\partial \xi^i} \log \sigma. \tag{2.15}$$

The dual projective structure is studied by Burdet and Perrin [1996].

3. Dually flat manifolds

When $\{S, g, \nabla, \nabla^*\}$ is torsion- and curvature-free, the manifold S is said to be dually flat. The Riemannian (Levi-Civita) curvature does not vanish in general. When it also vanishes, S is proved to be self-dual and flat, so that it reduces to the Euclidean space locally. Therefore, a dually flat manifold is a generalization of the Euclidean space. The dually flat structure inherits some of the Euclidean properties. See Nagaoka and Amari [1982], Amari [1995], Amari and Nagaoka [1994].

Since ∇ and ∇^* are flat, we have two local affine coordinate systems $\theta = (\theta^i)$ and $\eta = (\eta_i)$ such that the coordinate curves are ∇- and ∇^*-geodesics and the natural basis fields $\{\mathrm{e}_i\}$ and $\{\mathrm{e}^{*i}\}$ are ∇- and ∇^*-parallel, that is,

$$\nabla_{\mathrm{e}_i}\mathrm{e}_j = 0, \qquad \nabla_{\mathrm{e}^*_i}\mathrm{e}^*_j = 0, \tag{3.1}$$

respectively, where

$$\mathrm{e}_i = \frac{\partial}{\partial \theta^i}, \qquad \mathrm{e}^{*i} = \frac{\partial}{\partial \eta_i}. \tag{3.2}$$

They are determined to within affine transformations. We can choose a pair of ∇- and ∇^*-affine coordinates such that they are biorthogonal everywhere,

$$\langle \mathrm{e}_i, \mathrm{e}^{*j} \rangle = \delta_i^j. \tag{3.3}$$

The following theorem holds in this coordinate system.

THEOREM 3.1. *In a dually flat manifold, there exist two convex potential functions $\psi(\theta)$ and $\varphi(\eta)$ locally, such that the two coordinates are connected by the Legendre transformation,*

$$\psi(\theta) + \varphi(\eta) - \sum \theta^i \eta_i = 0 \tag{3.4}$$

$$\theta^i = \frac{\partial}{\partial \eta_i}\varphi(\eta), \quad \eta_i = \frac{\partial}{\partial \theta^i}\psi(\theta). \tag{3.5}$$

The metric tensors $g_{ij} = \langle \mathbf{e}_i, \mathbf{e}_j \rangle$ and $g^{ij} = \langle \mathbf{e}^{*i}, \mathbf{e}^{*j} \rangle$ are mutually inverse matrices and are given by

$$g_{ij} = \frac{\partial^2}{\partial \theta^i \partial \theta^j}\psi(\theta), \quad g^{ij} = \frac{\partial^2}{\partial \eta_i \partial \eta_j}\varphi(\eta). \tag{3.6}$$

On the contrary, when a differentiable convex function $\psi(\theta)$ exists in a manifold M with coodinates θ, we can introduce the following dually flat structure

$$\{M, g, \nabla, \nabla^*\}$$

given by

$$g_{ij}(\theta) = \frac{\partial^2}{\partial \theta^i \partial \theta^j}\psi(\theta),$$
$$\Gamma_{ijk}(\theta) = 0$$

and

$$\overset{*}{\Gamma}{}^{ijk}(\eta) = 0$$

in the ∇^*-affine coordinates

$$\eta_i = \frac{\partial}{\partial \theta^i}\psi(\theta).$$

Therefore, the dually flat structure is regarded as the natural geometry of the Legendre transformation.

A dually flat manifold has moreover the following structure which is not known by the theory of the Legendre transformation. Let us define a local function $D : S \times S \to \mathrm{R}$ in a dually flat S. This is a function $D(p : q)$ of two points p, q of S, and is defined by

$$D(p : q) = \psi(\theta_p) + \varphi(\eta_q) - \sum \theta_p^i \eta_{qi}, \tag{3.7}$$

where $\theta_p = (\theta_p^i)$ and $\eta_q = (\eta_{q^i})$ are the ∇- and ∇^*-affine coordinates of p and q, respectively.

It is easy to prove that

$$D(p : q) \geq 0,$$

where the equality holds if $p = q$, and, when $q = p + dp$ is infinitesimally close to p, it is expanded as

$$D(p : p + dp) = \frac{1}{2}\sum g_{ij} d\theta^i d\theta^j.$$

Hence, D is a generalization of a half of the square of the Riemannian distance. The symmetry condition

$$D(p : q) = D(q : p)$$

does not in general hold nor the triangular inequality. The function D is called the divergence function. When the dually flat manifold is self-dual, it is Euclidean and the divergence reduces to a half of the squared Euclidean distance between p and q. It is symmetric in this case. The divergence in general satisfies the following generalized Pythagoras theorem.

THEOREM 3.2. *Let p, q and r be three points in a dually flat S such that the ∇^*-geodesic connecting p and q is orthogonal at q to the ∇-geodesic connecting q and r. Then,*

$$D(p:q) + D(q:r) = D(p:r). \tag{3.8}$$

An important consequence of this theorem is the following projection theorem.

THEOREM 3.3. *Let V be a closed subset of a dually flat S that has a smooth boundary hypersurface ∂V. Let p be a point outside V. Then, the point $\hat{p} \in V$ that minimizes $D(p:q)$, $q \in V$, is the ∇^*-geodesic projection of p to V, that is, when $D(p:\hat{p})$ is the minimum of $D(p:q)$, $q \in V$, $p \in \partial V$ and the ∇^*-geodesic connecting p and $\hat{p}$ is orthogonal to ∂V at $\hat{p}$.*

There is another property which is important for applications. For dual local coordinate systems $\theta = (\theta^i)$ and $\eta = (\eta_i)$, we partition the coordinates into two parts,

$$\theta = (\theta_1, \theta_2), \qquad \eta = (\eta_1, \eta_2)$$

where the θ_1 and η_1 parts have k-dimensions and the remaining θ_2 and η_2 have n - k dimensions. Let us consider a family of k-dimensional hyper-surfaces parametrized by n - k real parameters $c = (c_{k+1}, \cdots, c_n)$,

$$A^k(c) = \{\theta \mid \theta_2 = c\}$$

Then, A^k is a ∇-flat hypersurface, and $A^k(c)$'s form a (local) foliation of S,

$$A^k(c) \bigcap A^k(c') = \phi$$

when $c \neq c'$ and

$$\bigcup_c A^k(c) \supset S.$$

Dually to the above, we have another foliation,

$$B^{n-k}(d) = \{\eta \mid \eta_1 = d\}$$

where $d = (d_1, \cdots, d_k)$ is the parameter to specify $(n-k)$-dimensional submanifolds. Any $B^{n-k}(d)$ is ∇^*-flat.

THEOREM 3.4. *For a dual pair of foliations $\mathcal{A} = \{A^k(c)\}$ and $\mathcal{B} = \{B^{n-k}(d)\}$, a point p is determined uniquely as the intersection of a $A^k(c)$ and a $B^{n-k}(d)$. Hence, (c, d) can be used as a local coordinate system, called the mixed coordinate system. Let $p = (c_p, d_p)$ and $q = (c_q, d_q)$ in the mixed coordinates. Then,*

$$D(p:q) = D(p:\hat{p}) + D(\hat{p}:q), \tag{3.9}$$

$$D(q:p) = D(q:\hat{q}) + D(\hat{q}:p), \tag{3.10}$$

where $\hat{p}$ is the ∇^*-projection of p to $A^k(c_p)$ and $\hat{q}$ is the ∇-projection of q to $B^{n-k}(d_q)$.

We have discussed the properties of dually flat manifolds. When S is a manifold of dually constant curvatures, it also enjoys some nice properties expressed by transformations and potential functions. They are studied by Kurose [1993].

4. Some Unsolved Mathematical Questions

The dual structure originated from information sciences, especially statistics. Some special cases of the dual structure is given rise to by the affine differential geometry (Blashkke [1923], Nomizu and Sasaki [1994], Nomizu and Pinkall [1987], Simon et al. [1991]) and also by the Hessian manifold of Shima [1995]. However, there remain many problems to be studied from the mathematical point of view. We show some of them.

1. Embedding problem : It is known that an n-dimensional Riemannian manifold is embedded isometrically in an $n(n+1)/2$-dimensional flat (i.e. Euclidean) manifold. Is it possible to embed an n-dimensional manifold M of the dual structure in an m-dimensional dually flat manifold S such that the dual structure of M is induced from that of S? How large m is necessary? It is conjectured from statistics that $m = \infty$ is in general necessary. This is because a general statistical model $M = \{p(x, \xi)\}$ is not a curved exponential family. When and only when M has a finite-dimensional sufficient statistics for repeated observations, it is a curved exponential family which is embedded in a dually flat exponential family. Then, what is the characterization of a manifold which is embedded in a finite-dimensional dually flat manifold compatibly?
2. Affine embedding problem : Let M be an n-dimensional submanifold in an $(n+1)$-dimensional affine space. Then, it is known from affine differential geometry that a dual structure is induced in M. However, it produces only a special type of the dual structure. More precisely, when and only when a dual structure is conformally flat, it can be realized as a hypersurface of an $(n+1)$-dimensional affine space (Nomizu and Sasaki [1994], Kurose [1990; 1991]). Then, what type of the dual structure is induced in an n-dimensional submanifold of an m-dimensional affine space? See Lauritzen [1987]. In other words, what is the characterization of the dual structure which is realized as a submanifold of a finite-dimensional affine space?
3. Dual flatness : Given a Riemannian manifold $\{M, g\}$ it is sometimes possible to introduce a dual flat structure when one can find T such that $\{M, g, T\}$ is dually flat. Is this always possible? If possible, is T unique? If it is not always, what is the characterizaton of a manifold having such a T?
4. Global structure : We have so far studied only local properties of the dual structure. How are local structures related to global properties? Some preliminary studies are found in Shima [1995]. This is an interesting mathematical problem.
5. Quantum probability distributions : The manifold of quantum probability has attracted much attention in relation with problems of quantum observation and information transmission by quantum states. The non-commutativity plays a fundamental role, and the related manifold has torsion structure. Construction of non-commutative information geometry is an important problem. See Petz [1994a, b], Nagaoka [1989; 1994], Hasegawa [1993] and others.

6. Infinite-dimensional statistical manifolds : We need mathematical foundations on the infinite-dimensional manifold of probability distributions (Friedrich [1991], Lafferty [1988], Kambayashi [1994]). Here, the topology of the manifold is different from that derived from the Fisher information metric. It is interesting to know the statistical role of the symplectic structure induced in it (Friedrich [1991], Lafferty [1988]; see also Barndorff-Nielsen and Jupp [1989]). Infinite dimensional structures are related to nonparametric and semiparametric statistical inference (Amari and Kawanabe [1996]).
7. Dual flatness and completely integrable dynamical systems : There are discussions concerning intrinsic relations between the dually flat structure and complete-integrability of dynamical systems. It is pointed out that the affine projection method of Linear Programming can be formulated as a completely integrable dynamical system (Bayer and Lagarias [1989]). On the other hand, the trajectory of the solution by the inner method is proved to be a ∇^*-geodesic of the related dually flat manifold. Such relations to information geometry are found in the Brockett flow and in the Lax form. See Nakamura [1993], [1994] and Fujiwara and Amari [1995].
8. Finslerian structure : When the probability distributions $\{p(x,\xi)\}$ are statistically singular such as they do not have finite Fisher information, we cannot introduce a Riemannian structure. A Finslerian structure was suggested in such a case (Amari). This is related to asymptotics of stable distributions instead of the central limit theorem and asymptotic Gaussian distributions.

5. Manifold of Probability Distributions

We touch upon the invariant structure of the manifold of probability distributions from which the dual geometry originated. See Amari [1985], Amari et al. [1987], Barndorff-Nielsen [1988], Barndorff-Nielsen, Cox and Reid [1986], Murray and Rice [1993], Kass [1989], etc. Let S be a set of the probability distributions

$$S = \{p(x,\xi)\}$$

of a random variable x parametrized by an n-dimensional real vector parameter $\xi = (\xi^1, \cdots, \xi^n) \in \mathrm{R}^n$. Here, $p(x,\xi)$ is a probability density with respect to a common measure $d\mu$ of the sample space $\mathcal{X}$. Under a certain regularity conditions, S is regarded as an n-dimensional manifold with a local coordinate system ξ. We give some examples.

1. Discrete distributions S_n.

   Let $\mathcal{X}$ be the finite sample space $\mathcal{X} = \{0, 1, 2, \cdots, n\}$ where x takes on $n+1$ values $0, 1, \cdots, n$. The probability of x is written as

$$p(x) = \sum_{i=0}^{n} \delta_i(x) p_i, \tag{5.1}$$

   where $\delta_i(x) = 1$ when $x = 1$ and 0 otherwise and

$$p_i = \mathrm{Prob}\{x = i\}. \tag{5.2}$$

   Hence, the set of all such distributions is parametrized by $\eta = (\eta_i)$,

$$\eta_i = p_i, \quad i = 1, \cdots, n$$

because of $\sum_{i=0}^{n} p_i = 1$. Hence, this is an n-dimensional manifold denoted by S_n.

2. Exponential family.

When the probability density function is written as

$$r(x,\theta) = \exp\left\{\sum_{i=1}^{n} \theta^i k_i(x) - m(x) - \psi(\theta)\right\}, \tag{5.3}$$

the family of such distributions are said to be of the exponential type. Here, $\theta = (\theta^1, \cdots, \theta^n)$ is called the natural parameter to specify distributions. By introducing a vector $x = (x_i)$, $x_i = k_i(x)$, and by taking a common measure

$$d\mu(x) = \exp\{m(x)\}dx,$$

it is rewritten in the simple form

$$p(x,\theta) = \exp\{\sum \theta^i x_i - \psi(\theta)\}, \tag{5.4}$$

where $p(x,\theta)$ is the probability density with respect to the measure $d\mu(x)$. A family $S = \{p(x,\theta)\}$ is said to be an exponential family when it can be represented in the form (5.4). The manifold S_n of discrete distributions is an exponential family because it is rewritten as

$$p(x,\theta) = \exp\{\sum \theta^i x_i - \psi(\theta)\}$$

where

$$x_i = \delta_i(x),$$
$$\theta^i = \log \frac{p_i}{p_0}.$$

A smooth submanifold M of an exponential family is said to be a curved exponential family. Let $u = (u^1, \cdots, u^m)$, $m < n$, be local coordinates of M, and is represented by the equation

$$M:\ \theta = \theta(u).$$

Then, the probability distributions of M are parametrized by u in the form

$$p(x,u) = \exp\{\sum \theta^i(u) x_i - \psi[\theta(u)]\}. \tag{5.5}$$

This is not an exponential family unless $\theta(u)$ is linear.

3. Infinite-dimensional space.

Let us consider the set of all the probability distributions which have smooth non-zero densities with respect to the Lebesgue measure,

$$p(x) > 0.$$

Such a set $S = \{p(x)\}$ does not form a manifold unless we introduce some conditions in the behaviors of $p(x)$ at $x \to \pm\infty$. The set $\mathcal{P}$ of all the smooth distributions over the one-dimensional sphere (circle) S^1 was studied by Friedrich [1991] and Lafferty [1988], see also Kambayashi [1994]. Friedrich and Lafferty showed that $\mathcal{P}$ is homeomorphic to the manifold of all the smooth deformations of S^1, while Kambayashi used the manifold of paths. Geometry of the manifolds of Gaussian and Poissonian processes were studied by Takahashi and Sekine.

What is the natural geometrical structure to be introduced in the manifold of probability distributions? It was Rao [1945] who proposed the Riemannian structure by using the Fisher information matrix. The matrix $g = (g_{ij})$ defined by

$$g_{ij}(\xi) = E\left[\frac{\partial \log p(x,\xi)}{\partial \xi^i}\frac{\partial \log p(x,\xi)}{\partial \xi^j}\right]$$

is an important quantity known as the Fisher information, where E denotes the expectation with respect to the distribution $p(x,\xi)$. Chenstov [1972] proved that the Fisher information is the unique metric under the Markovian morphisms among $\{S_m\}$, $m = 1, 2, \cdots$ in the discrete distributions (see also Campbell [1985]). He also introduced a family of affine connections called the α-connections (Amari [1982], [1985]) defined by

$$\Gamma^{(\alpha)}_{ijk}(\xi) = \{ij;k\} - \frac{\alpha}{2}T_{ijk}, \tag{5.6}$$

$$T_{ijk}(\xi) = E\left[\frac{\partial \log p(x,\xi)}{\partial \xi^i}\frac{\partial \log p(x,\xi)}{\partial \xi^j}\frac{\partial \log p(x,\xi)}{\partial \xi^k}\right]. \tag{5.7}$$

The duality of the affine connections was formulated by Nagaoka and Amari [1982] (see also Amari [1985], Murray and Rice [1993]).

Under a certain regularity conditions, the α-structure given by $\{S, g, \alpha T\}$ or $\{S, g, \nabla^{(\alpha)}, \nabla^{(-\alpha)}\}$ is the unique geometrical structure under the transformation of random variable x to a sufficient statistics $y = f(x)$. A typical case of such f is a bijective transformation. (See Amari and Nagaoka [1994], see also Li [1993] for regularity conditions). A variable $y = f(x)$ is called a sufficient statistics, when the probability distribution $p(x,\xi)$ is of the form

$$p(x,\xi) = h(y;\xi)r(y)$$

where r does not depend on ξ. In this case, y carries all the necessary information concerning the underlying probability distributions.

6. Applications of Information Geometry

6.1. Statistical inference. Because information geometry originated from the manifold of probabilities, it gives naturally a mathematical foundation of statistical inference. The basic structures will be explained here intuitively. To this end, let us consider a curved exponential family $M = \{p(x,u)\}$, which is embedded in an exponential family $S = \{q(x,\theta)\}$,

$$q(x,\theta) = \exp\{\sum \theta^i x_i - \psi(\theta)\}, \tag{6.1}$$

$$p(x,u) = q\{x,\theta(u)\} = \exp\{\sum \theta^i(u)x_i - \psi(\theta)\}. \tag{6.2}$$

The manifold M is a curved manifold in S, where $u = (u^a)$ is a coordinate system in M.

Let $x_1, x_2, \cdots, x_N$ be N independent observations from an unknown probability distribution $p(x,u)$. From the observed sample, we need to estimate the true value u by $\hat{u} = \hat{u}(x_1, \cdots, x_N)$. In the framework of the exponential family S, we take the η-coordinate system in it given by

$$\eta_i = \frac{\partial}{\partial \theta^i}\psi(\theta) = E_\theta[x_i]. \tag{6.3}$$

Then, the natural estimator $\hat{\eta}$ in S is given by replacing the expectation by the sample average,

$$\hat{\eta}_i = \frac{1}{N}\sum x_i. \tag{6.4}$$

This is the maximum likelihood estimator in S and the $\hat{\eta}$ summarizes all the Fisher information included in N observations. Let $\hat{p}$ be the point in S whose η-coordinates are given by $\hat{\eta}$. We call $\tilde{p}$ the observed point. However, $\hat{p}$ does not in general belong to M so that it does not give an estimator $\hat{u}$. Geometrically speaking, an estimator e is a mapping from S to M,

$$e : S \to M \ ; \ \hat{p} \mapsto \hat{u},$$

where $\hat{u}$ is the parameter representing $p(x, \hat{u})$ in M. Let

$$e^{-1}(u) = \{p|\ e(p) = u,\ p \in S\}. \tag{6.5}$$

We call $e^{-1}(u)$ the ancillary manifold of u associated with the estimator. When and only when the observed $\hat{p}$ belongs to $e^{-1}(u)$, the estimated value is u.

In order that an estimator is consistent, that is, $\hat{u}$ converges to the true u in probability as N tends to infinity, it is necessary and sufficient that

$$e^{-1}(u) \ni p(x, u). \tag{6.6}$$

Now, let us consider the expected estimation error given by the covariance matrix

$$C^{ab} = \frac{1}{N}E[(\hat{u}^a - u^a)(\hat{u}^b - u^b)]. \tag{6.7}$$

When N is large, the law of large numbers guarantees that $\hat{p}$ is in an $(1/\sqrt{N})$-neighborhood of the true distribution. Hence, in order to evaluate the performance of a consistent estimator, we can linearize the estimation problem in the tangent space $T_S(u)$ and $T_M(u)$ of S and M at $p(x, u)$. The linear analysis gives the following well known universal result of the asymptotic theory: When and only when $e^{-1}(u)$ is orthogonal (with respect to the Fisher information metric) to M, a consistent estimator has the minimum expected error given by

$$C^{ab} = g^{ab} + O\left(\frac{1}{N}\right) \tag{6.8}$$

where $g^{-1} = (g^{ab})$ is the inverse of the Fisher metric of M.

When N is not so large, we need to evaluate the term $O\left(\frac{1}{N}\right)$. In this case, the linear approximation in the tangent space T_S is not sufficient, and we need to evaluate the curvatures which show how T_S changes. For a bias-corrected efficient estimator, we have the following universal result:

$$C^{ab} = g^{ab} + \frac{1}{2N}\left(2\left[H_M^{(e)}\right]^{2ab} + \left[H_A^{(m)}\right]^{2ab} + \left[\Gamma_M^{(m)}\right]^{2ab}\right) + O\left(\frac{1}{N^2}\right), \tag{6.9}$$

where $\left[H_M^{(e)}\right]^2$, $\left[H_A^{(m)}\right]^2$ and $\left[\Gamma_M^{(m)}\right]^2$ are the squares of the e-curvature of M, m-curvature of $e^{-1}(u)$ and the m-connection of M, respectively. Hence, an estimator is second-order efficient when the associated $e^{-1}(u)$ has vanishing m-curvature. The maximum-likelihood estimator attains this.

We can similarily construct the higher-order asymptotic theory of testing hypothesis and interval estimation by using information geometry. The theory is extended to general non-curved exponential cases where the fibre bundle structure is used (Amari [1987a], Barndorff-Nielsen and Jupp [1989]), to sequential estimation where conformal information geometry is used (Okamoto, Amari and Takeuchi [1991]), and to semiparametric cases where the Hilbert bundle structure is introduced (Amari [1982], Amari and Kawanabe [1996]).

6.2. Linear systems and time series. Let us consider a time series $\{x_t\}$, $t = 0, \pm 1, \pm 2, \cdots$, generated by white Gaussian noises $\{z_t\}$ as

$$\sum_{i=0}^{p} a_i x_{t-i} = \sum_{i=1}^{q} b_i z_{t-i}. \tag{6.10}$$

The probability distribution of $\{x_t\}$ is specified by the parameter $\mathrm{u} = (a_0, \cdots, a_p; b_1, \cdots, b_q)$ and is written symbolically as $p(\{x_t\}, \mathrm{u})$. The transformation from $\{z_t\}$ to $\{x_t\}$ is called a linear system, so that we can study the set of the linear systems of the form (6.10) by information geometry. See Amari [1987b].

A control system modifies its performance by adding a feedback loop from the output. The set of linear feedback systems also form a differentiable manifold, where the information-geometric dual structure is introduced. It plays an important role in control systems theory. See Ohara and Amari [1994]; Ohara, Suda and Amari [1995].

6.3. Neural networks and nonlinear systems. Let us consider a nonlinear input-output system

$$\mathrm{x} = \mathrm{f}(\mathrm{z}; \mathrm{u}) \tag{6.11}$$

parametrized by parameter u, where z is an input signal and x is the corresponding output signal. The set of such nonlinear systems forms a differentiable manifold where u is a (local) coordinate system. When z is a random input, (6.11) defines a probability distribution $p(\mathrm{x}, \mathrm{u})$ specified by u. Hence, the dual structure is introduced in the manifold of nonlinear systems.

Sometimes, the input-output relation is stochastic, for example,

$$\mathrm{x} = \mathrm{f}(\mathrm{z}; \mathrm{u}) + \mathrm{n} \tag{6.12}$$

where n is a random noise. In this case, the behaviour of a system is specified by a conditional probability distribution $p(\mathrm{x}|\mathrm{z}; \mathrm{u})$ of x conditioned on z. This also introduces the dual structure of information geometry. It can be extended to the nonlinear dynamical systems

$$\mathrm{x}_{t+1} = \mathrm{f}(\mathrm{x}_t; \mathrm{u}). \tag{6.13}$$

Neural networks are nonlinear systems which include both signal transformations (feedforward networks) and dynamics (recurrently connected networks). They are regarded as neural manifolds, where the synaptic connection weights play the role of a coordinate system. The neural information processing is studied by information geometry. See Amari [1991], [1995]; Amari, Kurata and Nagaoka [1992].

6.4. Information theory and statistics. Information theory studies problems of information transmission where signals are generated from an information source which is subject to a probability distribution. Hence, the dual structure is naturally introduced when we consider a family of information sources. Information geometry plays a fundamental role in the multiterminal statistical inference problem where statistical inference is discussed under multiterminal information transmission. The dual foliation of the underlying manifold plays a fundamental role, because it decomposes the marginal structures from correlational structures. See Amari [1989], Amari and Han [1989].

6.5. Linear programming, convex analysis and completely integrable dynamical systems. A dually flat manifold is naturally constructed from a convex function and the related Legendre transformation. For example, we take the linear programming problem. Let M be a convex region in R^n,

$$M = \{\theta | \theta \in \mathrm{R}^n, \ \sum_{i=1}^{n} A_i^\mu \theta^i - b^\mu \geq 0, \ \mu = 1, \cdots, m\}. \tag{6.14}$$

The linear programming problem is to find a point $\boldsymbol{\theta}$ that minimizes a linear function

$$f(\theta) = \sum c_i \theta^i. \tag{6.15}$$

We can introduce a convex function ψ in the inner region of M by

$$\psi(\theta) = -\sum_{\mu=1}^{m} \log \left(\sum A_i^\mu \theta^i - b^\mu \right). \tag{6.16}$$

This automatically defines the dually flat structure in M. It was shown by Fujiwara and Amari [1995] that the Karmarkar affine projection method is the inner method of obtaining the optimal θ along the ∇^*-geodesic. It was also shown by Bayer and Lagarias [1989] that the method can be reformulated as a completely integrable dynamical system. This is also related to the Brockett flow and solitons. The relation between the dually flat structure and the completely integrable dynamical systems is an important unsolved problem which attracts much attention. See Nakamura [1993], [1994], Fujiwara and Amari [1995].

6.6. Quantum information geometry. Information geometry originated from probability distributions. Quantum mechanics defines another type of probability, the complex probability, where non-commutative operators play a fundamental role. It is an important problem to construct non-commutative information geometry to understand the quantum observation problem. Information transmission via quantum states, and quantum-statistical estimation of states are also an important problem. There are lots of efforts along this line. See Petz [1994a, b], Nagaoka [1989, 1994], Fujiwara and Nagaoka, Hasegawa [1993].

7. Conclusions

Information geometry is a field of mathematical sciences which includes wide areas of applications. It also suggests mathematics a new but classic concept, in its nature, of the dual affine connections, which should be studied further in the modern context. The present paper reviews the fundamental concepts of information geometry and surveys its applications briefly. The field is developing rapidly, and there are lots of problems to be solved. It is hoped that the present review ignites cooperation between mathematical scientists and pure mathematicians.

References

[1] AMARI, S., *Differential geometry of curved exponential families — curvature and information loss*, Ann. Statist. **10**, pp. 357–385 (1982)

[2] AMARI, S., *Differential-Geometrical Methods in Statistics*, Springer lecture Notes in Statistics **28**, (1985).

[3] AMARI, S., *Differential geometry of a parametric family of invertible linear systems — Riemannian metric, dual affine connections and divergence*, Mathematical systems Theory **20**, pp. 53–82 (1987b).

[4] AMARI, S., *Fisher information under restriction of Shannon information*, AISM **41**, pp. 623–648 (1989).

[5] AMARI, S., *Dualistic Geometry of the Manifold of Higher-Order Neurons*, Neural Networks **4**, pp. 217–227 (1989)

[6] AMARI, S., BARNDORFF-NIELSEN, O. E., KASS, R. E., LAURITZEN, S. L. and RAO, C. R,, *Differential Geometry in Statistical Inferences*, IMS Lecture Notes Monograph Series **10**, Hayward California IMS (1987).

[7] AMARI, S. and HAN, T. S., *Statistical inference under multi-terminal rate restrictions — a differential geometrical approach*, IEEE Trans. on Information Theory **IT-35**, pp. 217–227 (1989).

[8] AMARI, S. and KAWANABE, M., *Information geometry of semiparametric statistical models and estimating functions*, to appear (1996).

[9] AMARI, S., KURATA, K. and NAGAOKA, H., *Information Geometry of Boltzmann Machines*, IEEE Trans. on Neural Networks **3** No.2, pp. 260–271 (1992).

[10] AMARI, S. and NAGAOKA, H., *Method of Information Geometry*, Iwanami Lecture Series in Applied Mathematics (in Japanese, to be translated into English by AMS and Oxford University Press) (1994).

[11] BARNDORFF-NIELSEN, O., *Parametric Statistical Models and Likelihood*, Springer Lecture Notes in Statistics **50**, (1988).

[12] BARNDORFF-NIELSEN, O., COX, R. D. and REID, N., *The role of differential geometry in statistical theory*, Int. Staatist. Rev. **54**, pp. 89–96 (1986).

[13] BARNDORFF-NIELSEN, O. and JUPP, P. E., *Approximating exponential models*, AISM **41**, pp. 247–267 (1989).

[14] BAYER, D. A. and LAGARIAS, J. C., *The nonlinear geometry of linear programming*, Trans. AMS. **314**, pp. 499–527 (1989)

[15] BLASCHKE, W., *Vorlesungen über Differentialgeometrie II*, Affine Differentialgeometrie, Springer Berlin (1923).

[16] BURDET, G. and PERRIN, M., *Conformal implementation of the projective equivalence over statistical manifolds*, Indagationes Mathematical, to appear (1996).

[17] CHENTSOV, N. N., *Statistical Decision Rules and Optimal Inference* (in Russian), Nauka Moscow; translated in English, (1982), AMS Rhode Island (1972).

[18] FRIEDRICH, V. T., *Die Fisher-information und symplektische Strukturen*, Math. Nachr. **153**, pp. 273–296 (1991).

[19] FUJIWARA, A. and AMARI, S., *Gradient systems in view of information geometry*, Physica D **80**, pp. 317–327 (1995).

[20] HASEGAWA, H., *α-divergence of the non-commutative information geometry*, Rep. Math. Phys. **33**, pp. 87–93 (1993).

[21] KAMBAYASHI, T., *Statistical manifold of infinite dimension*, Transactions of the Japan Society for Industrial and Applied Mathematics **4** No3, pp. 211–228 (in Japanese) (1994).

[22] KASS, R. E., *The geometry of asymptotic inference* (with discussions), Statistical Science **4**, pp. 188–234 (1989).

[23] KUROSE, T., *On the divergence of 1-conformally flat statistical manifold*, Tohoku Math.Ê J. **46**, pp 427–433 (1994).

[24] KUROSE, T., *On the realization of statistical manifolds in affine space*, (1994).

[25] KUROSE, T., *Dual connections and affine geometry*, Math. Z. **203**, pp. 115–121 (1990).

[26] LAFFERTY, J. D., *The density manifold and configuration space quantization*, Trans. AMS **305**, pp. 699–741 (1988)

[27] LAURITZEN, S. L., *Conjugate connection in statistical theory*, in Dodson, C. T. J. ed., Geometrization of Statistical Theory, ULDM Publications, pp. 33–51 (1987).

[28] LI, H., *Determination of geometrical structures on statistical experiments, probability and statistics*, Proceedings of the Wuhan Meeting (eds. A. Badrkian, P-A Meyer, J-A Yan), World Scientific, pp. 133–1148 (1993).

[29] MURRAY, M. K. and RICE, J. W., *Differential Geometry and Statistics*, Chapman & Hall (1993).

[30] NAGAOKA, H., *On the relation between Kullback divergence and Fisher information — from classical systems to quantum systems*, an outline of IEICE Tech. Rep. **89**, pp. 9–14 (1989).

[31] NAGAOKA, H., *Differential geometrical aspects of quantum state estimation and relative entropy*, METR94-**14** (1994).

[32] NAGAOKA, H. and AMARI, S., *Differential geometry of smooth families of probability distributions*, METR **82-7**, University of Tokyo (1982).

[33] NAKAMURA, Y., *Completely integrable gradient systems on the Manifolds of Gaussian and multinomial distributions*, Japan Journal of Industrial and Applied Mathematics **10** No.2, pp. 179–189 (1993).

[34] NAKAMURA, Y., *A Tau-function for the finite Toda molecule and information spaces*, Contemporary Mathematics **179** (1994).

[35] NOMIZU, K. and PINKALL, O., *On the geometry of affine immersions*, Math. Z. **195**, pp. 165–178 (1987).

[36] NOMIZU, K. and SASAKI, T., *Affine Differential Geometry*, Cambridge Trans. in Mathematics (1994).

[37] OHARA, A. and AMARI, S., *Differential geometric structures of stable state feedback systems with dual connections*, Kybernetica **30**, pp. 369–386 (1994).

[38] OHARA, A., SUDA, N. and AMARI, S., *Dualistic differential geometry of positive definite matrices and its applications to related problems*, Linear Algebra Appl. (to appear).

[39] OKAMOTO, I, AMARI, S. and TAKEUCHI, *Asymptotic theory of sequential estimation procedures for curved exponential families*, Ann. Statist. **19**, pp. 961–981 (1991).

[40] PETZ, D., *Geometry of canonical correlation on the state space of a quantum system*, J. Math. Phys. **35**, pp. 780–795 (1994a).

[41] PETZ, D., *Monotone metric on matrix spaces*, Linear Algebra Appl. (1994b).

[42] RAO, C. R., *Information and accuracy attainable in the estimation of statistical parameters*, Bull. Calcutta. Math. Soc. **37**, pp. 81–91 (1945).

[43] SHIMA, H., *Homogeneous Hessian Manifolds*, Ann. Inst. Fourier Grenoble **30**, pp. 91–128 (1980).

[44] SHIMA, H., *Vanishing theorems for compact Hessian manifolds*, Ann. Inst. fourier Grenoble **36** 3, pp. 183–205 (1986).

[45] SHIMA, H., *Hessian manifolds of constant Hessian sectional curvature*, J. Math. Soc. Japan **47** 4, pp. 735–753 (1995).

[46] SIMON, U., SCHWENK-SCHELLSCHMIDT, A. and BIESEL, H., *Introduction to the Affine Differential Geometry of Hypersurfaces*, Lecture Notes, Science U. of Tokyo (1991).

DEPARTMENT OF MATHEMATICAL ENGINEERING AND INFORMATION PHYSICS, UNIVERSITY OF TOKYO, TOKYO 113, JAPAN; RIKEN FRONTIER RESEARCH PROGRAM, BRAIN INFORMATION PROCESSING

Current address: RIKEN FRP, Wako-shi, Hirosawa, Saitama 351-01, Japan

E-mail address: amari@sat.t.u-tokyo.ac.jp

Contemporary Mathematics
Volume **203**, 1997

An Example of Dynamical Behaviour of the Relative Entropy

G. BURDET †, H. NENCKA ‡, M. PERRIN †

ABSTRACT. The differential geometry content of a particular statistical model, namely the Gaussian manifold, is used to exhibit a dynamical aspect of the relative entropy. According to the value of the parameter which characterizes the Chentsov-Amari connection, elliptic or hyperbolic comportments are attained.

1. Introduction

Suppose we have a statistical problem involving a n-parameter familly of probability functions, since 1945, [**5**], differential geometrical considerations have been introduced on the parameter space by endowing it with a Riemannian metric in terms of the Fisher information matrix. Thereafter, an affine-metric point of view was set by defining a one-parameter family of linear connections [**1**][**2**][**4**]. This framework is now called *information geometry*, and is applicable not only to statistics but also to more wide areas of information sciences such as control system theory or time-series analysis, information theory, neural networks,...

Another quantity is often to be taken into account, namely the *relative entropy* which expresses the mean information for discrimination in favor of a distribution probability against another one. The relative entropy is, à priori, connection independent, but it is known that it is intimately related to a particular one, whence the idea to introduce a family of divergences, in one to one relation with the family of connections and generalizing the relative entropy. However the interest of such a generalization is not obvious, so in this paper we want to adopt another gait, namely to study the behaviour of the relative entropy along auto-parallel curves in terms of the parameter characterizing the connection family. The investigation will be achieved on a skinny, but almost fully explicit, example, the one-dimensional Gaussian distribution.

1991 *Mathematics Subject Classification.* 94 A 17 ; 53 C 99.

2. Self-parallel curves of the Gaussian manifold

Let us consider the gaussian law $N(\mu, \sigma)$ with $\mu \in \mathbb{R}$ and $\sigma \in]0, +\infty[$ an exponential family, the probability function of which being written in the familiar form

$$p(x) = \frac{1}{\sigma\sqrt{2\pi}} \exp\left[-\frac{(x-\mu)^2}{2\sigma^2}\right] \tag{1}$$

with respect to the Lebesgue measure over $\mathbb{R}$. Following a well established formalism (see Ref [2]), a 2-dimensional *statistical* manifold $\mathcal{N}$ can be associated, the *Gaussian manifold* defined by a single one atlas with coordinate system $\{\mu, \sigma\}$, so $\mathcal{N}$ is an half-plane. Both fully symmetric covariant tensor fields, denoted by g and t, are realized as expectation of tensorial products of the differential of the Gaussian log-likehood, one gets

$$g = \frac{1}{\sigma^2}\ [\,\mathrm{d}\mu \otimes \mathrm{d}\mu\ + 2\ \mathrm{d}\sigma \otimes \mathrm{d}\sigma] \tag{2}$$

$$t = \frac{2}{\sigma^3}\ [\,\mathrm{sym}(\mathrm{d}\mu \otimes \mathrm{d}\mu \otimes \mathrm{d}\sigma) + 4\ \mathrm{d}\sigma \otimes \mathrm{d}\sigma \otimes \mathrm{d}\sigma\,]\,. \tag{3}$$

The first, g, endows $\mathcal{N}$ with a Riemannian structure. From the second, t, called the tensor of connection deficiency, a 1-parameter ($\alpha \in \mathbb{R}$) family of torsionless, non-metric linear connections is constructed satisfying the relation $\overset{\alpha}{\nabla} g = \alpha t$. Corresponding to the expressions (2-3), the non-zero coefficients of the connection family are given by

$$\Gamma^{\sigma}_{\mu\mu} = \frac{1-\alpha}{2\sigma}\,, \quad \Gamma^{\mu}_{\mu\sigma} = \Gamma^{\mu}_{\sigma\mu} = -\frac{1+\alpha}{\sigma}\,, \quad \Gamma^{\sigma}_{\sigma\sigma} = -\frac{1+2\alpha}{\sigma}\,, \tag{4}$$

and for the Riemann and Ricci tensor components one gets

$$(\overset{\alpha}{Riem})^{\mu}_{\sigma\mu\sigma} = (\overset{\alpha}{Ric})_{\sigma\sigma} = 2(\overset{\alpha}{Riem})^{\sigma}_{\mu\sigma\mu} = 2(\overset{\alpha}{Ric})_{\mu\mu} = \frac{\alpha^2-1}{\sigma^2}\,. \tag{5}$$

Moreover the scalar α-curvature of $(\mathcal{N}, g, t)$ is constant and given by

$$\overset{\alpha}{R} =: g^{-1}(\overset{\alpha}{Ric}) = \alpha^2 - 1\,, \tag{6}$$

As noticed by Amari, the statistical Gaussian manifold $(\mathcal{N}, g, t)$ is diffeomorphic to the Poincaré half-plane for $\alpha = 0$.

Self-parallel curves $t \to C(t), t \in I \subset \mathbb{R}$, in a statistical manifold are such that the vector field $\dot{C} =: \frac{\mathrm{d}C}{\mathrm{d}t}$ is parallel translated along C with respect to the α-connection, equivalently they are curves of acceleration zero $\ddot{C} =: \overset{\alpha}{\nabla}_{\dot{C}}\dot{C} = 0$.

COROLLARY. *In the choosen coordinate system $\{\mu, \sigma\}$ a curve in $(\mathcal{N}, g, t)$ is self-parallel if and only if its coordinate functions $\{\mu, \sigma\} \circ C$ satisfy*

$$\ddot{\mu} - 2\frac{1+\alpha}{\sigma}\ \dot{\mu}\dot{\sigma} = 0 \quad , \quad \ddot{\sigma} + \frac{1-\alpha}{2\sigma}\dot{\mu}^2 - \frac{1+2\alpha}{\sigma}\ \dot{\sigma}^2 = 0 \tag{7}$$

using a common abbreviation for the coordinate functions.

Then the existence and uniqueness theorem for ordinary differential equations implies there exists a unique self-parallel curve starting at $\{\mu_0, \sigma_0\} = C(0)$ with initial velocity $\{\dot{\mu}_0, \dot{\sigma}_0\} = \dot{C}(0)$.

Let us recall the well known Riemannian case ($\alpha = 0$), there are two kinds of *geodesics* - all "vertical" lines described by the solutions $\{\mu = \mu_0\ ,\ \sigma = \sigma_0\ e^t\}$, - all semicircles (with respect to the Euclidean metric conformally related to g) depicted by $\{\sigma = \sigma_0/\mathrm{ch}(t)\ ,\ \mu = \mu_0 + \sigma_0\sqrt{2}\ \mathrm{th}(t)\}$ where $\sigma_0 \in\ ]0,+\infty[$ and $\mu_0, t \in \mathbb{R}$; hence $(\mathcal{N}, g)$ is geodesically complete.

Let us now turn to the general case with the following first remark. The differential system (7) remains invariant by translations on μ and so admits the solutions

$$C_V = \left\{\mu = \mu_1\ ,\ \sigma = \sigma_1 t^{-\frac{1}{2\alpha}},\ t > 0\right\}$$

i.e all "vertical" lines, but with respect to which the α-connection family is not complete.

To show the existence of self-parallel curves generalizing the Riemannian semicircles, the notion of cUp *-related statistical manifolds*, introduced in [**3**], has to be used. Here the symbol cUp refers to a conformal implementation of the projective geometry in which the properties of both geometries are preserved, namely angle and self-parallel curves, up to a reparametrization.

Let us consider a family of metrics $\tilde{g}$ conformally related to the initial metric following $\tilde{g} = \sigma^\alpha g$ and a family of symmetric tensor fields $\tilde{t} = \sigma^\alpha(t - \mathrm{sym}(g \otimes \mathrm{d}\ (\ \log\sigma)))$. From the results of Ref [3] one gets the following

PROPOSITION. *For any $\alpha \neq 0$, there is a unique connection $\tilde{\nabla}$ projectively related to $\overset{\alpha}{\nabla}$ through the 1-form $\alpha\ \mathrm{d}\ (\log\sigma)$, which is such that $\tilde{\nabla}\tilde{g} = \alpha\tilde{t}$.*

The coefficients of $\tilde{\nabla}$ are given by

$$\tilde{\Gamma}^{\sigma}_{\mu\mu} = \frac{1-\alpha}{2\sigma}\ ,\quad \tilde{\Gamma}^{\mu}_{\mu\sigma} = \tilde{\Gamma}^{\mu}_{\sigma\mu} = \tilde{\Gamma}^{\sigma}_{\sigma\sigma} = -\frac{1}{\sigma}\ , \tag{8}$$

and the differential system defining the self-parallel curves ($s \to \tilde{C}(s)$) is the following

$$\mu'' - 2\mu'\ \frac{\sigma'}{\sigma} = 0\quad ,\quad \frac{\sigma''}{\sigma} - \left(\frac{\sigma'}{\sigma}\right)^2 + \lambda\ \left(\frac{\mu'}{\sigma}\right)^2 = 0 \tag{9}$$

with $\lambda = \frac{1}{2}(1-\alpha)$, and $'$ denoting the derivation with respect to the new affine parameter denoted by s.

The system (9) can be explicitly solved, so one gets the following

PROPOSITION. *Half-conics, of the Euclidean half-plane, of all possible shapes and sizes are solutions of Eq.(9) according to the sign of λ, more exactly*

$$\tilde{C}_+ = \Big\{ half - ellipses\ centered\ in\ (\mu = m\ , \sigma = 0)$$
$$horizontal\ half - axis\ b\sqrt{2/\lambda},$$
$$vertical\ half - axis\ b, \forall \lambda > 0 \Big\}$$

and

$$\tilde{C}_- = \Big\{ half - branchs\ of\ hyperbolas :$$
$$asymptotes\ \sigma - \epsilon\sqrt{-\lambda}(\mu \quad m)\ ,$$
$$vertex\ \sigma = b\ in\ (\mu = m, \sigma = 0),\ \forall \lambda < 0 \Big\}.$$

where $m = \mu_o + \frac{\sigma_o \sigma'_o}{\lambda \mu'_o}$ *and* $b = \sigma_o \sqrt{1 + \frac{1}{\lambda}(\frac{\sigma'_o}{\mu'_o})^2}$ *are expressed in terms of the initial conditions* $\sigma_o > 0, \mu'_o \in \mathbb{R} \backslash 0,\ \sigma'_o, \mu_o \in \mathbb{R}$ *for* $s = o$. *Both types of conics verify* $\lambda(\mu - m)^2 + \sigma^2 = b^2$,

$$\tilde{C}_0 = \Bigg\{ horizontal\ lines\ \mu - \mu_o = s,\ \sigma = \sigma_o\ ,$$
$$half - parabolas\ \mu - \mu_o = \frac{\sigma_o \mu'_o}{2\sigma'_o}\left(\left(\frac{\sigma}{\sigma_o}\right)^2 - 1\right)\ ,\ for\ \lambda = 0 \Bigg\}.$$

Respective affine parametrizations are given by

$$\lambda > 0 : \qquad \sigma = \frac{\sigma_o}{\mathrm{ch}(\gamma s) + \beta\,\mathrm{sh}(\gamma s)}\ ,$$
$$\mu - \mu_o = \frac{\sigma_o \sigma'_o}{\lambda \mu'_o}\left(1 - \frac{\mathrm{ch}(\gamma s) + \beta^{-1}\,\mathrm{sh}(\gamma s)}{\mathrm{ch}(\gamma s) + \beta\,\mathrm{sh}(\gamma s)}\right) \tag{10.a}$$

$$\lambda = 0 : \qquad \sigma = \sigma_o\ ,$$
$$\mu - \mu_o = \ \mu'_o\ s\ ,$$
$$et : \sigma = \sigma_o\ \exp\left(\frac{\sigma'_o}{\sigma_o} s\right)\ ,$$
$$\mu - \mu_o = \frac{\sigma_o \mu'_o}{2\sigma'_o}\left(\exp\left(2\frac{\sigma'_o}{\sigma_o} s\right) - 1\right) \tag{10.b}$$

$$\lambda < 0 : \qquad \sigma = \frac{\sigma_o}{\cos(\gamma s) + \beta \sin(\gamma s)}\ ,$$
$$\mu - \mu_o = \frac{\sigma_o \sigma'_o}{\lambda \mu'_o}\left(1 - \frac{\cos(\gamma s) + \beta^{-1} \sin(\gamma s)}{\cos(\gamma s) + \beta \sin(\gamma s)}\right) \tag{10.c}$$

where $\gamma = \frac{1}{\sigma_o}\sqrt{|\sigma_o'^2 + \lambda\, \mu_o'^2|}$, $\beta\gamma = -\frac{\lambda}{|\lambda|}\frac{\sigma_o'}{\sigma_o}$, *and, for* $\lambda < 0$, *the following constraints* $\sigma_o'^2 < |\lambda|\, \mu_o'^2$ *et* $\gamma s \in]-\frac{\pi}{2}, +\frac{\pi}{2}[$.

Then, from projective properties included into the notion c ∪ p-related statistical manifolds, one gets the following

PROPOSITION. *Self-parallel curves of the Gaussian manifold* $N(\mu, \sigma^2)$, *endowed with the associated* α*-connections and so solutions of Eq.(9), are the vertical lines* C_V *the curves* C_+, C_- *and* C_0, *with* $C = \tilde{C} \circ s$, *described as functions of an affine parameter* t *related to the previous parameter* s *by*

$$\frac{dt}{ds} = k\sigma^{-2\alpha}(s) \tag{11}$$

where $k \in \mathbb{R} \setminus \{0\}$.

For all λ the parameter t is proportional to the primitive $\int \sigma^{-2\alpha}(s)ds$ which is null at $s = 0$, but two remarks are in order

i) primitives which give $t(s)$ are explicitly known for $2\alpha \in \mathbb{Z}$ only,
ii) only some particular cases can be inverted to give $s(t)$, which are

$$\begin{array}{ll} \alpha = -1 : s = \mathrm{Argth}(t) , & \alpha = -\frac{1}{2} : s = \mathrm{Argsh}(tgt) \\ \alpha = \frac{1}{2} \;\;\; : s = \mathrm{Argsh}(t) , & \alpha = 1 \;\;\; : s = \frac{1}{2}\log(t+1) \end{array}$$

It is worth noticed that for $\alpha = \frac{1}{2}$ (and o), $t \in \mathbb{R}$, and for $\alpha = 1$, $t \in [0, +\infty[$. On the other hand, for all integer values of 2α, t belongs to an open interval I strictly contained into $\mathbb{R}$, this means that self-parallel curves are covered in a finite time.

3. Relative entropy between two same limit point curves

The *relative entropy* between two probabilities P et P^* gives a measure for discrimination in favor of P against P^*, it is also denamed *Kullback information* and is given by $K(P; P^*) = E_P[\ Log\ \frac{p}{p^*}\ ]$ in terms of probability densities . In fact this notion is more usefull than the Riemannian distance which do not reflect any stochastic or statistical properties, so is quite useless. It is easy to show that the relative entropy between two Gaussian distributions is given by

$$K(N; N^*) = \frac{1}{2}\left\{ \ln\left(\frac{\sigma^*}{\sigma}\right)^2 - \left(\frac{\sigma}{\sigma^*}\right)^2 + \left(\frac{\mu - \mu^*}{\sigma^*}\right)^2 - 1 \right\}. \tag{12}$$

Let us call *limit points* any couples (μ, σ) for which $\sigma \to 0$ when s goes to $+\infty$ ou $-\infty$. Therefore $\tilde{C}_+$ half-ellipses have two limit points, whereas $\tilde{C}_0$ half-parabolas have only one's.

Then, interesting *dynamical* properties can be exhibited by selecting, out of any same limit point sheaf into the $\tilde{C}_+$ family, couples of $finite\ limit\ relative\ entropy\ self-parallel\ curves$, half-ellipses here.

From relations (10.a) and (12) it is shown that they correspond to the following initial conditions $\beta = \beta^*, \gamma = \gamma^*$, again, by using (12), one gets the

PROPOSITION. *The relative entropy of two Gaussian distributions evolving over two $\tilde{C}_+$-self parallel curves and remaining finite at the common limit-point (γs tending to $-\infty$), is given by*

$$K(N;N^*) = \frac{1}{2}\Bigg\{\log\left(\frac{b^*}{b}\right)^2 - \left(\frac{b}{b^*}\right)^2 - 1 + \frac{1}{\lambda}\left(\frac{b-b^*}{b^*}\right)^2 \frac{1+\beta}{1-\beta}\exp(2\gamma s)\Bigg\}. \tag{13}$$

If the common limit point is obtained with $\gamma s \to +\infty$, the relative entropy is given by

$$K(N;N^*) = \frac{1}{2}\Bigg\{\log\left(\frac{b^*}{b}\right)^2 - \left(\frac{b}{b^*}\right)^2 - 1 + \frac{1}{\lambda}\left(\frac{b-b^*}{b^*}\right)^2 \frac{1-\beta}{1+\beta}\exp(-2\gamma s)\Bigg\}. \tag{14}$$

Rel.(13-14) are characteristic of an *hyperbolic* comportment, self-parallel curves move away exponentially in one direction and draw closer in the same manner in the opposite one. The importance of this gap is gauged by the relative entropy.

Let us note that this relative entropy unstability phenomenon has nothing to do with the scalar α-curvature shift of sign for $\alpha = -1$, see Rel.(6), without repercussion on the $\tilde{C}_+$-family comportment.

Concerning the $\tilde{C}_0$-family, finite limit relative entropy pair of parabolic self-parallel curves are selected by keeping $\sigma_o'/\sigma_o = \sigma_o^{*\prime}/\sigma_o^*$, then, from Rel.(10.b) and (12), one establishes the following

PROPOSITION. *The relative entropy of two Gaussian distributions evolving over two $\tilde{C}_0$-self parallel curves and remaining finite at the common limit-point ($\sigma_o's/\sigma_o$ tending to $-\infty$), is given by*

$$K(N;N^*) = \frac{1}{2}\Bigg\{\ln\left(\frac{\sigma_o^*}{\sigma_o}\right)^2 - \left(\frac{\sigma_o}{\sigma_o^*}\right)^2 - 1 + \frac{(\mu_o' - \mu_o^{*\prime})^2}{(2\sigma_o^{*\prime})^2}\exp 2(\sigma_o'/\sigma_o)s\Bigg\}. \tag{15}$$

This shows that the $\tilde{C}_0$–family is also concerned with an exponential type unstability when $\sigma_o's/\sigma_o$ tends to $+\infty$ i.e. produces an hyperbolic type comportment.

There is no limit-point on half- hyperbolic curves, but it is interesting to consider two branchs whose abcissas of minima coincide which is obtained with initial conditions such that $\beta = \beta^*, \gamma = \gamma^*$, then, by taking into account for Rel.(2.3.c) and (3.1) one gets the

PROPOSITION. *The relative entropy of two Gaussian distributions evolving over two $\tilde{C}_-$-self parallel curves whose abcissas of minima coincide is given by*

$$K(N; N^*) = \frac{1}{2}\left\{ log\left(\frac{b^*}{b^*}\right)^2 - \left(\frac{b}{b^*}\right)^2 - 1 \right.$$
$$\left. - \frac{1}{\lambda}\left(\frac{b-b^*}{b^*}\right)^2 \sin^2(\gamma s + \text{Arctg } \beta)\right\}. \tag{16}$$

In this case,it should be noticed that the relative entropy stays finite (and periodic) all along the curves, so an other evolution type is displayed which, curiously, is reminiscent that of *elliptic* zones observed in bi-dimensional fluids turbulent runnings.

Now, it is worth noticing that all the above described relative entropy properties are carried over the self-parallel curves C of the Gaussian manifold in the right affine parametrization, according to Rel.(11), the comportment speed only can be modified, for instance

- in Rel.(15), the divergence governed by $\exp 2(\sigma_o'/\sigma_o)s$ for the value $\alpha = 1$, becomes $(t+1)^{\sigma_o'/\sigma_o}$,

- in Rel (14), with $\alpha = \frac{1}{2}$, the $\exp(2\gamma s)$ divergence transforms into $(t + \sqrt{t^2+1})^\gamma$,

- finally, for $\alpha = o$, the exponential divergence is preserved. In that case, the relative entropy divergence is identical with the one's affecting horocycle segment lengths of the Poincaré half-plane geodesics, the hyperbolic system parangon.

4. Conclusion

Hence, relative entropy dynamical properties on the Gaussian manifold have been described, at least qualitatively, but mainly shown up through the notion of c∪p *related statistical manifolds*.

In other words, a *dynamical system* aspect of statistical models has been divulged by this very simple example, in which the Riemannian distance is replaced by the relative entropy.

Moreover, the type of the dynamical comportment depends upon a real parameter i.e. on the choice of a connection in the α-family.

By widely extrapolating the above results, one can imagine the fundamental part that has to play the α parameter in various fields

- control parameter to warrant stability and avoid chaotic phenomena in statistical inference and in algorithmic self-parallel curve based methods,

- behavioural control parameter in cognitive function modelization.

References

1. S-i.Amari, *Differential Geometrical Methods in Statistics*, Lecture Notes in Statistics., vol. 28, Springer-Verlag, New York, 1985.
2. S-i. Amari, O.E. Barndorff-Nielsen, R.E. Kass, S.L. Lauritzen and C.R. Rao, *Differential Geometry in Statistical Inference*, IMS Lecture Notes-Monographs Series 10 (S.S. Gupta, eds.), Amer. Math. Soc., Providence, R. I., 1987.
3. G. Burdet, M. Perrin,, *Conformal Implementation of the Projective Equivalence over Statistical Manifolds*, Prétirage CPT 94/ P.3001. To be published in Indagationes Mathematicae..
4. N.N. Chentsov, *Statistical Decision and Optimal Inference*, "Nauka'", Moscow, 1972 (Russian); English transl., American Mathematical Society., vol 53, Amer. Math. Soc., Providence, R. I., 1982..
5. C.R. Rao, *Information and the accuracy attainable in the estimation of statistical parameters*, Bull.Calcutta Math.Soc. **37** (1945), 81–91.

† Centre de Physique Théorique, CNRS-Luminy, Case 907, F-13288 Marseille, France

E-mail address: perrin@cptsu2.univ-mrs.f r

‡ Universidad da Madeira, Colegio dos jesuitas, Largo do colegio, PT-9000 Funchal, Madeira, Portugal

E-mail address: nencka@dragoeiro.uma.pt, nencka@cptsu2.univ-mrs.fr

Contemporary Mathematics
Volume **203**, 1997

Information Geometry and Learning in Formal Neural Networks

P. COMBE AND H. NENCKA

ABSTRACT. Applications of the statistical manifolds are considered in the framework of stochastic neural networks. Information geometry provides a natural geometrical interpretation for the learning processes of such networks. The learning rules are discussed on some examples.

1. Introduction

A formal neural network is a device the architecture of which is given by the topology of a finite connected directed labeled graph $\mathcal{G} = (\mathcal{V}, \mathcal{L})$. The graph is composed by a set of vertices $\mathcal{V} = \{x_1, \ldots, x_n\}$ being the set of automata (the formal neurons) and the set of directed edges $\mathcal{E} = \{e_1, \ldots, e_m\}$ with $m > n$, describing the connections between neurons (the synaptic connections). The labels (the synaptic weights) are defined as a function $W : \mathcal{E} \to \mathbb{R}$. In the following we limit ourself to simplicial graphs that is graphs with no multiple edges or loops. Then between two vertices $x_i \neq x_j$ there is at most one edge $e = e_{ij} = (x_i, x_j)$ such that the origin of e_{ij} is $\partial_0 e_{ij} := x_j$ and the extremity $\partial_1 e_{ij} := x_i$. It follows that in this case the synaptic weights can be seen as a function on a subspace of $\mathcal{V} \times \mathcal{V}$, $W(e_{ij}) := W(x_i, x_j) := W_{ij}$. A second family of internal parameters, corresponding to some thresholds of individual neurons, can be introduced as a function $b : \mathcal{V} \to \mathbb{R}$. In the following for a sake of simplicity we disgard these parameters.

The states of the network are functions $u : \mathcal{V} \to \mathcal{S}$, where $\mathcal{S} \subseteqq \mathbb{R}$ is the set of possible outputs of an individual neuron. In the case of a stochastic network the states are just random variables $u \in \mathcal{S}^{\mathcal{V}}$ and the dynamics of the network is specified by transition probabilities between possible states.

The learning of a formal neural network consists in a modification of the internal parameters of the system, in view to enable the network to accomplish

1991 *Mathematics Subject Classification.* Primary 53 C 99, 68 T 05 ; Secondary 94 A 17.

a prescribed task. In the case of stochastic networks the learning processes are based on the discrimination of the probability distributions of the possible outputs of the network with respect to expected one [**1**]. This approach has a natural geometrical interpretation in terms of affine dually flat connections for a statistical manifold of discrete probabilities [**3,7,11**]. In section 2 we discuss and illustrate briefly the geometrical notion introduced by Amari in his lecture [**5**] in the case of discrete probabilities. The examples of sequential and parallel Boltzmann machines [**1,2,8**] are considered in section 3. Finally in the last section an example of a perceptron working as an encoder [**9,1**] is considered.

2. Geometry of discrete probabilities

Let $(\Omega, \mathcal{P}(\Omega), P)$ be a probability space where $\Omega = \{\omega^1, \dots, \omega^m\}$ is a finite space and $\mathcal{P}(\Omega)$ the algebra of all subsets of Ω. This algebra is generated by the family $\{\{\omega^1\}, \dots, \{\omega^m\}\}$ of one-point subsets of Ω. A discrete probability measure P on the measurable space $(\Omega, \mathcal{P}(\Omega))$ is fully defined by its distribution function p:

$$p(\omega^k) = P[\{\omega^k\}], \quad \forall\ k = 1, \dots, m. \tag{1}$$

The distribution p appears as a random variable such that:

$$p(\omega^k) = p^k; \qquad 0 \leq p^k \leq 1, \quad \sum_{k=1}^{m} p^k = 1. \tag{2}$$

and can be written as

$$p = \sum_{k=1}^{m} p^k \delta_{\omega^k}, \quad \text{where} \quad \delta_{\omega^k}(\omega) = \begin{cases} 1 & \text{if } \omega = \omega^k, \\ 0 & \text{otherwise.} \end{cases} \tag{3}$$

Using the $m-1$ coordinates $\eta^k = p^k, \quad (k = 1, \dots, m-1)$ the distribution of probability p can be rewritten, in this privileged coordinate system $\mathcal{N}$, as

$$p = \sum_{k=1}^{m-1} \eta^k \delta_{\omega^k} + (1 - \sum_{k=1}^{m-1} \eta^k) \delta_{\omega^m}. \tag{4}$$

The set $\mathcal{S}$ of the probabilities absolutely continuous with respect to the uniform probability is an (m-1)-open simplex $\{0 < p^k < 1, \forall\ k = 1, \dots, m-1\}$. This simplex $\mathcal{S}$ can be seen as a differentiable manifold defined by an atlas containing only one chart $\varphi : \mathcal{S} \to \mathcal{N}$, the set of privileged coordinates. The application which associates to the probability $P \in \mathcal{S}$ its distribution p is called the distribution representation or along Amari [**3**] the (-1)–representation of $\mathcal{S}$.

Many other representations of the manifold $\mathcal{S}$ can be defined in terms of random variables. In particular there exists an affine representation called the information representation (or (+1)–representation) which is associated to the information quantity. This representation L is defined by

$$\begin{aligned} L(P) = \ln p &= \sum_{k=1}^{m} \ell_k \delta_{\omega^k}, \qquad \ell_k = \ln p^k,\ k = 1, \dots m, \\ &= \sum_{k=1}^{m-1} \theta_k \delta_{\omega^k} - \psi(\theta), \quad \theta_k = \ln \frac{p^k}{p^m},\ k = 1, \dots, m-1, \end{aligned} \tag{5}$$

with

$$\psi(\theta) = \ln(1 + \sum_{k=1}^{m-1} e^{\theta_k}). \tag{6}$$

Taking the expectation of (5) with respect to the measure P we obtain the Legendre transformation between the coordinates η and θ

$$\begin{aligned} \psi(\theta) - H(\eta) - \sum_{k=1}^{m-1} \theta_k \eta^k = 0, \quad H(\eta) = -\mathbb{E}_P[\ln p], \\ \eta^k = \mathbb{E}[\delta_{\omega^k}] = \frac{\partial \psi}{\partial \theta_k}, \quad \theta_k = -\frac{\partial H}{\partial \eta^k}, \end{aligned} \tag{7}$$

where H is the entropy of distribution p.

These probabilistic considerations take a new light in the framework of the Information Geometry. As it has been shown [11] equipped with the Fisher information metric g, the manifold $\mathcal{S}$ becomes a Riemannian manifold. This metric is defined through a 1-1-correspondance $\Lambda(A) = A_\eta$ between the tangent space $T_P(\mathcal{S})$ at point $P \in \mathcal{S}$ and the linear subspace of the space T_η with $\eta = \varphi(P)$ of real random variables defined for any η by

$$T_\eta = \left\{ A_\eta \in \mathbb{R}^\Omega : A_\eta = \sum_{k=1}^{m-1} A^k(\eta) \left(\frac{\partial \ell}{\partial \eta^k} \right)_\eta, \quad \ell(\omega, \eta) = (L(\varphi^{-1}(\eta))(\omega) \right\}, \tag{8}$$

where $\left(\frac{\partial \ell}{\partial \eta^k}\right)_\eta$ is the partial derivative of ℓ as a function of the η's and of ω and the coefficients $A^k(\eta)$ are the components of A with respect to the privileged basis ($A = \sum_k A^k(\eta) \frac{\partial}{\partial \eta^k}$). This linear space, under the assumption of existence of second moment, is Euclidean space equipped with the following scalar product

$$< A_\eta, B_\eta >= \mathbb{E}_P[A_\eta B_\eta], \tag{9}$$

which allows to define a metric (the Fisher information metric) g on $\mathcal{S}$ by

$$g(A, B) = \mathbb{E}_P[A_\eta B_\eta], \tag{10}$$

or in the privileged coordinates system

$$g_{ij} = \frac{\delta_{ij}}{\eta^i} + \frac{1}{\eta^m} = -\frac{\partial^2 H}{\partial \eta^i \partial \eta^j}, \quad \text{with } \eta^m = (1 - \sum_{k=1}^{m-1} \eta^k). \tag{11}$$

The inverse matrix with entries denoted by g^{ij} has the following expression:

$$g^{ij} = \eta^i\delta_{ij} - \eta^i\eta^j = \frac{\partial^2\psi}{\partial\theta_i\partial\theta_j}, \tag{12}$$

where δ_{ij} is the Kronecker symbol.

The skewness tensor $t(A,B,C) = \mathbb{E}_P[A_\eta B_\eta C_\eta],\ \ A,B,C,\in T_P$ has the coordinate components

$$t_{ijk} = \frac{1}{(\eta^m)^2} + 2\left(1 - \frac{1}{\eta^m}\right)\frac{\delta_{jk}}{\eta^k} - \frac{\delta_{ij}\delta_{jk}}{{\eta^j}^2}, \tag{13}$$

and

$$t^h_{ij} = g_{jk}\eta^k - \frac{\delta_{hj}\delta_{jk}}{\eta^j}. \tag{14}$$

From the relation

$$\overset{\alpha}{\nabla} g = \alpha t, \tag{15}$$

between the α-covariant derivative of the metric tensor g and skewness tensor t we can introduce on the statistical manifold $(\mathcal{S}, g, t)$ a family of affine connections labeled by α, the α-connections which are non-Riemannian if $\alpha \neq 0$. Moreover the couples $(\alpha, -\alpha),\ \ \alpha \neq 0$ define dual connections with respect to the metric g,

$$Ag(B,C) = g(\overset{\alpha}{\nabla} B, C) + g(B, \overset{-\alpha}{\nabla} C), \quad A,B,C \in T_P. \tag{16}$$

The symbols of the α-connections are given by

$$\overset{\alpha}{\Gamma}{}^i_{jk} = \frac{1+\alpha}{2} t^i_{jk}, \tag{17}$$

The following tensors

$$\overset{\alpha}{R}_{hijk} = \frac{1-\alpha^2}{4}(g_{hj}g_{ik} - g_{hk}g_{ij}), \quad \overset{\alpha}{R}{}^h_{ijk} = \frac{1-\alpha^2}{4}(\delta^h_j g_{ik} - \delta^h_k g_{ij}), \tag{18}$$

are called the Riemann α–curvatures. The contraction of the Riemann α–curvature gives the Ricci α–curvature with components $\overset{\alpha}{R}_{ij} = \overset{\alpha}{R}{}^k_{ijk}$ with the usual summation convention,

$$\overset{\alpha}{R}_{ij} = \frac{1-\alpha^2}{4}(m-2)g_{jk}. \tag{19}$$

The contraction of the Ricci α–curvature with the metric $\overset{\alpha}{R} = g^{ij}\overset{\alpha}{R}_{ij}$ gives the scalar α–curvature,

$$\overset{\alpha}{R} = \frac{1-\alpha^2}{4}(m-1)(m-2). \tag{20}$$

In the case of discrete probability the Riemann α-curvature (18) vanishes only for $\alpha = \pm 1$, which corresponds to the cases described in (4) and (5). We have seen that the (± 1)–flat manifolds are dual in the sense that there exists at least one bi-orthonormal basis in T_P which defines a couple of dual coordinate systems

(η, θ), related by the Legendre transformation (7) where ϕ and φ are generating potentials of the Fisher metric: $\phi(\eta)$ and $\psi(\theta)$ such that:

$$g_{ij} = \frac{\partial^2 \phi}{\partial \eta^i \partial \eta^j} = \frac{\partial \theta_i}{\partial \eta^j}, \quad g^{ij} = \frac{\partial^2 \psi}{\partial \theta_i \partial \theta_j} = \frac{\partial \eta^i}{\partial \theta_j}. \tag{21}$$

In the case of a probability on a set of cardinality m there exists at least one family $\{X_j,\ j = 1, \dots, m\}$ of random variables such that

$$\ell(P) = \ln p = \sum_{i=1}^{m-1} \theta_i X_i - \psi(\theta), \tag{22}$$

$\psi(\theta)$ being determined by the normalization condition $P[\Omega] = 1$. The dual variables of θ^i's are given by $\eta_i = \mathbb{E}_P[X_i]$ and $\phi(\eta) = -H(\eta)$ the entropy of p.

On such dually (± 1)–flat manifold the existence of the Legendre transform allows to define a measure of the discrimination of Q in favor of P by introduction of the non-negative function $D : \mathcal{S} \times S \to \mathbb{R}$ defined by

$$D(P; Q) = \phi(\theta(P)) + \psi(\eta(Q)) - \sum_{i=1}^{m-1} \eta^i(P)\theta_i(Q) = \mathbb{E}_P\big[\ln \frac{p}{q}\big], \tag{23}$$

where $\theta(P)$ and $\eta(P)$ are dual coordinate systems in T_P. The so-called divergence $D(P; Q)$ is just the Kullback-Liebler information quantity of Q relatively to P.

$$D(P; Q) = \sum_{i=1}^{m} p_i \ln \frac{p_i}{q_i}. \tag{24}$$

Finally we have a Pytagorian theorem and a projection property:

THEOREM. *Let P, Q and R three points in $\mathcal{S}$ such that the $(+1)$-self-parallel curve connecting P and Q is orthogonal (for the Fisher metric) at Q to the (-1)-self-parallel curve connecting Q and R. Then*

$$D(P; Q) + D(Q; R) = D(P; R). \tag{25}$$

This theorem provides a natural notion of (-1)–orthogonal projection on a $(+1)$-submanifold. Finally,

PROPOSITION. *Let $\mathcal{E}$ be a smooth submanifold of a (-1)–flat manifold $\mathcal{S}$, a probability measure $Q_P \in \mathcal{E}$ is the best approximation in $\mathcal{E}$ of a probability measure $P \in \mathcal{S}$, in the sense of the divergence, $Q_P = \min_{Q \in \mathcal{E}} D(P, Q)$, if Q_P is the (-1)-orthogonal projection of P on $\mathcal{E}$. Moreover, Q_P is the unique when $\mathcal{E}$ is $(+1)$-convex submanifold, i.e.*

$$L(G, t) = (1 - t)L(G_1) + tL(G_2)$$

is in $\mathcal{E}$ for all $t \in [0, 1]$ if G_1 and G_2 are in $\mathcal{E}$.

3. Symmetric Boltzmann Machines

A symmetric Boltzmann machine is a probabilistic formal neural network, defined by a non-directed simplicial graph $\mathcal{G} = (\mathcal{V}, \mathcal{E})$ which is not necessarily complete but where each vertex x_i has a high degree of connectivity. Because the graph is non-directed the weights are symmetric $W_{ij} = W_{ji}$ where $W_{ij} = W(e_{ij}) = W(x_i, x_j)$. Neurons are probabilistic binary automata, that is the state space of each neuron is $\{0,1\}$ then a configuration of the network[1] is given by a point $\omega = \{\omega_1, \omega_2, \ldots, \omega_n\}$ in $\Omega = \{0,1\}^n$ where $n = \mid \mathcal{V} \mid$ the number of vertices of $\mathcal{G}$ and $|\Omega| = 2^n$.

It can be easily seen that any power $(\omega_i)^r$ of the random variable ω_i is equal to ω_i, a function on $\{0,1\}^n$ can be written as a polynomial of order $\le 2^n$. The logarithm of a finite probability distribution p depends only on $2^n - 1$ parameters.Then we have the following expression:

$$\ln p(\omega) = \sum_{i_1=1}^{n} \theta^1_{i_1}\omega_{i_1} + \sum_{i_1<i_2=1}^{n} \theta^2_{i_1 i_2}\omega_{i_1}\omega_{i_2} + \cdots + \theta^n_{i_1 i_2 \ldots i_n}\omega_{i_1}\omega_{i_2}\ldots\omega_{i_n} - \Psi(\theta), \quad (26)$$

with $i_1 < i_2 < \cdots < i_n$ and where $\Psi(\theta)$ is the normalisation constant. The θ's define a new parametrization of $\mathcal{S}$ related to the privileged one (4,5) by:

$$\begin{aligned} \ell_k &= \ln p(\omega^k), \qquad \omega^k = \{\omega^k_{i_1}, \omega^k_{i_2}, \ldots, \omega^k_{i_n}\}, \quad 1 \le k \le 2^n, \\ \theta^r_{i_1 i_2 \ldots i_r} &= 2^{-n} \sum_{k=1}^{2^n} \ell_k \omega^k_{i_1}\omega^k_{i_2}\ldots\omega^k_{i_r}, \qquad r \le n. \end{aligned} \quad (27)$$

The dual coordinates of the θ's are given by:

$$\eta_r^{i_1 i_2 \cdots r} = \frac{\partial \Psi}{\partial \theta^r_{i_1 i_2 \ldots i_r}} = \mathbb{E}_P[\omega_{i_1}\omega_{i_2}\ldots\omega_{i_r}]. \quad (28)$$

Neurons are updated according to the state of the connected neurons. An individual neuron at vertex x_i is, at time $t+1$, in the state 0 or 1 according to the probability law:

$$\omega_i(t+1) = \begin{cases} 1 & \text{with probability } p^i_\beta = [1 + \exp(-\beta h_i(t))]^{-1} \\ 0 & \text{with probability } q^i_\beta = 1 - p^i_\beta = [1 + \exp(+\beta h_i(t))]^{-1} \end{cases} \quad (29)$$

where β is a characteristic parameter of the neurons and $h_i(t+1)$ is an effective field defined by

$$h_i(t) = h_i(\omega(t)) = \sum_{\substack{j=1 \\ j \ne i}}^{n} W_{ij}\omega_j(t), \quad (30)$$

to simplify, the threshold of neurons is chosen to be at zero ($b_i = 0,\ \forall i$).

The individual probability law (29) does not define the dynamics of the networks, we need to precise how are updated the state of the netwetwork. In the

[1]We assume that there is no hidden neurons and that each neuron is an input and an output neuron.

following we consider two extreme cases where the updating is sequential or parallel respectively. In a sequential Boltzmann machine one neuron is updated at each step of time, this neuron is chosen along the uniform law but in a parallel Boltzmann machine all neurons are updated simultaneously. In both cases the dynamics is given by a Markov chain with an unique stationnary probability measure.

Let us remark that the uniqueness of the stationary state implies that the Boltzmann machine has an output probability independent of the input. However, such machines are able, in the limit $\beta \to \infty$, to make some pattern recognition. Indeed, the stationary distribution tends to the uniform distribution on the global minima of a Lyapunov function in the sequential case and to a Manhattan function in the parallel case. A Manhattan function is such a Lyapunov function where the dependence in the state of the system is relaxed.

3.1. Sequential Symmetric Boltzmann Machine. In sequential Boltzmann machine [**1,2,4,8**] neurons are updated in a sequential way according to a uniform random trial. The fired neuron is updated along the probabilistic law (28). This procedure determines a Markov chain $\omega(t)$ on the set of configuration $\Omega = \{0,1\}^n$ and is defined by the transition probability:

$$P[\omega(t+1) = v \mid \omega(t) = u] = \begin{cases} \frac{1}{n} P_k(u) & \text{if } v = u^{(k)}, \\ 1 - \sum\limits_{i=1}^{n} P_i(u) & \text{if } v = u, \\ 0 & \text{otherwise,} \end{cases} \tag{31}$$

The configuration $u^{(k)} = \{u_1, \ldots, 1-u_k, \ldots, u_N\}$, $u_i \in \{0,1\}$ is obtained from $u = \{u_1, \ldots, u_k, \ldots, u_N\}$ by the change of the state of the neuron x_k alone. The transition probability of this change is given by

$$P_k(u) = P[\omega(t+1) = u^{(k)} \mid \omega(t) = u] = \frac{1}{1 + e^{\beta \Delta E_k(u)}}, \tag{32}$$

with $\Delta E_k(u) = E(u^{(k)}) - E(u)$ where $E(u)$ is the energy function

$$E(u) = -\frac{1}{2} \sum_{\substack{i,j=1 \\ i \neq j}}^{n} W_{ij} u_i u_j = - \sum_{i<j=1}^{n} W_{ij} u_i u_j. \tag{33}$$

The transition probability (31) defines a Markov chain with a unique stationary probability, the Gibbs probability measure on Ω

$$G_\beta(u) = \frac{e^{-\beta E(u)}}{Z(\beta)}, \quad Z(\beta) = \sum_{u \in \{0,1\}^n} e^{-\beta E(u)}, \tag{34}$$

which has the simple $(+1)$–representation

$$\ln G_\beta(u) = \beta \sum_{i<j=1}^{n} W_{ij} u_i u_j - \ln Z(\beta). \tag{35}$$

The family of all possible states of the Boltzmann machine is just the (+1)–linear submanifold of $\mathcal{S}$ defined by the $2^n - 1$ linear equation:

$$\begin{cases} \theta^2_{ij} & = \beta W_{ij}, \\ \theta^m_{i_1 i_2 \dots i_m} & = 0, \qquad m \neq 2. \end{cases} \tag{36}$$

We denote this Gibbs submanifold by $\mathfrak{G}$ of the Boltzmann machine. Because of the 1–linear the Gibbs submanifold inherits of the dually flat structure of $\mathcal{S}$. Moreover $\mathfrak{G}$ being by itself a 1–flat manifold with coordinate system $\theta^2_{ij} = W_{ij}$, the dual coordinate system in $\mathfrak{G}$ is the η_2 part of the dual coordinate system η in $\mathcal{S}$.

$$\eta_2^{ij} = \beta \mathbb{E}_{P_{\mathfrak{G}}}[u_i u_j] = \beta \mathbb{E}_P[u_i u_j], \tag{37}$$

where $P_{\mathfrak{G}}$ is the probability in $\mathfrak{G}$ induced by P by the relations $\theta^m_{i_1 i_2 \dots i_m} = 0$ for all $m \neq 2$. Let us remark that the $\mathbb{E}_G[u_{i_1} u_{i_2} \dots u_{i_m}]$ are not necessarily 0 but non linear function of the η_2^{ij}. Then the dual coordinate system in $\mathfrak{G}$ does not define a (−1)–linear submanifold of $\mathcal{S}$. However, $\mathfrak{G}$ is (+1)-flat submanifold of $\mathcal{S}$ then there exists a unique (−1)–self parallel ortogonal projection G_p on $\mathfrak{G}$ of the probability P. The projection G_p is the best approximation of P by a probability in $\mathfrak{G}$. By the pytagorian theorem if G is a Gibbs measure:

$$D(P;G) = D(P;G_P) + D(G_P;G). \tag{38}$$

Then the minimal value for the divergence is fixed by the projection property and is given by $D(P;G_P)$.

Then the learning method proposed by Ackley, Hilton and Sejnowski [**2**] which consists to minimize the divergence $D(P;G)$ is reduced to a motion of the current point G toward G_P in $\mathfrak{G}$. This learning procedure is implemented by the steepest descent gradient method which consists in the modification of the synaptic weight W_{ij} in $W_{ij} + \delta W_{ij}$ with

$$\delta W_{ij} = -\epsilon \sum_{\{kl\}} g^{ij,kl} \frac{\partial D(P;G)}{\partial W_{kl}} = \epsilon\beta \sum_{\{k\ell\}} g^{ij,k\ell} \{\mathbb{E}_P[u_k u_\ell] - \mathbb{E}_G[u_k u_\ell]\}, \tag{39}$$

where ϵ is a sufficiently small parameter. Under this infinitesimal transformation $\eta_2^{ij} = \beta\mathbb{E}_G[u_i u_j]$, the dual variables of W_{ij} in $\mathfrak{G}$ is changed in $\eta_2^{ij} + \delta\eta_2^{ij}$,

$$\delta\eta_2^{ij} = \sum_{\{k\ell\}} g_{ij,k\ell} \delta W_{k\ell}. \tag{40}$$

It follows the following proposition

PROPOSITION.. *The gradient learning method for sequential symmetric Boltzman machine consists to follows the* (+1)*–self parallel curve in* $\mathfrak{G}$ *from the current point* G *to the* (−1)*–self parallel ortogonal projection* G_P *of* $P \in \mathcal{S}$ *on* $\mathfrak{G}$. *This method provides a minimization algorithm of the divergence* $D(P;G)$ *in* $\mathfrak{G}$.

3.2. Parallel Symmetric Boltzmann Machine. In a parallel Boltzmann machine [**1,2,6**]the dynamics is parallel, all the neurons are updated simultaneously, then, for thresholds taken at zero, the transition probability takes the form:

$$\Pi[\omega(t+1) = v \mid \omega(t) = u] = \prod_{i=1}^{n} \frac{1}{1 + e^{\beta(1-2v_i)h_i(u)}}, \quad h_i(u) = \sum_{\substack{j=1 \\ j\neq i}}^{n} W_{ij} u_j. \quad (41)$$

In this case the Markov chain defined by this transition probability has also a unique stationary probability measure:

$$\Pi_\beta(u) = \frac{1}{N(\beta)} \sum_{v\in\{0,1\}^n} e^{-2\beta\nu(v,u)}, \quad \nu(v,u) = -\frac{1}{2}\sum_{i=1}^{n} v_i h_i(u), \quad (42)$$

where $N(\beta)$ is the normalization factor

$$N(\beta) = \sum_{u\in\{0,1\}^n} \sum_{v\in\{0,1\}^n} e^{-2\beta\nu(v,u)}. \quad (43)$$

This stationary probability can be written

$$\Pi_\beta(u) = \frac{1}{N(\beta)} \prod_{i=1}^{n} \left(1 + e^{\beta h_i(u)}\right), \quad (44)$$

Taking the logarithm

$$\ln \Pi_\beta(u) = \sum_{i=1}^{n} \ln\left(1 + \exp\{\beta \sum_{\substack{j=1 \\ i\neq j}} W_{ij} u_j\}\right) - \ln N(\beta). \quad (45)$$

This last equation can be written in the form of (26) where the $\theta^k_{i_1 i_2 \cdots K}$ are non linear functions of the W_{ij}. Then the submanifold $\mathfrak{P}$ of the output stationary measures of a given parallel Boltzmann machine is not a (+1)–linear submanifold of $\mathcal{S}$ and moreover not a (+1)-convex submanifold in the sense that (+1)-self parallel curve joining two elements of $\mathfrak{P}$ is in $\mathfrak{P}$. The uniqueness of the dual (−1)–projection is not automatically insured and needs a study of the (±1)–self parallel curves.

A learning algorithm based on the minimization by the gradient method of the Kullback divergence

$$D(P;\Pi_\beta) = \sum_{u\in\{0,1\}^n} P(u) \ln \frac{P(u)}{\Pi_\beta(u)}, \quad (46)$$

leads to

$$\begin{aligned}\delta W_{ij} &= -\epsilon \sum_{\{kl\}} g^{ij,kl} \frac{\partial D(P;\Pi_\beta)}{\partial W_{kl}} \\ &= \epsilon\beta \sum_{\{k\ell\}} g^{ij,k\ell} \{\mathbb{E}_P[u_k(t)u_\ell(t-1)] - \mathbb{E}_\Pi[u_k(t)u_\ell(t-1)] \\ &\qquad + \mathbb{E}_P[u_k(t-1)u_\ell(t)] - \mathbb{E}_\Pi[u_k(t-1)u_\ell(t)]\}, \qquad (47)\end{aligned}$$

which does not depend on the present configuration but rather on the trajectory, then the geometrical interpretation is no longer simple and needs a more precise characterization of the learning trajectory .

4. Information transfert by a noisy integrator neuron

The Boltzmann machines are able to learn a probabilistic surrounding and to reproduce it with a reasonable approximation but are not able to give information on an input signal, because there is a unique stationary probability measure. Nevertheless, in the limit $\beta \to 0$, obtained through a simulated anneling algorithm, a Boltzmann machine becomes a deterministic machine able to carry out a pattern recognition [**1,2**]. In this section we are interested in formal neural networks able to process an input signal ξ emitted by a source with a probability $p_s(\xi)$ to produce an output code ζ. The mathematical description of such encoder is given by its joint probability $p(\xi, \zeta)$. The source and code probability distributions are just the marginal distributions

$$\begin{aligned}p_s(\xi) &= \sum_\zeta p(\xi,\zeta), \\ p_c(\zeta) &= \sum_\xi p(\xi,\zeta) = \sum_\xi p(\zeta \mid \xi) p_s(\xi),\end{aligned} \qquad (48)$$

where $p(\zeta \mid \xi)$ is the transition probability (transfert function) of the system. A characterization of the information about the source that the system convey can be evaluated by a comparison between the joint probability $p(\xi, \zeta)$ and the product of the marginal distribution $p_s(\xi)p_c(\zeta)$ which corresponds to the case where the code is independent of the signal. The code contains the maximum of information on the input signal when the correlation between input and output signals is maximum. Then the determination of the network which gives the best code reduces to an inference problem which can be solved by maximalization of the Kullback relative entropy:

$$\mathcal{K}(p, p_s p_c) = \sum_{\xi,\zeta} p(\xi,\zeta) \ln \frac{p(\xi,\zeta)}{p_s(\xi)p_c(\zeta)} = \sum_\xi p_s(\xi) \sum_\zeta p(\zeta \mid \xi) \ln \frac{p(\zeta \mid \xi)}{p_c(\zeta)}. \qquad (49)$$

In this section we illustrate the geometrical aspect of this learning process on the simplest example of a continuous perceptron consisting in a single neuron with n input channels connected to sensory units and a noisy output channel.

At each instant of time some signal $\xi = \{\xi_1, \xi_2, \ldots, \xi_n\}$ activates the sensory units, the information transfered by the neuron is assumed to be processed by an integration process followed by the action of a gain function f. Then the output takes the form:

$$\zeta = f(h) + \nu, \qquad h = \sum_{i=1}^{n} W_i \xi_i, \tag{50}$$

where ν is an additive output noise independent of the input with a probability law π with bounded entropy and f is assumed to be taken in a class of non-decreasing derivable fuction such that $0 \leq f(u) \leq 1$.

In the case where the W_i are fixed constants the only degree of freedom are those of f and the extrema of $\mathcal{K}(q, q_h p_c)$ defined by $\mathcal{K}(p, p_s p_c)$ are the extrema of

$$\mathcal{K}(q, q_h p_c) = \int du d\zeta q_h(u) q(\zeta \mid h) \ln \frac{q(\zeta \mid u)}{p_c(\zeta)}. \tag{51}$$

where $q_h(u)$ is the probability distribution density of the random variable h, $q(h, \zeta)$ the joint probability of h and ζ and

$$q(u \mid \zeta) = \pi(\zeta - f(u)), \tag{52}$$

the conditional density of probability. However,

$$\mathcal{K}(q, q_h p_c) = H(q_c) - H(\pi), \tag{53}$$

where the entropy $H(\pi)$ of the distribution π and the entropy of the distribution q_h are given by:

$$H(q_c) = \int du q_c(u) \ln q_c(u) = \int du q_h(u) \ln \frac{q_h(u)}{f'(u)}, \tag{54}$$

we have used expressions (48) and (50). Then the maximum of $\mathcal{K}(p, p_s p_c)$ is given by the minimum of the divergence

$$D(q_h; f') = \int du q_h(u) \ln \frac{q_h(u)}{f'(u)}, \tag{55}$$

It follows that the adaptation of the gain function of a formal neuron to have a maximun of information conveyed by the output ζ about the input ξ can be seen as a geometrical problem.

Concluding remarks

The previous examples have exhibited some difficulties for a geometrical interpretation of the gradient learning based on the Kullback information quantity. These difficulties exist in particular in the case where the submanifolds defined by the machine are not (± 1)–flat even if the discrete probability manifold is α–flat for any α. A geometrical approach of learning processes needs a more detailed studies of the α–self-parallel curves. Moreover, such a study allows to control the approximation and the stability of the algorithm used in the simulations.

REFERENCES

[1]. E. Aarts, J. Korst, *Simulated annealing and Boltzmann machines. A stochastic approach to combinatorial optimization and neural computing*, J. Wiley and sons, 1989.

[2]. D.Ackley, G. Hilton, T. Sejnowski, *A learning algorithm for Boltzmann machines*, Cognitive Science **9** (1985), 147–169.

[3]. S-I.Amari,, *Differential Geometrical Methods in Statistics*, Lecture Notes in Statistics., vol. 28, Springer-Verlag, New York, 1985.

[4]. ———, *Information Geometry of Boltzmann Machines*, IEEE Transactions on Neural Networks **3 7** (1992), 260–271.

[5]. ———, *Information Geometry*, Contemporary Mathematics **this issue**.

[6]. B. Apolloni, D. de Falco, *Learning by parallel Boltzmann machines*, IEEE Transaction on Information Theory **37** (1991), 1162.

[7]. N.N. Chentsov, *Statistical Decision and Optimal Inference*, "Nauka", Moscow, 1972 (Russian); English transl., American Mathematical Society., vol 53, Amer. Math. Soc., Providence, R. I., 1982..

[8]. G. Hilton, T. Sejnowski, *Learning and relearning in Boltzmann machines*, Parallel Distributed Processing: Exploration in Microstructures of Cognition., Vol I, Ch 7 (L.L. McClelland, D.E. Rumelhart, eds.), MIT Press, Cambridge, 1986, pp. 147–169.

[9]. J.-P. Nadal, N. Parga, *Information processing by perceptron in an unsupervised learning task.*, Networks **4** (1993), 295–312.

[10].———, *Nonlinear neurons in the low-noise limit: a factorial code maximizes information transfer*, Networks **5** (1994), 565–581.

[11].C.R. Rao, *Information and accurency attainable in the estimation of statistical parameters*, Bull.Calcutta Math.Soc. **37** (1945), 81–91.

CENTRE DE PHYSIQUE THÉORIQUE, CNRS-LUMINY, CASE 907, F-13288 MARSEILLE, FRANCE

E-mail address: combe@cptsu2.univ-mrs.fr

UNIVERSIDADE DA MADEIRA, COLEGIO DOS JESUITAS, LARGO DO COLEGIO, PT-9000 FUNCHAL, MADEIRA, PORTUGAL

E-mail address: nencka@dragoeiro.uma.pt, nencka@cptsu2.univ-mrs.fr

Contemporary Mathematics
Volume **203**, 1997

Statistical Dynamics and Information Geometry

R. F. STREATER

ABSTRACT. We relate the techniques used in statistical dynamics to the Fisher information metric and Amari's dual affine connections.

In these lectures, given at the Clifford Conference, Madeira, August, 1995, and at the Luminy Workshop on Dynamics, Stochastics and Complexity, (September, 1995), I shall describe a non-linear projection operator, useful in non-equilibrium thermodynamics, in terms of information geometry. This geometry is the subject of Prof. Amari's lecture. Our maxim (MAXIM = polite slogan) is that dissipation is loss of information, not loss of energy. In all its forms, including heat, energy is conserved. Loss in information is recognised physically as gain in entropy. Our aim is to add dissipation to reversible dynamics in such a way that both the first and the second law of thermodynamics are obeyed. Part of the dissipation will be due to our using a reduced description in the sense of Ingarden [**1, 2**], and part will be due to our choosing a random dynamics; thus the theory is necessarily probabilistic. Ingarden's "level of description" is fixed by choosing a list of random variables, (the "slow variables"), say $\mathcal{X} = \text{Span}\,\{X_1, X_2, \dots, X_n\}$, among which are the sure function 1 and the energy. The larger n is, the more detailed is our description. We shall present a systematic method of coarse-graining which leads to an autonomous system of equations of motion for the means of $X_1, X_2, \dots, X_n$. I call this method "statistical dynamics" [**3**], and this is briefly described in the next two sections.

1. Random dynamics

In a probability theory there is a sample space Ω, which in statistical physics consists of the possible configurations of the system. We assume here that Ω is countable, and furnish it with the trivial measurable structure, in which every subset of Ω is measurable. We specify n "slow" random variables which span a subspace $\mathcal{X}$; among these is the energy

$$\mathcal{E} : \Omega \to \mathbf{R},$$

which will be assumed to be non-negative. In this way, Ω becomes split up into disjoint sets, Ω_E, the energy-shells:

$$\Omega_E = \{\omega \in \Omega : \mathcal{E}(\omega) = E\}.$$

1991 *Mathematics Subject Classification.* Primary 53C99, 82C05; Secondary 62B10.

Then $\Omega = \sqcup_E \Omega_E$. To guarantee thermodynamic stability we assume that the partition function is finite for each $\beta > 0$:

$$Z(\beta) = \sum_{\omega} \exp(-\beta \mathcal{E}(\omega)).$$

This enables us to define the canonical state

$$p_\beta(\omega) = Z(\beta)^{-1} \exp(-\beta \mathcal{E}(\omega)),$$

called *thermal equilibrium* with temperature β^{-1}. Later we shall need to suppose that the variance of the energy in any canonical state is finite; this puts a mild condition on the growth of the density of states as a function of energy. For many systems the canonical state is the limiting distribution for large times, but in the presence of conservation laws other states, called grand canonical states, are obtained. The stability condition implies that $|\Omega_E| < \infty$ for each E. This allows us to define the uniform distribution

$$p_E(\omega) = \begin{cases} |\Omega_E|^{-1} & \text{if } \omega \in \Omega_E \\ 0 & \text{otherwise} \end{cases}$$

This is called the microcanonical state of energy E. We shall see that any mixture of microcanonical states is a fixed point of random dynamics which conserve energy. This means the following.

Let Γ_E be the permutation group of Ω_E, and put $\Gamma = \prod_E \Gamma_E$. We say that Γ is the group of energy-conserving permutations. A *reversible dynamical law* is prescribed by choosing $\gamma \in \Gamma$; we write the group action on the right: $\omega \mapsto \omega\gamma$ is the microscopic change in one time-step. The orbit through ω under time-evolution, taken to be the discrete group $\mathbf{Z}$, is the set

$$\{\dots, \omega\gamma^{-2}, \omega\gamma^{-1}, \omega, \omega\gamma, \omega\gamma^2, \dots\}.$$

By our choice of Γ this lies entirely inside the energy-shell containing ω. We say that the action of Γ_E is *transitive* if the orbit through any point of Ω_E is Ω_E itself. This implies that the action is ergodic (any invariant set is either empty or the whole of Ω_E).

It is convenient to identify the vector space, consisting of real functions on Ω having finite support, with $\operatorname{Span}\Omega$, the set of finite formal sums of elements of Ω with real coefficients. Thus the function f, which takes values f_j at ω_j, $j = 1, 2, \dots$ and zero elsewhere, is written

$$f = \sum_{j=1}^{n} f_j \omega_j.$$

In particular, the point $\omega_0 \in \Omega$ represents the function that is 1 at ω_0 and zero elsewhere. The linear span of Ω with positive coefficients is called the positive cone, $\operatorname{Span}\Omega_+$. We can complete $\operatorname{Span}\Omega$ in the $\|\bullet\|_1$-norm

$$\|f\|_1 = \sum_{\omega} |f(\omega)|$$

to obtain the space $\ell_1(\Omega)$ of finite signed measures. The closure of $\operatorname{Span}\Omega_+$ in $\ell_1(\Omega)$ is the space of finite measures, and those of norm 1 form the state-space Σ:

$$\Sigma(\Omega) = \{p \in \ell_1(\Omega) : \|p\|_1 = 1, \quad p(\omega) \geq 0\}.$$

This is a bounded closed convex set, whose extreme points are the point measures.

The dual space of ℓ_1 is $\ell_\infty(\Omega)$, the W^*-algebra $\mathcal{A}$ of all bounded real-valued functions on Ω; that is, the set of all bounded random variables. These are the microscopic observables of the theory. We shall write

$$p.f = \sum_\omega p(\omega) f(\omega)$$

This is the mean (or expectation) of f when the system is in the state p. Since $\ell_1(\Omega) \subseteq \ell_\infty(\Omega)$ this identifies ℓ_1 with part of its dual, and, indeed, $\operatorname{Span}\Omega$ can be made into an algebra by pointwise multiplication. This is a dangerous game: the time-evolution of the states (Schrödinger picture) is opposite to that of the observables (Heisenberg picture). We avoid ambiguity by regarding $\operatorname{Span}\Omega$ as a Γ-bimodule, writing $p \mapsto p\gamma$ for the Schrödinger picture, and $f \mapsto \gamma f$ for the Heisenberg picture. The latter action, the left action, is defined by duality, which is expressed as associativity:

$$p\gamma.f = p.\gamma f.$$

Thus if $p = \sum_\omega p(\omega)\omega$, then $p\gamma = \sum_\omega p(\omega)\omega\gamma$, and if $f = \sum_\omega f(\omega)\omega$, then

$$\gamma f = \sum_\omega f(\omega)\gamma\omega = \sum_\omega f(\omega)\omega\gamma^{-1}.$$

We note that both these maps are isometries in either norm, and that $\operatorname{Span}\Omega$ is, in a natural way, a bimodule for the group algebra $\operatorname{Span}\Gamma$.

The specification of $\mathcal{E}$ selects an important class of states, those with finite (mean) energy:

$$p.\mathcal{E} = \sum_\omega p(\omega)\mathcal{E}(\omega) < \infty.$$

All states in $\operatorname{Span}\Omega$ have finite energy. The permutations $\gamma \in \Gamma$ leave $\mathcal{E}$ invariant: $\mathcal{E}(\omega) = \mathcal{E}(\omega\gamma)$, and so we get the first law of thermodynamics in average form: if p has finite energy, so has $p\gamma$, and $p.\mathcal{E} = p\gamma.\mathcal{E}$.

The discrete nature of Ω gives us a satisfactory entropy of a state

$$S(p) = -\sum_\omega p(\omega)\log p(\omega),$$

which is ambiguous up to an overall factor (depending on the base used for the logarithm). If the sum diverges we take $S(p) = +\infty$ by convention. The entropy is preserved under any $\gamma \in \Gamma$: such a law of dynamics is *reversible.* To get a non-trivial second law (in which entropy increases with time) we can use *random dynamics.* Thus, let $\lambda_j \geq 0,\ \ j = 1, \ldots, N$, and $\sum_j \lambda_j = 1$. Let $\gamma_1, \ldots, \gamma_N \in \Gamma$, and consider the element

$$\tau = \sum_{j=1}^N \lambda_j \gamma_j$$

of the positive cone in $\operatorname{Span}\Gamma$. Since $\sum \lambda_j = 1$, we say that τ lies in the cap of the cone. The right action of τ on the state-space Σ:

$$p \mapsto p\tau = \sum_{j=1}^N \lambda_j p\gamma_j.$$

is said to be the random dynamics given by τ. Since τ is a convex mixture of permutations, it is bistochastic. The dual map acting on the left of $\mathrm{Span}\,\Omega \subseteq \mathcal{A}$, defined by $f \mapsto \tau f$, where

$$p\tau.f = p.(\tau f) \text{ for all } p \in \Sigma$$

is consistent with the $\mathrm{Span}\,\Gamma$-bimodule structure of $\mathrm{Span}\,\Omega$. Then we get both laws of thermodynamics:

THEOREM 1.1. *For each τ in the cap of the positive cone in SpanΓ,*

$$\begin{aligned} p\tau.\mathcal{E} &= p.\mathcal{E} \\ S(p\tau) &\geq S(p). \end{aligned}$$

The microcanonical state p_E for any energy E, and any mixture of these, is a fixed point of any such τ. Under suitable ergodic conditions, one can show that for any initial $p \in \Sigma(\Omega_E)$,

$$p\tau^n \longrightarrow p_E \text{ as } n \longrightarrow \infty;$$

this is called the theorem on the equipartition of energy. Since $p_E\tau = p_E$ we do not get convergence to the canonical state (except in the trivial cases where the initial state is distributed so that the probability of the event Ω_E is already $p_\beta(\Omega_E)$ for some β).

One way to mix the energy shells, and to guarantee convergence to the canonical state, is to leave the class of bistochastic maps, replacing τ by a stochastic map obeying "detailed balance". Typical of this approach is the "Glauber dynamics" of the Ising model; in this, transitions occur between the energy-shells; the probabilities of transition from Ω_E to $\Omega_{E'}$ and the reverse are taken to be in the ratio $e^{\beta(E-E')}$. The canonical state p_β is a fixed point of the Glauber dynamics, and it is easy to arrange that the dynamics take any initial state to this state at large times. These models do not describe isolated systems, but are isothermal: mean energy is not conserved, and entropy might decrease as well as increase. The value of β is built into the dynamics, and is not determined by the initial state. We shall see that isothermal dynamics is the Legendre transform of the isolated dynamics, and that these are dual in the geometric sense as introduced by Amari [**4**].

To describe isolated dynamics, we keep the bistochastic random dynamics, and introduce more randomness, which mixes the energy-shells, in a way that keeps both laws of thermodynamics. We borrow an idea from Jaynes [**5**] and Ingarden [**2**], which turns out to be close to the theory of estimation used in statistics. In the next section we give this idea, and in later sections we relate it to information geometry.

2. Reduced Description and the map Q

Given Ω and an energy $\mathcal{E}$ obeying the condition of thermodynamic stability, then p_β is the unique state of maximum entropy among all states with the same value

$$\overline{\mathcal{E}} = p_\beta.\mathcal{E} = Z_\beta^{-1} \sum_\omega e^{-\beta\mathcal{E}(\omega)}\mathcal{E}(\omega)$$

of the mean energy. Gibbs knew this. The proof in finite dimensions uses two Lagrange multipliers Z, β and also the strict concavity of the entropy. Shannon interpreted entropy as missing information, which led Jaynes [**5**] to interpret p_β as the best choice to make for the state, if $\overline{\mathcal{E}}$ is the only information to hand: it is the most neutral way to express our lack of other knowledge.

Jaynes uses this argument to explain the ubiquity of the canonical state– he says that p_β arises so often because usually we only measure $\overline{\mathcal{E}}$. I am not convinced. We find p_β in many situations where we measure not only $\overline{\mathcal{E}}$ but the means of other random variables as well. A famous example of this is the background cosmic radiation, which was found to be distributed according to Planck's law (which is p_β for a particular choice of Ω, $\mathcal{E}$) even though the means of many powers of $\mathcal{E}$, and not only $\overline{\mathcal{E}}$, were measured. Indeed, the very act of verifying that the distribution is p_β involves measuring other moments of $\mathcal{E}$. The logical extension of Jaynes's own argument was made by Ingarden [**2**]. He was led to the concept of "higher order temperature". Suppose that we measure, for example, $\overline{\mathcal{E}}$ and $\overline{\mathcal{E}^2}$ in a series of measurements. This is equivalent to measuring the mean and the variance of the energy. The state of maximum entropy, given the mean and variance, has the form

$$p_{\beta_1,\beta_2} = Z(\beta_1,\beta_2)^{-1}\exp\{-\beta_1\mathcal{E} - \beta_2\mathcal{E}^2\}. \tag{2.1}$$

Here, β_2^{-1} is called (by Ingarden) a second-order temperature. This state is not Gibbsian, except in the accidental case when $\beta_2 = 0$. The theory does not allow us to predict β_1 or β_2: they are to be found by measurements. In practice states very close to a canonical state arise very often, even though we measure the variance, as we do when we make a note of the fluctuations. Ingarden, in trying to explain this, says that for large systems there is not much difference between Eqn. (2.1) and a canonical state. But this is only true if β_2 is small, and this cannot be the explanation of the common occurrence of small β_2.

In contrast to the ideas of Jaynes, in statistical dynamics the ubiquitous nature of the canonical state finds its explanation in that many systems are at or close to equilibrium. Others not in equilibrium are not obliged to be close to canonical states, even if we only measure their mean energies. We now adapt Ingarden's development of Jaynes's theory to non-equilibrium situations. Suppose we measure the means $\{\overline{X}\}$ of the random variables $X \in \mathcal{X}$. We assume that 1 and the energy are in $\mathcal{X}$. Since the energy is unbounded in many models, we must allow that some of the X_j are unbounded. To allow us to proceed smoothly, we assume the following:

DEFINITION 2.1. We say that $\mathcal{X}$ obeys the extended stability condition if, for each n-tuple $\beta = \{\beta_1, \dots, \beta_n\}$ of positive numbers, the sum

$$Z = \sum_\omega \exp\{-\sum \beta_j X_j(\omega)\}$$

is finite, and the first two moments (and cross moments) of X_j, $j = 1, \dots, n$ in the state p_β, defined in Eqn. (2.2) below, are finite.

If $\mathcal{X}$ obeys the extended stability condition, then the state of maximum entropy with the means $\overline{X}_j$ is found ([**3**], Theorem 8.6)) to be

$$p_\beta = p_{\beta_1,\dots,\beta_n}(\omega) = Z^{-1}_{\beta_1,\dots,\beta_n}\exp\{-\sum_j \beta_j X_j\} \tag{2.2}$$

where $\beta = (\beta_1, \dots, \beta_n) \in \mathbf{R}^n$ is the set of generalised (inverse) temperatures, determined by

$$Z = \sum_\omega \exp\{-\sum_j \beta_j X_j(\omega)\}$$

and

$$\overline{X}_j = Z^{-1} \sum_\omega X_j(\omega) \exp\{-\sum_j \beta_j X_j(\omega)\}.$$

The state p_β is the least prejudiced choice of state, given these means. States of the form Eqn. (2.2) form an "exponential family". The set $\mathcal{X}$ defines the level of description provided by this family. If $\mathcal{X}$ is not dense in the whole algebra of random variables, the family can give only a reduced description of a state. The $X_1, \dots, X_n$ are taken to be the *slow varibles*, whose (mean) dynamics we wish to describe. Roughly speaking, random variables not included in $\mathcal{X}$ are the fast variables, which reach equilibrium in one time-step. This leads us the define the map Q, which replaces the current state p of the system by pQ (writing the action on the right) the state of maximum entropy having the same means as p for each $X \in \mathcal{X}$:

$$pQ.X = p.X, \qquad X \in \mathcal{X}.$$

Thus pQ lies in the exponential family. There is no need for these means to have been measured experimentally for the map Q to be applied – although, being slow variables, they could have been; we use Q to implement the moving to equilibrium of the fast variables. The map Q is related to the method of maximum likelihood, used in statistics to find the "best" approximation to a given distribution p from among members of a family of distributions. In neural nets, the state pQ is sometimes called the "inference" made by the brain, when the input causes the neurons to adopt the state p.

To get the dynamics, we start with p in the exponential family; then one time-step is given by

$$p \mapsto p\tau Q$$

where τ is an energy-conserving random dynamics. To enable us to give a geometrical picture of the dynamics, we shall require that τ is such that $p' = \tau p$ has finite first moments for all the elements of $\mathcal{X}$ for any p in the exponential family. Since mean energy is preserved by Q, and Q is entropy-increasing, we see that our map τQ obeys both laws of thermodynamics. The law of dynamics gives an orbit entirely within the exponential family defined by $\mathcal{X}$. The models got in this way are non-linear, and generally mix up the energy-shells. In many cases, the limit state is a canonical state. This is a satisfactory desciption of an isolated system. We should stress that the fast variables, omitted from the description, are not necessarily fast when they follow the dynamics given by the random dynamics τ. They are made into fast variables by the map Q itself; we imagine that there are some further forces and effects that are not included in the dynamics τ, which have the effect of randomising the variables omitted from $\mathcal{X}$. An obvious example of a variable that is slow relative to an energy-conserving dynamics is the indicator function χ of any energy-shell; this is more than slow – it is invariant. We cannot include all such functions in $\mathcal{X}$ if we are to arrive at dynamics that mix the energy-shells. In the Ising model the function χ is a rather non-local function of the spin variables, and the extra randomness achieved by omitting such functions from $\mathcal{X}$ expresses the sensitivity of long-range many-body correlations to outside noise.

The reduced description and the map Q were used by Kossakowski [**6**] in a general theory of non-equilibrium thermodynamics. This theory concerns quantum theory, but otherwise is more special than ours, in that the microscopic dynamics in [**6**] is taken to be irreversible. The use of the exponential family to describe non-equilibrium thermodynamics was also used by other authors, for example [**7**].

A well-known example of the map Q is Boltzmann's Stosszahlansatz; this is not so much a hypothesis or ansatz as a procedure, which replaces the state of two particles after each scattering by the tensor product of the marginal probabilities of each particle. More abstractly, let $\Omega = \Omega_1 \times \Omega_2$ and let $\mathcal{A}_1$, $\mathcal{A}_2$ be the algebras of all bounded random variables on Ω_1 and Ω_2 respectively. Put $\mathcal{A} = \mathcal{A}_1 \otimes \mathcal{A}_2$. We choose the slow variables to be any element of $\mathcal{A}_1$ or $\mathcal{A}_2$. The marginal map $\mathcal{M}_1 : \Sigma(\Omega) \to \Sigma(\Omega_1)$ is defined to be

$$(pM_1)(\omega_1) = \sum_{\omega_2} p(\omega_1, \omega_2); \tag{2.3}$$

similarly the marginal map $\mathcal{M}_2 : \Sigma(\Omega) \to \Sigma(\Omega_2)$ is defined. Then the "Boltzmann map" is the map

$$p \mapsto pQ = \mathcal{M}_1 p \otimes \mathcal{M}_2 p.$$

This map is known to increase the entropy (unless p is a fixed point), and indeed pQ is the state of maximum entropy with the same means as p for all random variables in $\mathcal{X} = \text{Span}\,\{\mathcal{A}_1, \mathcal{A}_2\}$. If the components of p are positive, the same is true of Qp and the exponential family in question is the set of states of the form $\{\exp(B_1 + B_2) : B_1 \in \mathcal{A}_1,\, B_1 \in \mathcal{A}_2\}$.

3. The Fisher Metric and Amari Connections

The map Q has a neat description in terms of information geometry. This geometry has two components, the Riemannian metric and the affine geometry defined by the choice of an affine connection. The metric on the manifold $\Sigma(\Omega)$ was introduced by Fisher [**8**] and developed by Amari, who also introduced a family of affine connections [**4**]. These topics are treated in Amari's lecture here, so I shall be very brief.

Given a faithful state it can be written as $p_\beta \in \Sigma(\Omega)$ in some exponential family with parameters β; we can define a scalar product on $\mathcal{A}(\Omega)$ by

$$\langle f, g\rangle_\beta = \sum_\omega p_\beta(\omega) f(\omega) g(\omega).$$

This defines a norm $\| \bullet \|_\beta$ on $\mathcal{A}$. We denote by $\mathcal{A}_\beta$ the Hilbert space of random variables with finite $\| \bullet \|_\beta$-norm. This is the Hilbert space of the Gelfand-Naimark-Segal representation determined by p_β, and it contains some unbounded random variables in general. The dual norm on $\text{Span}\,\Omega$ is

$$\|p\|_{-\beta} = \sup_f \{p.f : \|f\|_\beta = 1\} = \sum_\omega p_\beta^{-1}(\omega)(p(\omega))^2.$$

This is a Hilbert norm, coming from the scalar product

$$\langle p, q\rangle_{-\beta} = \sum_\omega (p_\beta(\omega))^{-1} p(\omega) q(\omega)$$

We denote by $\mathcal{A}_{-\beta}$ the Hilbert space completion of $\operatorname{Span}\Omega$ with finite $\|\bullet\|_{-\beta}$-norm; it does not include all of Σ in general.

Now suppose we have an exponential family S_n, parametrised by the intensive variables $\beta \in \mathbf{R}^n$; this is embedded in the manifold $\Sigma(\Omega)$. Now restrict $\langle\bullet,\bullet\rangle_{-\beta}$ to the tangent space of S_n at the point β, which is an affine hypersurface in $\mathcal{A}_{-\beta}$ denoted by $T_\beta(S_n)$. This gives a scalar product on $T_\beta(S_n)$, which is what is meant by a Riemannian structure on S_n. This is the Fisher information metric. Amari shows that

$$\langle\frac{\partial}{\partial\beta_i},\frac{\partial}{\partial\beta_j}\rangle(\beta) = p_\beta.\left(\frac{\partial}{\partial\beta_i}(\log p_\beta)\frac{\partial}{\partial\beta_j}(\log p_\beta)\right). \tag{3.1}$$

We now turn to the other geometrical object, the affine connection. A linear space such as $\operatorname{Span}\Omega$ (with $|\Omega| = n < \infty$) is an affine space, that is, a space on which the affine group $\mathrm{Aff}(n)$ acts. This group can act on the same space in different ways, so an affine space is really a pair, the underlying space (here $\operatorname{Span}\Omega$), and a given action of $\mathrm{Aff}(n)$. We may define an action as follows. Let $(A,b) \in \mathrm{Aff}(n) = GL(n,\mathbf{R}) \bowtie \mathbf{R}^n$, (the semidirect product); the action of $\mathrm{Aff}(n)$ on $\mathbf{R}^n$ is the map $c = \{c_1,\dots,c_n\} \mapsto c' = Ac+b$. This can be carried to $\operatorname{Span}\Omega$ by

$$\sum_1^n c_j\omega_j \mapsto \sum_1^n c'_j\omega_j. \tag{3.2}$$

Affine coordinates in an affine space are those, like $\{c_j\}_{j=1,\dots,n}$, which transform in this way under the given group action on the space. Alternatively, if an action is not already given and we deem a set of coordinates to be affine, then this determines a unique action of $\mathrm{Aff}(n)$ on the space.

A *connection* on a manifold S is an isomorphism between $T_\beta(S)$ and $T_{\beta'}(S)$ when $\beta' = \beta + d\beta$ is close to β. This isomorphism transports a vector ∂ in $T_\beta(S)$ to one in $T_{\beta'}(S)$; it is also called parallel transport. When S is embedded in an affine space (as S_n is embedded in Σ) we can use the affine coordinates to define a connection: let $\beta = \{\beta_1,\dots,\beta_n\}$ be our coordinates at T_β, taking the values $\beta' = \{\beta'_1,\dots,\beta'_n\}$ at $T_{\beta'}$. Define the linear isomorphism between T_β and $T_{\beta'}$ to be the linear extension of the map $\partial_i \mapsto \partial'_i$. In this case, we have an *affine* connection. In affine coordinates the covariant derivative ∇ is the ordinary partial derivative. In other coordinates the covariant derivative is obtained from this by the usual chain rules. Conversely, a connection is determined by the covariant derivative ∇ it defines; we shall then speak of the connection ∇. If, as our notation suggests, $\{\beta_1,\dots,\beta_n\}$ are the intensive variables associated with the slow random variables $X_1,\dots,X_n$, in that $p \in S_n$ has the form p_β as in Eqn. (2.2), then the affine connection defined by this construction is called the $(+1)$-connection by Amari. On the other hand, if $p \in S_n$ is parametrised by its extensive variables (also called the expectation coordinates)

$$\eta_1 = p_\beta.X_1, \quad \eta_2 = p_\beta.X_2, \quad \dots, \quad \eta_n = p_\beta.X_n, \tag{3.3}$$

the connection defined by taking these as affine coordinates is called the (-1)-connection by Amari. Since the expectation coordinates are linear combinations of the coordinates c of Eqn. (3.2), the affine structures using c_j and η_j are the same. Both the (±1)-connections are torsionless and curvature-free (or flat). Given any connection, we can use it to transport a tangent vector ∂ at β along a curve C joining β and β', by steps of $d\beta$, from T_β to $T_{\beta'}$. Let us denote this isomorphism:

$T_\beta \longrightarrow T_{\beta'}$ by Π_C. The (± 1)-connections are not metric connections; when we transport a pair of vectors ∂_1 and ∂_2 in T_β along C using either of these connections, the scalar product changes:

$$\langle \partial_1, \partial_2 \rangle(\beta) \neq \langle \Pi_C \partial_1, \Pi_C \partial_2 \rangle(\beta').$$

We say a connection is trivial if Π_C is independent of C, as is the case for the (± 1)-connections. Amari defines a family, the (α)-connections, $\alpha \in \mathbf{R}$, of which only the (0)-connection is metric: it is the Levi-Civita connection derived from the Fisher metric.

The (± 1) connections are dual in the sense of Amari; if $\Pi_C^{\pm}$ are the corresponding transports along a curve C, we have

$$\langle \partial_1, \partial_2 \rangle(\beta) = \langle \Pi_C^{+} \partial_1, \Pi_C^{-} \partial_2 \rangle(\beta'). \tag{3.4}$$

More generally, the $(\pm\alpha)$-connections are in duality, and the (0)-connection is self-dual (and so is metric).

Geodesics are defined in terms of a connection, not the metric. A geodesic on S_n is a curve $\{\beta(t),\ 0 \leq t \leq 1\}$ obtained by parallel transport along the curve. That is, $\dot\beta(t)$ is parallel to $\dot\beta(t')$. Equivalently, a geodesic for a flat connection is a straight line in affine coordinates. This is illustrated by the map $p \mapsto p' = pQ$, in which the expectation coordinates $\eta_1, \ldots, \eta_n$ of Eqn. (3.3) are unchanged. We can obtain p' by drawing a straight line in Σ (as seen in the coordinates c_j) joining p and p', and this can be described as a (-1)-geodesic. We shall show in the next section that this line is the (-1)-geodesic that cuts S_n at right angles (in the Fisher metric on the ambient space, Σ). Instead of using (-1)-geodesics, we could define the dual to Q, using straight lines in the canonical (or intensive) coordinates cutting S_n at right angles. We shall show that these are appropriate in isothermal and isopotential dynamics.

4. Dual Flat Connections and Potentials

Amari shows that when a Riemannian manifold S possesses a pair of dual torsionless flat connections, ∇, ∇^*, then there exists a pair (θ, η) of dual coordinate systems such that θ is ∇-affine and η is ∇^*-affine. Moreover, putting

$$\partial_i = \frac{\partial}{\partial \theta^i},\ \partial^j = \frac{\partial}{\partial \eta_j},$$

$$\langle \partial_i, \partial_j \rangle = g_{ij},\ \ \langle \partial^i, \partial^j \rangle = g^{ij},$$

the matrices g_{ij} and g^{ij} are inverses, and act as raising and lowering operators for vector fields, thus: $\partial^i = g^{ij}\partial_j$. He also shows that there exists a pair of *potential functions* $\psi(\theta), \phi(\eta)$, such that

$$g_{ij} = \partial_i \partial_j \psi(\theta),\ g^{ij} = \partial^i \partial^j \phi(\eta). \tag{4.1}$$

These potentials are related by the Legendre transform:

$$\theta^i = \partial^i \phi(\eta), \qquad \eta_i = \partial_i \psi(\theta) \tag{4.2}$$

$$\psi(\theta) + \phi(\eta) - \sum_i \theta^i \eta_i = 0. \tag{4.3}$$

For example, for an exponential family, say $S_n = \{p_\beta\}$, the coordinates β are affine for $\nabla^{(+1)}$, and writing p_β as in Eqn. (2.2), we find that $\psi(\beta) = \log Z(\beta)$. This is the cumulant generating function, so that the dual coordinates are

$$\eta_i = \partial_i\psi(\theta) = p_\theta.X_i, \qquad \text{the expectation coordinates.} \tag{4.4}$$

Also, the metric at p_β is

$$g_{ij} = \partial_i\partial_j\psi(\beta) = p_\beta.\left(X_i - \overline{X}_i\right)\left(X_j - \overline{X}_j\right). \tag{4.5}$$

If $|\Omega| = \infty$ we see the need for the extended stability condition. One then shows that

$$-\phi(\eta) = S(p_\eta), \text{ the entropy, and} \qquad \psi(\theta) = -S + \sum_i \theta^i\eta_i, \text{ the free energy.}$$

When ∇^+ and ∇^- are dual flat connections, we can define the *divergence* D between two points P_1 and P_2 on S_n to be

$$D(P_1, P_2) = \psi(\theta_1) + \phi(\eta_2) - \theta_1.\eta_2.$$

The divergence is some measure of the separation between P_1 and P_2, but is not a metric. It obeys

1. $D(\theta, \theta') \geq 0$, with equality only when $\theta = \theta'$.
2. $\partial_i D(\theta, \theta') = \partial_i' D(\theta, \theta') = 0$ at $\theta = \theta'$.
3. $\partial_i\partial_j D(\theta, \theta') = g_{ij}(\theta)$.
4. D is strictly convex.

It is not symmetric: $D(\theta, \theta') \neq D(\theta', \theta)$, and it violates the triangle inequality, being more like the square of the distance than the distance. But it does obey a form of Pythagoras theorem: let P, P', P'' be three points in Σ such that the $\nabla^{(-)}$-geodesic (P, P') from P to P' is perpendicular to the $\nabla^{(+)}$-geodesic (P', P'') from P' to P''; then

$$D(P, P'') = D(P, P') + D(P', P'').$$

In our case, D is the relative information of Kullback.

We now return to the map Q, from a faithful state p to the point $pQ \in S_n$. Suppose that $|\Omega| = N < \infty$; we can supplement the slow variables $X_1, \ldots, X_n$ by $N - n$ further random variables, say $X_{n+1}, \ldots, X_N$, such that $\log p$ can be written as

$$\log p = \sum_{i=1}^{N} \beta^i X_i + \text{const.}$$

Then the family S_n can be expressed in terms of the intensive variables $\beta^1 \ldots, \beta^N$ as the hyperplane

$$S_n = \{\beta : \beta^{n+1} = 0, \beta^{n+2} = 0, \ldots, \beta^N = 0\}.$$

Thus S_n is isomorphic to $\mathbf{R}^n$, and a $\nabla^{(+)}$-geodesic in S_n is a straight line in the coordinates $\beta^1, \ldots, \beta^n$. The tangent space at $p_\beta \in S_n$ is spanned by

$$\frac{\partial}{\partial\beta^1}, \ldots, \frac{\partial}{\partial\beta^n}.$$

Now define the expectation coordinates for any $p \in \Sigma$ by $\eta_j = p.X_j, j = 1, \ldots, N$. The map Q is, as we saw, obtained as the $\nabla^{(-)}$- geodesic (pp') joining p to the point $p' \in S_n$ having the same values of the expectation coordinates $\eta_1, \ldots, \eta_n$. Since these are constant along the geodesic, the tangent to (pp') is a sum of

$$\frac{\partial}{\partial \eta_{n+1}}, \ldots, \frac{\partial}{\partial \eta_N}.$$

It follows that (pp') is orthogonal to S_n, since

$$\langle \frac{\partial}{\partial \eta_i}, \frac{\partial}{\partial \beta^j} \rangle = g^{ik} \langle \frac{\partial}{\partial \beta^k}, \frac{\partial}{\partial \beta^j} \rangle = g^{ik} g_{jk} = \delta^i_j = 0$$

if $1 \le i \le n$ and $n+1 \le j \le N$. Similar results hold if Ω is countable, under the extended condition of thermodynamic stability.

In summary, the map Q is obtained by constructing the $\nabla^{(-)}$-geodesic through p which cuts the family S_n at right-angles at p'; then $pQ = p'$.

5. Isothermal and Isopotential Dynamics

In [**3**] we describe isothermal dynamics, which is constructed using a heat-particle as follows. To the system Ω_c with energy $\mathcal{E}_c$ we add a heat-particle, with sample space Ω_γ and heat-energy $\mathcal{E}_\gamma$; the total sample space is $\Omega = \Omega_c \times \Omega_\gamma$, and a point $\omega \in \Omega$ is thus a pair $(\omega_c, \omega_\gamma)$. We usually assume that $\Omega_\gamma = \{0, 1, 2, \ldots\}$, ("counting space") where the integer counts the quanta present. The total energy is the random variable $\mathcal{E}$ on Ω:

$$\mathcal{E}(\omega) = \mathcal{E}_c(\omega_c) + \mathcal{E}_\gamma(\omega_\gamma).$$

$\mathcal{E}_\gamma(\omega_\gamma)$ is usually taken to be proportional to ω_γ. We assume that $\mathcal{E}$ is thermodynamically stable. As in isolated dynamics, the random dynamics τ is taken to be a mixture of permutations of Ω preserving the energy-shells of $\mathcal{E}$. Let

$$p_{\gamma\beta} = Z^{-1} \exp\{-\beta \mathcal{E}_\gamma(\omega_\gamma)\}$$

with $\beta \in \mathbf{R}$. We start with a state $p = p_c \otimes p_{\gamma\beta}$ and apply the map τ followed by the marginal map onto the first factor, as in Eqn. (2.3), thus:

$$p_c \otimes p_{\gamma\beta} \mapsto \mathcal{M}_c\left((p_c \otimes p_{\gamma\beta})\tau\right) \otimes p_{\gamma\beta}.$$

This is the isothermal dynamics at beta β determined by τ.

It might be that there are other conserved quantities in a theory besides the energy; in atomic and chemical reactions there is usually at least one other quantity, the atomic number. Let $\mathcal{N}(\omega_c)$ be an integer-valued random variable, not a function of $\mathcal{E}_c$. Then $(\mathcal{E}, \mathcal{N})$ divide Ω into *energy-number shells*

$$\Omega_{E,N} = \{\omega \in \Omega : \mathcal{E}(\omega) = E, \;\; \mathcal{N}(\omega_c) = N\}.$$

A reversible isolated dynamics γ that conserves energy and atomic number is a permutation of each $\Omega_{E,N}$, and a random mixture τ of such permutations conserves the mean energy and the mean atomic number. To preserve these conservation laws in the reduced description it is only necessary to include $\mathcal{N}$ as well as $\mathcal{E}$ in the set of slow variables. The map Q, using the $\nabla^{(-)}$-geodesic perpendicular to S_n, will then conserve these mean values. The isolated dynamics determined by τ is then the map $p \mapsto p\tau Q$. We must modify the concept of ergodicity in the presence of further conserved quantities. Let us say that τ is $(\mathcal{E}, \mathcal{N})$-ergodic if τ conserves $\mathcal{E}$ and $\mathcal{N}$ and for each E and N, the uniform distribution on $\Omega_{E,N}$ is the only fixed point of

τ. Then with care (a sufficient condition is that $|\Omega| < \infty$) the map $p \mapsto p(\tau Q)^n$ converges to the grand canonical state

$$p_{\beta,\mu}(\omega) = Z_{\beta,\mu}^{-1} \exp\{-(\mathcal{E}(\omega) - \mu\mathcal{N}(\omega_c))\} \tag{5.1}$$

as $n \to \infty$.

The construction of isopotential, isothermal dynamics is similar to that of isothermal dynamics. The total system is taken to be $\Omega = \Omega_c \times \Omega_\gamma \times \Omega_n$, where both Ω_γ and Ω_n are counting space. We start with both Ω_γ and Ω_n is equilibrium, in the grand canonical state

$$p_{\beta,\mu} = Z_{\beta,\mu}^{-1} \exp\{-\beta(\mathcal{E}_\gamma(\omega_\gamma) - \mu\mathcal{N}(\omega_n))\}. \tag{5.2}$$

We express that p is in thermal equilibrium and at a uniform chemical potential by taking the initial state to be

$$p = p_c \otimes p_{\beta,\mu}(\omega_\gamma, \omega_n).$$

Having decided on a bistochastic map τ that preserves energy and atomic number, we take one time-step of the isothermal, isopotential dynamics to be

$$p \mapsto (\mathcal{M}_c p\tau)(\omega_c) \otimes p_{\beta,\mu}(\omega_\gamma, \omega_n) = p'_c \otimes p_{\beta,\mu}.$$

This law of dynamics is linear in p_c and reduces the Gibbs free energy, which turns out to be simply related to the relative entropy of the state p and the fixed point of the Markov chain; the latter represents thermal and chemical equilibrium. One time-step will take us out of the exponential family of states defined by the slow variables. The dynamics must be reduced by projecting the new point p' onto S_n. Since in isothermal, isopotential dynamics the values of the mean energy and atomic number are not conserved, there is no reason to use the $\nabla^{(-)}$-geodesic, which conserves these means. Indeed, all parts of the system are at the same beta, so the projection should preserve this fact. Similarly, in isopotential dynamics the projection back onto S_n should preserve the chemical potential of all parts of the system. This suggests that, for isothermal, isopotential dynamics, the projection should be the $\nabla^{(+)}$-geodesic cutting S_n at right angles. This is the dual to projection used in the isothermal case. More generally, we could use as affine coordinates certain extensive variables $\eta_1, \dots, \eta_k$ and the remaining ones could be intensive variables $\beta_{k+1}, \dots$. This type of model has not been explored yet.

6. The Dynamic Ising Model

In this section we show the results of applying these ideas to the Ising model, which is usually taken as entirely static. We shall need only the case of a finite system. Let $(\Lambda, \mathcal{L})$ be a finite graph, with vertices $x \in \Lambda$ and edges (representing bonds) $b \in \mathcal{L}$. Let Ω_x, the same for all x, be the sample space of configurations at x. For example, $\Omega_x = \{0, 1\}$ for the lattice gas, and $\Omega_x = \{-1, +1\}$ for the magnet. These are obviously the same from an abstract point of view. Put $\Omega = \prod_x \Omega_x$, so that a configuration $\omega \in \Omega$ is a finite sequence $\omega = (\omega_1, \omega_2, \dots, \omega_{|\Lambda|})$ with $\omega_j \in \Omega_j$. So for these models, $|\Omega_x| = 2$ and we have $|\Omega| = 2^{|\Lambda|}$ field configurations. Then

$$\operatorname{Span}\Omega = \mathbf{R}^{2^{|\Lambda|}}.$$

The energy in the Ising model is the random variable

$$\mathcal{E}(\omega) = -J \sum_{\substack{b \in \mathcal{L} \\ b=(x,y)}} \omega_x \omega_y = \sum_b \mathcal{E}_b(\omega)$$

where $J > 0$. The total spin is the random variable

$$\mathcal{S}(\omega) = \sum_{x \in \Lambda} \omega_x = \sum_{x \in \Lambda} \mathcal{S}_x(\omega)$$

where $\mathcal{S}_x(\omega) = \omega_x$. These, and their means, are called extensive variables, because they are sums of local contributions.

We (C. Connaughton and I) [**9**] choose $\{\mathcal{E}_b, \mathcal{S}_x\}$ as our slow variables (accessible variables, variables of interest) and so these define our level of description. Other random variables such as $\omega_x\omega_y$, where $(x, y) \notin \mathcal{L}$, or $\omega_x\omega_y\omega_z$, are taken as the fast variables (or nuisance variables, as they are known in estimation theory). We noted already that the theory does not require that these are fast variables in the linear dynamics $p \mapsto pQ$; we make them fast in the step Q, which replaces the current state $p\tau$ by one in which the nuisance variables are absent, i.e. they are as uniformly distributed as is consistent with the given values of the means. This step thus models randomising effects not present in τ. It is not necessary that all the variables in $\mathcal{X}$ are slow variables relative to the dynamics. If they are fast, this will simply mean that we are treating exactly some variables that might have been treated statistically. The set $\mathcal{X}$ does contain some genuine slow variables, namely, the sums of local energies over a set $\mathcal{L}_0$ of bonds in $\mathcal{L}$; call this $\mathcal{E}(\mathcal{L}_0)$. If the size of the boundary of $\mathcal{L}_0$ is small compared with the size of $\mathcal{L}_0$, then the rate of change of $\mathcal{E}(\mathcal{L}_0)$ will usually be small relative to the size of $\mathcal{E}(\mathcal{L}_0)$, and so for large systems these are slow variables. This will be a general feature of any extensive variable whose total value is conserved, and which satisfies a local conservation law. This explains why many macroscopic laws of physics and chemistry are expressed in terms of densities of globally conserved quantities.

In the model in question [**9**], the number of slow variables is $|\Lambda| + |\mathcal{L}| \ll 2^{|\Lambda|}$ for large systems, so our choice of slow variables leads to a big reduction in the level of description. We have worked out the case when $|\Lambda| = 3$, $|\mathcal{L}| = 3$. This represents a molecule of three atoms in a triangle, with two states for each atom and three bonds of equal strength. We have $|\Omega| = 8$, so the general state is described by seven parameters. Since we use six slow variables, there has not been much reduction in the level of description; the reduction consists in omitting the mutual correlation between the three spins.

The sample space Ω is divided into two energy-shells, with $E = J$ and $E = -3J$:

$$\begin{aligned} \Omega_J &= \left\{ \begin{pmatrix} + & & + \\ & - & \end{pmatrix}, \begin{pmatrix} - & & + \\ & + & \end{pmatrix}, \begin{pmatrix} - & & - \\ & + & \end{pmatrix}, \right. \\ & \qquad \left. \begin{pmatrix} + & & - \\ & - & \end{pmatrix}, \begin{pmatrix} + & & - \\ & + & \end{pmatrix}, \begin{pmatrix} - & & + \\ & - & \end{pmatrix} \right\} \\ \Omega_{-3J} &= \left\{ \begin{pmatrix} + & & + \\ & + & \end{pmatrix}, \begin{pmatrix} - & & - \\ & - & \end{pmatrix} \right\} \end{aligned}$$

The ground state is therefore a doublet. The dynamics τ is to be a mixture of permutations mapping each energy-shell to itself. The non-trivial permutation of the ground states in Ω_{-3J} involves three spin-flips, and some permutations of Ω_J involve double spin-flips. Let us suppose that these do not enter the dynamics, which we take to be an uniform mixture of the six two-cycles γ_j that flip one spin only. This expresses the locality of the interaction. The six permutations are

$$\gamma_1 = \begin{pmatrix} + & + & - \\ + & - & - \end{pmatrix}, \quad \gamma_2 = \begin{pmatrix} - & + & + \\ - & + & - \end{pmatrix}, \quad \gamma_3 = \begin{pmatrix} + & - & + \\ - & - & + \end{pmatrix},$$
$$\gamma_4 = \begin{pmatrix} - & + & + \\ - & - & + \end{pmatrix}, \quad \gamma_5 = \begin{pmatrix} + & + & - \\ - & + & + \end{pmatrix} \quad \gamma_6 = \begin{pmatrix} + & - & + \\ + & - & - \end{pmatrix}.$$

Other energy-conserving cycles of order 6 involve collaboration at different sites, and are omitted. Thus we choose

$$\tau = (1 - \lambda)1 + \lambda/6 \sum_{j=1}^{6} \gamma_j .$$

We seek dynamics within the exponential family S_6:

$$p_{\beta,\mu} = Z^{-1} \exp\left(\mu_1 \mathcal{S}_1 + \mu_2 \mathcal{S}_2 + \mu_3 \mathcal{S}_3 - \beta_{12}\mathcal{E}_{12} - \beta_{23}\mathcal{E}_{23} - \beta_{31}\mathcal{E}_{31}\right). \tag{6.1}$$

The intensive variables $\beta \in \mathbf{R}^3$ are the inverse temperatures of the bonds, and $\mu \in \mathbf{R}^3$ are the chemical potentials of the spins (up to conventional factors). These are the $\nabla^{(+)}$-affine coordinates of our problem.

To get the isolated dynamics we start with a given energy-density field $\{p.\mathcal{E}_b\}_{b\in\mathcal{L}}$ and a given spin-density field $\{p.\mathcal{S}_x\}_{x\in\Lambda}$; we then construct the unique state $p_{\beta,\mu} \in S_6$ in the exponential family having these mean values. We then find $p_{\beta,\mu}\tau$; it is not in the exponential family S_6, but we can compute new mean values, $p\tau.\mathcal{E}_b,\ \ b \in \mathcal{L}$ and $p\tau.\mathcal{S}_x,\ \ x \in \Lambda$. We then apply Q, which is the projection onto S_6 given by the $\nabla^{(-)}$-geodesic, through $p\tau$, that cuts S_6 at right angles. This is one time-step. Proceed; we get an orbit $\{p(t)\}$ through S_6, which we call the isolated dynamics of the Ising model.

To get some idea of what this law of dynamics implies, we let the time-step become infinitesimal, replacing λ, the probability of one spin-flip in one time-step, by λdt, where now λ is the rate. Let $S_x(t) = p(t).\mathcal{S}_x$ and $E_b(t) = p(t).\mathcal{E}_b$. We get differential equations of the Onsager type, but non-linear:

$$\frac{d}{dt} S_x = F(x, \beta(t), \mu(t)) \tag{6.2}$$
$$\frac{d}{dt} E_b = G(b, \beta(t), \mu(t)) \tag{6.3}$$

where $(\beta, \mu) \in \mathbf{R}^6$ are implicit functions of S_x and E_b at time t. The functions F and G can be found by Mathematica [**9**].

We show that if the initial energy is sufficiently small then the point $\mu = 0$, $\beta = (\beta_0, \beta_0, \beta_0)$ is the unique global attractor, where $\beta_0 > 0$ is determined by the initial energy. The usual perturbation theory about this fixed point leads to linear equations e.g. for $S_2(t)$:

$$3\frac{dS_2}{dt} = \lambda\kappa[(S_3 - S_2) - (S_2 - S_1)] - \frac{4\lambda}{3e^{4\beta J} + 1} S_2(t) \tag{6.4}$$

where $\kappa = (e^{4\beta J} - 1)/(3e^{4\beta J} + 1) > 0$. The first term represents diffusion of spins, being an approximation to the second space derivative of S_2. This term conserves total spin. The second term leads to the decay of the spins at a rate equal to $R = 4\lambda/(9e^{4\beta J} + 3)$, which converges to $\lambda/3$ as $\beta \to 0$, and vanishes as $\beta \to \infty$, at zero temperature. The decay of the spin is thus seen to be induced by the heat present.

We also find a heat equation (in the linear approximation) for β and for its inverse, in which the conductivity is determined by λ and J [**9**].

Study of this model, and similar ones with larger lattices, allows us to formulate the concept of metastability for the Ising model. While no finite system can show a phase transition in the accepted sense, we can calculate the stability of the Ising state with, say, positive values for S_x, under the perturbation caused by the random dynamics. Obviously, in the linear approximation the decay rate R, say, will be proportional to λ. Let us define the coefficient of metastability (relative to perturbations given by τ) to be λ/R. This, which has the units of time, measures how long the spins survive when random dynamics τ of unit strength is added to the static Ising model. For small temperatures the life-time of the spins is exponentially long, whereas for temperatures of the order of $4J$ this time is $O(1)$, and this does not change much from there to infinite temperatures. So the random dynamics acts as a sort of probe, seeking evidence of a phase transition. It is to be expected that as $|\Lambda|$ becomes large, and the lattice approximates the two-dimensional square lattice, this coefficient will become very small below a certain value β_c of beta, but will be roughly constant above β_c. We have no results, as yet, in this direction.

References

[1] R. S. Ingarden, Information theory and variational principles in statistical theories, *Bull. Acad. Polon. Sci. Ser. Math.-Astro.-Phys.* **11**, 541-547 (1963).

[2] R. S.Ingarden, Towards mesoscopic thermodynamics: small systems in higher-order states, *Open Systems and Information Dynamics*, **1** 75-102, (1992). See also: R. S. Ingarden and T. Nakagomi, The second order extension of the Gibbs state, *Open Systems and Information Dynamics*, **1**, 243-258 (1992).

[3] **Statistical Dynamics**, 275 pp, Imperial College Press, 1995.

[4] S.-I. Amari, **Differential Geometric Methods in Statistics**, *Lecture Notes in Statistics*, **28**, Springer-Verlag, New York, 1985.

[5] E. T. Jaynes, Information Theory and Statistical Mechanics, *Physical Review* **106**, 620-630; and II, *ibid*, **108**, 171-190 (1957).

[6] A. Kossakowski, On the Quantum Informational Thermodynamics, *Bulletin de l'académie polonaise des sciences*, **17**, 263-267 (1969).

[7] F. Schlögl, Stochastic measures in non-equilibrium thermodynamics, *Phys. Rep.*, **62**, 267-380, (1980).

[8] R. A. Fisher, Theory of statistical estimation, *Proc. Camb. Phil. Soc.* **22**, 700 (1925).

[9] C. Connaughton and R. F. Streater, The Dynamic Ising Model, in preparation.

DEPT. OF MATHEMATICS,, KING'S COLLEGE,, STRAND,, LONDON, WC2R 2LS
E-mail address: R.STREATER@kcl.ac.uk

Part IV

NONCOMMUTATIVE GEOMETRY

Contemporary Mathematics
Volume **203**, 1997

On Finite Differential Calculi

D. KASTLER, J. MADORE, and T. MASSON

ABSTRACT. In the case of matrix algebras, a comparison is made between the differential calculus based on derivations and that based on the Dirac operator. In particular the appropriate Dirac operator is given which reproduces the differential calculus based on derivations.

1. Introduction

The extension to noncommutative algebras of the definition of a differential calculus has been given both without [**6**] and with [**14**] the use of the derivations of the algebra. The former is of course a more general procedure since it can be used even in the case of algebras which have no derivations, for example the algebras of continuous or measurable functions on a differential manifold. At least in the finite-dimensional case which we shall consider here a comparison between the two approaches is readily made. In particular the differential calculus of an algebra based on derivations can be easily put in the form of the differential calculus introduced by Connes & Lott [**10**] which is based on a generalization of the Dirac operator.

In Section 2 we shall give a brief description of the differential calculus based on derivations. In Section 3 we shall do the same for the differential calculus which is based on a generalization of the Dirac operator. In Section 4 we show the relation between the two. In Section 5 we briefly recall how physical theories can be constructed by considering in conjunction with these finite calculi the ordinary differential calculus of space-time.

2. Derivations

A smooth global vector field on a manifold V is equivalent to a derivation of the algebra of smooth functions $\mathcal{C}(V)$ defined on V [**29**]. As such it has a natural extension to noncommutative geometry [**6, 7, 8**]. The algebras M_n of $n\times n$ complex matrices furnish finite analogues of compact parallelizable manifolds. Let $\mathrm{Der}(M_n)$ be the vector space of all derivations of M_n. It is an elementary fact of algebra that all derivations of M_n are inner. This means that every element X of $\mathrm{Der}(M_n)$ is of the form $X = \mathrm{ad}\, f$ for some f in M_n. The derivation X is real if and only if f is

1991 *Mathematics Subject Classification.* Primary 47C10; Secondary 14A22 .
Supported in part by the CNRS.

anti-hermitian. The main difference with the commutative case lies in the fact that $\mathrm{Der}(M_n)$ is not an M_n-module; if $X \in \mathrm{Der}(M_n)$ and $f \in M_n$ then in general fX is not in $\mathrm{Der}(M_n)$. It is however a free module over the center of M_n, that is, over the complex numbers; it is a vector space of dimension $n^2 - 1$. This is the analogue of the fact that the global vector fields on a parallelizable manifold V form a free $\mathcal{C}(V)$-module.

Let λ_a, for $1 \leq a \leq n^2 - 1$, be an anti-hermitian basis of the Lie algebra of the special unitary group SU_n. The product $\lambda_a \lambda_b$ can be written in the form

$$\lambda_a \lambda_b = \frac{1}{2} C^c{}_{ab} \lambda_c + \frac{1}{2} D^c{}_{ab} \lambda_c - \frac{1}{n} g_{ab}. \tag{2.1}$$

The g_{ab} are the components of the Killing metric, a positive, nondegenerate inner product on the Lie algebra. We have suppressed the unit element of M_n which should appear in the above formula as a factor of g_{ab}. The structure constants $C^c{}_{ab}$ are real. The Killing metric can be defined in terms of them by the equation

$$g_{ab} = -\frac{1}{2n} C^c{}_{ad} C^d{}_{bc}. \tag{2.2}$$

We shall lower and raise indices with g_{ab} and its inverse g^{ab}. The tensor C_{abc} is completely antisymmetric and D_{abc} is completely symmetric and trace-free. We shall normalize the λ_a so that $g_{ab} = \delta_{ab}$.

The set of λ_a is a set of generators of M_n. It is not a minimal set but it is convenient because of the fact that the derivations

$$e_a = \mathrm{ad}\, \lambda_a \tag{2.3}$$

form a basis over the complex numbers for $\mathrm{Der}(M_n)$. We recall that the adjoint action is defined by

$$e_a f = \mathrm{ad}\, \lambda_a(f) = [\lambda_a, f], \qquad f \in M_n. \tag{2.4}$$

Any element X of $\mathrm{Der}(M_n)$ can be written as a linear combination $X = X^a e_a$ of the e_a where the X^a are complex numbers.

The vector space $\mathrm{Der}(M_n)$ has a Lie-algebra structure. In particular the derivations e_a satisfy the commutation relations

$$[e_a, e_b] = C^c{}_{ab}\, e_c. \tag{2.5}$$

We define the algebra of forms $\Omega^*(M_n)$ over M_n just as we did in the commutative case, as the graded algebra of multilinear, completely antisymmetric maps of the derivations into the algebra. The product is defined also using the same formula [**14**]. First we set $\Omega^0(M_n)$ to be equal to M_n. Then we define df for $f \in M_n$ by the equation

$$df(e_a) = e_a(f). \tag{2.6}$$

This means in particular that

$$d\lambda^a(e_b) = [\lambda_b, \lambda^a] = -C^a{}_{bc} \lambda^c. \tag{2.7}$$

We define the set of 1-forms $\Omega^1(M_n)$ to be the set of all elements of the form fdg with f and g in M_n. So $\Omega^1(M_n)$ is a left M_n-module. We could have also defined $\Omega^1(M_n)$ to be the set of all elements of the form $(dg)f$ which would define $\Omega^1(M_n)$ as a right module. Since M_n is not a commutative algebra one has to distinguish between left and right multiplication. Although fdg and $(dg)f$ are not equal the two definitions of $\Omega^1(M_n)$ coincide as a bimodule because of the relation $d(fg) = f(dg) + (df)g$. A differential form of order p or p-form is defined exactly as in the commutative case, as are the product and the differential d. As in the commutative case we have the formulae

$$d(\alpha\beta) = (d\alpha)\beta + (-1)^p \alpha d\beta \tag{2.8}$$

as well as

$$d^2 = 0. \tag{2.9}$$

The algebra $\Omega^*(M_n)$ is a graded differential algebra. It is easy to see however that if $\alpha, \beta \in \Omega^*(M_n)$ then in general

$$\alpha\beta \neq \pm\beta\alpha. \tag{2.10}$$

The set of $d\lambda^a$ constitutes a system of generators of $\Omega^1(M_n)$ as a left or right module but it is not a convenient one. For example $\lambda^a d\lambda^b \neq d\lambda^b \lambda^a$. There is a better system of generators completely characterized by the equations

$$\theta^a(e_b) = \delta^a_b. \tag{2.11}$$

We have suppressed the unit matrix which should appear as a factor of δ^a_b on the right-hand side of this equation. The θ^a are related to the $d\lambda^a$ by the equations

$$d\lambda^a = C^a{}_{bc}\lambda^b\theta^c \tag{2.12}$$

and their inverse

$$\theta^a = \lambda_b\lambda^a d\lambda^b. \tag{2.13}$$

They form a basis of the 1-forms dual to the derivations 2.3. Equation 2.12 follows immediately from 2.7 and 2.11 but the proof of the inverse Equation 2.13 uses the identities [**30**]

$$\begin{aligned}
C^a{}_{bc}C^c{}_{de}C^e{}_{fa} &= -nC_{bdf}, \\
C^a{}_{bc}C^c{}_{de}D^e{}_{fa} &= -nD_{bdf}, \\
C^a{}_{bc}D^c{}_{de}D^e{}_{fa} &= -\frac{1}{n}(n^2-4)C_{bdf}
\end{aligned}$$

between the structure constants and the tensor D_{abc}.

Because of the relation 2.11 we have

$$\theta^a \wedge \theta^b = -\theta^b \wedge \theta^a, \qquad \lambda^a\theta^b = \theta^b\lambda^a. \tag{2.14}$$

The θ^a satisfy the same structure equations as the components of the Maurer-Cartan form on the special unitary group SU_n:

$$d\theta^a = -\frac{1}{2}C^a{}_{bc}\,\theta^b \wedge \theta^c. \tag{2.15}$$

The product on the right-hand side of this formula is the product in $\Omega^*(M_n)$. Although this product is not in general antisymmetric, the subalgebra $\bigwedge^*$ of $\Omega^*(M_n)$ generated by the θ^a is an exterior algebra. Formula 2.15 means that it is a differential subalgebra. Since the θ^a commute with the elements of M_n we can identify $\Omega^*(M_n)$ with the tensor product of M_n and $\bigwedge^*$:

$$\Omega^*(M_n) = M_n \otimes_{\mathbb{C}} \textstyle\bigwedge^*. \tag{2.16}$$

The interior product and the Lie derivative are defined using the same formulae as in the commutative case.

From the generators θ^a we can construct the 1-form

$$\theta = -\lambda_a \theta^a \tag{2.17}$$

in $\Omega^1(M_n)$. From 2.13 we see that it can also be written as

$$\theta = -\frac{1}{n}\lambda_a d\lambda^a = \frac{1}{n} d\lambda_a \lambda^a. \tag{2.18}$$

Using θ we can rewrite 2.13 as

$$\theta^a = C^a{}_{bc}\lambda^b d\lambda^c - n\lambda^a\theta = -C^a{}_{bc}\lambda^b\theta\lambda^c. \tag{2.19}$$

Apart from the second term on the right-hand side this equation is related to 2.12 by an interchange of $d\lambda^a$ and θ^a. From 2.12 and 2.15 one sees that θ satisfies the condition

$$d\theta + \theta^2 = 0. \tag{2.20}$$

It satisfies with respect to the algebraic exterior derivative the same equation which the Maurer-Cartan form satisfies with respect to ordinary exterior derivation on the group SU_n. It follows directly from the definitions that the exterior derivative df of an element of M_n can be written in terms of a commutator with θ:

$$df = -[\theta, f]. \tag{2.21}$$

This is not true however for an arbitrary element of $\Omega^*(M_n)$ and this accounts for the difference between the differential calculus we are here considering and that of the next section.

There is a map of the trace-free elements of M_n onto the derivations of M_n given by $f \mapsto X_f = i\,\mathrm{ad}\, f$. The factor i has been included to make X_f real when f is hermitian. The 1-form θ can be defined, without any reference to the θ^a, as the inverse map:

$$\theta(X_f) = -if. \tag{2.22}$$

The complete set of all derivations of M_n is the natural analogue of the space of all smooth vector fields $\mathrm{Der}\big(\mathcal{C}(V)\big)$ on a manifold V. If V is parallelizable then $\mathrm{Der}\big(\mathcal{C}(V)\big)$ is a free module over $\mathcal{C}(V)$ with a set of generators e_α which is closed under the Lie bracket and which has the property that if $e_\alpha f = 0$ for all e_α then f is a constant function. Consider the Lie algebra 2.5 with the e_a defined by 2.3. We have supposed that the λ_a are elements of the fundamental representation of $\underline{su}_n$, the Lie algebra of SU_n. We could equally well have assumed that they lie in an n-dimensional representation of $\underline{su}_m$ for $m < n$. If the representation is irreducible they will generate M_n as an algebra and the corresponding set of derivations $\mathrm{Der}_m(M_n)$ will be a Lie subalgebra of the Lie algebra $\mathrm{Der}(M_n)$ of all the derivations of M_n. This freedom of representation allows one to introduce an extra integer parameter into some of the theories which we shall mention in the last section.

3. The Dirac Operator

In the previous section we used derivations to define differential forms and the d operator but we then saw in that we could have defined d as acting on elements of M_n by the equation $df = -[\theta, f]$. There is here no explicit reference to the set of derivations. The algebra of forms $\Omega^*(M_n)$ has a $\mathbb{Z}_2$ grading and θ is an odd element. Let Ω_S be the matrix algebra M_n itself with a $\mathbb{Z}_2$ grading. That is, Ω_S

can be expressed as the direct sum $\Omega_S = \Omega_S^+ \oplus \Omega_S^-$ where Ω_S^+ (Ω_S^-) are the even (odd) elements of Ω_S. It is possible to consider Ω_S as an algebra of forms. One can define on Ω_S a graded derivation $\hat{d}$ by the formula

$$\hat{d}f = -[\eta, f], \tag{3.1}$$

where η is an arbitrary anti-hermitian odd element of Ω_S and the commutator is taken as a graded commutator. We find that $\hat{d}\eta = -2\eta^2$ and for any $\alpha \in \Omega_S$,

$$\hat{d}^2\alpha = [\eta^2, \alpha]. \tag{3.2}$$

The $\mathbb{Z}_2$ grading can be induced from a decomposition $\mathbb{C}^n = \mathbb{C}^l \oplus \mathbb{C}^{n-l}$ for some integer l. The elements of Ω_S^+ are diagonal with respect to the decomposition; the elements of Ω_S^- are off-diagonal. If n is even and $l = n/2$ then it is possible to impose the condition

$$\eta^2 = -1. \tag{3.3}$$

The unit on the right-hand side of this equation is the unit in Ω_S. From 3.2 we see that $\hat{d}^2 = 0$. The map 3.1 is a differential. In this case we shall write

$$\hat{d} = d. \tag{3.4}$$

When 3.3 is satisfied the algebra Ω_S is a differential algebra with a $\mathbb{Z}_2$ grading. Notice that 2.20 is not satisfied for η. We have instead

$$d\eta + \eta^2 = 1. \tag{3.5}$$

The reason for the difference is that initially a relation of the form 3.1 was valid only for elements of $\Omega^0(M_n)$. In defining the second differential d we have imposed it on all elements of Ω_S.

If n is not even or, in general, if η^2 is not proportional to the unit element of M_n then $\hat{d}^2$ given by 3.2 will not vanish and Ω_S will not be a differential algebra. Consider now an arbitrary grading of M_n and let η be an arbitrary odd element. It is possible to construct over M_n^+ an integer-graded differential algebra $\Omega_\eta^* = \Omega_\eta^*(M_n^+)$ based on Formula 3.1. Let $\Omega_\eta^0 = M_n^+$ and let $\Omega_\eta^1 \equiv \overline{d\Omega_\eta^0} \subset M_n^-$ be the M_n^+-bimodule generated by the image of Ω_η^0 in M_n^- under $\hat{d}$. Define

$$\Omega_\eta^0 \xrightarrow{d_\eta} \Omega_\eta^1 \tag{3.6}$$

using directly 3.1: $d_\eta = \hat{d}$. Let $\overline{d\Omega_\eta^1}$ be the M_n^+-module generated by the image of Ω_η^1 in M_n^+ under $\hat{d}$. It would be natural to try to set $\Omega_\eta^2 = \overline{d\Omega_\eta^1}$ and define

$$\Omega_\eta^1 \xrightarrow{d_\eta} \Omega_\eta^2 \tag{3.7}$$

using once again 3.1. Every element of Ω_η^1 can be written as a sum of elements of the form $f_0\hat{d}f_1$. If we attempt to define an application 3.7 using again directly 3.1,

$$d_\eta(f_0\hat{d}f_1) = \hat{d}f_0\hat{d}f_1 + f_0\hat{d}^2f_1, \tag{3.8}$$

then we see that in general d_η^2 does not vanish. To remedy this problem we eliminate simply the unwanted terms. Let $\mathrm{Im}\,\hat{d}^2$ be the submodule of $\overline{d\Omega_\eta^1}$ consisting of those elements which contain a factor which is the image of $\hat{d}^2$ and define Ω_η^2 by

$$\Omega_\eta^2 = \overline{d\Omega_\eta^1}/\mathrm{Im}\,\hat{d}^2. \tag{3.9}$$

Then by construction the second term on the right-hand side of 3.8 vanishes as an element of Ω^2_η and we have a well defined map 3.7 with $d^2_\eta = 0$. This procedure can be continued to arbitrary order by iteration. For each $p \geq 2$ we let $\mathrm{Im}\,\hat{d}^2$ be the submodule of $\overline{d\Omega^{p-1}_\eta}$ defined as above and we define Ω^p_η by

$$\Omega^p_\eta = \overline{d\Omega^{p-1}_\eta}/\mathrm{Im}\,\hat{d}^2. \tag{3.10}$$

Since $\Omega^p_\eta \Omega^q_\eta \subset \Omega^{p+q}_\eta$ the complex Ω^*_η is a differential algebra. The Ω^p_η need not vanish for large values of p. In fact if $\eta^2 \propto 1$ we see that $\mathrm{Im}\,\hat{d}^2 = 0$ and $\Omega^p_\eta \simeq M^+_n(M^-_n)$ for p even (odd). An equivalent construction of Ω^*_η can be given [**10, 25**] using what is known as the universal differential calculus [**6, 7, 8, 27**].

Let $n = 2$. Then $\Omega_S = M_2$ with a $\mathbb{Z}_2$ grading. The most general element of Ω_S can be expanded using the 4 matrices $\{\lambda_a, 1\}$ with λ_1 and λ_2 odd and λ_3 and the unit even. The most general possible form for η is a linear combination of λ_1 and λ_2 and it can be normalized so that 3.3 is satisfied. Then Ω_S is a differential algebra of dimension 4. The algebra of forms $\Omega^*(M_2)$ constructed using the differential of the previous section is a $4 \times 8 = 32$ dimensional algebra. Over M_2 we have constructed then two differential calculi, one with an algebra of dimension 32 and a second of dimension 4, although more properly the algebra Ω_S should be considered as an algebra of forms over its even elements M^+_2. Using Ω_S one can construct a differential calculus over $\mathcal{C}(\mathbb{R}^4) \times \mathcal{C}(\mathbb{R}^4)$, the algebra of smooth functions on a double-sheeted space-time. This is one of the models [**9**] which we shall mention in the last section.

Let $n = 3$. Then $\Omega_S = M_3$ with a $\mathbb{Z}_2$ grading defined by the decomposition $\mathbb{C}^3 = \mathbb{C}^2 \oplus \mathbb{C}$. The most general possible form for η is

$$\eta = \begin{pmatrix} 0 & 0 & a_1 \\ 0 & 0 & a_2 \\ -a_1^* & -a_2^* & 0. \end{pmatrix} \tag{3.11}$$

For no values of the parameters can 3.3 be satisfied. The general construction yields $\Omega^0_\eta = M^+_3 = M_2 \times M_1$ and $\Omega^1_\eta = M^-_3$ as in the previous example but after that the quotient by elements of the form $\mathrm{Im}\,\hat{d}^2$ reduces the dimensions. One finds $\Omega^2_\eta = M_1$ and $\Omega^p_\eta = 0$ for $p \geq 3$. This differential calculus is closely related to that used in one of the models [**10**] which we shall mention in the last section.

4. The comparison

Now we show that the differential calculus of Section 2 can be put in the form of the language of Section 3. Set $N = n^2 2^{n^2-1}$. Then $\Omega^*(M_n)$ is a complex N-dimensional vector space with a natural $\mathbb{Z}_2$ grading. The algebra M_N has therefore also a natural $\mathbb{Z}_2$ grading. There is an imbedding of $\Omega^*(M_n)$ in M_N by left multiplication. We designate the image of a form α in M_N by $\hat{\alpha}$. The differential d has an extension $\hat{d}$ to all elements of M_N. Let L be an element of M_N with parity $|L|$ and α an arbitrary element of $\Omega^*(M_n)$. Define $\hat{d}L$ by

$$(\hat{d}L)\alpha = d(L\alpha) - (-1)^{|L|} L d\alpha. \tag{4.1}$$

Using the graded commutator this can be written in the form

$$\hat{d}L = [d, L]. \tag{4.2}$$

That is, $\hat{d}$ is just the commutator of d considered as an element of M_N. In particular $\hat{d}\hat{\alpha} = [d, \hat{\alpha}] = \widehat{d\alpha}$. We have seen that when acting on the elements of $\Omega^0(M_n) = M_n$ the exterior derivative d can be expressed as a commutator with the element θ. Although in general it is not possible to write the right-hand side of 4.2 as the image of a commutator in $\Omega^*(M_n)$, there is a sense in which this can almost be done. That is for any $\alpha \in \Omega^*(M_n)$ we can write

$$\hat{d}\hat{\alpha} = -[\eta, \hat{\alpha}] \tag{4.3}$$

where η is almost the image of an element of $\Omega^*(M_n)$.

Let $c^a = \hat{\theta}^a$ be the element of M_N which corresponds to left-multiplication by the element θ^a of $\Omega^1(M_n)$ and let $\bar{c}_a$ be the element of M_N which corresponds to the interior product by e_a. Both c^a and $\bar{c}_a$ are odd elements of M_N and their (graded) commutation rules are

$$[c^a, c^b] = 0, \qquad [\bar{c}_a, \bar{c}_b] = 0, \qquad [\bar{c}_a, c^b] = \delta^b_a. \tag{4.4}$$

They commute with the elements $\hat{\lambda}^a$. Let $\hat{\theta} = -\hat{\lambda}_a c^a$ be the image of θ in M_N and define η as

$$\eta = \hat{\theta} + \frac{1}{2} C^a{}_{bc} c^b c^c \bar{c}_a. \tag{4.5}$$

It is easy to verify that 4.3 is satisfied. It follows immediately from the definitions that

$$\hat{d}c^a = -[\eta, c^a] = -\frac{1}{2} C^a{}_{bc} c^b c^c. \tag{4.6}$$

Let L_a be the Lie derivative with respect to e_a. It is interesting to note also that

$$\hat{d}\bar{c}_a = L_a, \qquad \hat{d}L_a = 0. \tag{4.7}$$

But because $\bar{c}_a$ is not in the image of $\Omega^*(M_n)$ in M_N the expression $\hat{d}\bar{c}_a + [\eta, \bar{c}_a]$ does not vanish. It commutes however with all elements of the image of $\Omega^*(M_n)$ in M_N. We have

$$\eta^2 = 0, \tag{4.8}$$

and therefore $\hat{d}^2 = 0$. Consider the algebra Ω^*_η constructed from the η defined in 4.5. There is a natural imbedding of differential algebras

$$\Omega^*(M_n) \overset{i}{\hookrightarrow} \Omega^*_\eta. \tag{4.9}$$

We construct the map i by iteration. First of all there is the imbedding of $\Omega^0(M_n) = M_n$ into $\Omega^0_\eta = M_N^+$ given by left multiplication. We extend i to $\Omega^1(M_n)$ by setting

$$i(df) = -[\eta, \hat{f}] = d_\eta i(f). \tag{4.10}$$

Since both $\Omega^*(M_n)$ and Ω^*_η are graded differential algebras 4.9 is uniquely defined. The notation in this example is adapted from the standard notation used in the application of ghosts in the quantization of gauge fields.

In the above example we imbedded a differential algebra in a larger algebra such that in the latter the differential was given by a commutator. The larger algebra was the algebra of homomorphisms of the original algebra considered as a vector space. This construction can be generalized to arbitrary algebras. There is also another construction which uses a different extension. Let $\mathcal{A}$ be any (graded) algebra and d a (graded) derivation. It is possible to imbed $\mathcal{A}$ into an algebra $\mathcal{B}$

such that the derivation on the image is given by a commutator. We construct the larger algebra by simply adding to the original one an (odd) element η and defining the product ηa and $a\eta$ such that the (graded) commutator is given by $da = -[\eta, a]$. In general d need not be defined on η. In the previous example we added in fact n elements $\bar{c}_a$ to the original algebra. The larger algebra constructed was contained in M_N.

Connes has introduced [**6, 7, 8**] the noncommutative generalization of a K-cycle or spectral triple $(\mathcal{A}, \mathcal{H}, D)$. This consists of an algebra $\mathcal{A}$, a representation of $\mathcal{A}$ on a graded Hilbert space $\mathcal{H}$ and an odd operator D which is a generalization of the Dirac operator. From the spectral triple a differential calculus can be constructed. We have defined two examples of spectral triples. The first, $(M_n, \mathbb{C}^n, \eta)$ was used to construct the differential calculus used in the Connes-Lott model [**9, 10**]. The second, $(M_N, \mathbb{C}^N, \eta)$ was implicit in the construction of Dubois-Violette *et al.* [**15, 16, 17, 18**].

5. Extended Space-Time

The first attempts to introduce extra dimensions in order to unify the gravitational field with electromagnetism were made by Kaluza and Klein. It was suggested by Einstein & Bergmann that at sufficiently small scales what appears as a point will in fact be seen as a circle. Later, with the advent of more elaborate gauge fields, it was proposed that this internal manifold could be taken as a compact Lie group or even as a general compact manifold [**1, 2**]. The disadvantage of these extra dimensions is that they introduce even more divergences in the quantum theory and lead to an infinite spectrum of new particles. One can take this as motivation for introducing a modification of Kaluza-Klein theory with an internal structure which is described by a finite noncommutative geometry [**16, 31, 3, 33**].

Closely related to Kaluza-Klein theory is the study of Maxwell's equations in higher-dimensions. As has been shown by Forgacs & Manton [**35, 21**] and others [**20, 4, 22**] if one considers the Maxwell action on a principle fibre bundle over space-time with group G then the resulting effective action is that of a Yang-Mills theory unified with a Higgs field which takes its value in the Lie algebra of G. The quartic potential appears naturally since the Yang-Mills action is quartic in the gauge potentials. Like all theories in higher dimensions, these theories include an infinite tower of massive states over each massless state. A rather artificial truncation procedure must be introduced to eliminate them. At least in the cases where the group G is one of the unitary groups U_n, equivalent theories can be reproduced without truncation by using a finite noncommutative extension of space-time based on the algebras M_n of complex $n \times n$ matrices [**15, 16, 17, 18**] and by using the differential calculus based on derivations [**14**] which we reviewed in Section 2. The resulting physical theories [**31, 32**] contain only one free parameter and are not realistic. In order to reproduce the standard model with a set of Higgs fields in the fundamental representation of the gauge group it is necessary to use the more abstract differential calculi which we reviewed in Section 3, either the $\mathbb{Z}_2$-based calculus [**9, 12, 13**] or its $\mathbb{Z}$-based extension [**10, 12, 23, 28**] For more details of this we refer to the numerous review articles on the subject [**27, 37, 24**].

6. Acknowledgement

The authors would like to thank M. Dubois-Violette for enlightening conversations.

References

[1] Appelquist T., Chodos A., Freund P.G.O. 1987, *Modern Kaluza-Klein Theory and Applications*, Benjamin-Cummings.

[2] Bailin D., Love A. 1987, *Kaluza-Klein theories*, Rep. Prog. Phys. **50** 1087.

[3] Chamseddine A.H., Felder G., Fröhlich J. 1993, *Gravity in Non-Commutative Geometry*, Commun. Math. Phys. **155** 205.

[4] Chapline G., Manton N. S. 1980, *The Geometrical Significance of Certain Higgs Potentials: An approach to grand unification*, Nucl. Phys. **B184** 391.

[5] Cho Y. M. 1985, *Geometric Symmetry Breaking*, Phys. Rev. Lett. **55** 2932.

[6] Connes A. 1986, *Non-Commutative Differential Geometry*, Publications of the Inst. des Hautes Etudes Scientifiques. **62** 257.

[7] Connes A. 1990, *Géométrie noncommutative*, InterEditions, Paris.

[8] Connes A. 1994, *Noncommutative Geometry*, Academic Press.

[9] Connes A., Lott J. 1990, *Particle Models and Noncommutative Geometry*, in 'Recent Advances in Field Theory', Nucl. Phys. Proc. Suppl. **B18** 29.

[10] Connes A., Lott J. 1992, *The metric aspect of non-commutative geometry*, Proceedings of the 1991 Cargèse Summer School, Plenum Press.

[11] Connes A., Rieffel M.A. 1987, *Yang-Mills for non-commutative two-tori*, Contemp. Math. **62** 237.

[12] Coquereaux R. 1989, *Noncommutative geometry and theoretical physics*, J. Geom. Phys. **6** 425.

[13] Coquereaux R., Esposito-Farèse G., Vaillant G. 1991, *Higgs Fields as Yang-Mills Fields and Discrete Symmetries*, Nucl. Phys. **B353** 689.

[14] Dubois-Violette M. 1988, *Dérivations et calcul différentiel non-commutatif*, C. R. Acad. Sci. Paris **307** Série I 403.

[15] Dubois-Violette M., Kerner R., Madore J. 1989, *Gauge bosons in a noncommutative geometry*, Phys. Lett. **B217** 485.

[16] Dubois-Violette M., Kerner R., Madore J. 1989, *Classical bosons in a noncommutative geometry*, Class. Quant. Grav. **6** 1709.

[17] Dubois-Violette M., Kerner R., Madore J. 1990, *Noncommutative differential geometry of matrix algebras*, J. Math. Phys. **31** 316.

[18] Dubois-Violette M., Kerner R., Madore J. 1990, *Noncommutative differential geometry and new models of gauge theory*, J. Math. Phys. **31** 323.

[19] Dubois-Violette M., Kerner R., Madore J. 1991, *Super Matrix Geometry*, Class. Quant. Grav. **8** 1077.

[20] Fairlie, D.B. 1979, *The Interpretation of Higgs Fields as Yang Mills Fields*, Geometrical and Topological Methods in Gauge Theories, Proceedings, Montreal, Lecture Notes in Physics **129** 45.

[21] Forgács P., Manton N.S. 1980, *Space-Time Symmetries in Gauge Theories*, Commun. Math. Phys. **72** 15.

[22] Harnad J., Shnider S., Tafel J. 1980, *Group actions on principal bundles and dimensional reduction*, Lett. in Math. Phys. **4** 107.

[23] Iochum B., Schücker T. 1994, *A Left-Right Symmetric Model à la Connes-Lott*, Lett. Math. Phys. **32** 153.

[24] Iochum B., Schücker T. 1994, *Yang-Mills-Higgs versus Connes-Lott*, Preprint, CPT-94/P.3090.

[25] Kalau W., Papadopoulos N.A., Plass J., Warzecha J.-M. 1995, *Differential Algebras in Non-Commutative Geometry*, Jour. of Geom. and Phys. (to appear).

[26] Kastler D. 1988, *Cyclic cohomology within the differential envelope*, Hermann, Paris.

[27] Kastler D. 1993, *A detailed account of Alain Connes' version of the standard model in non-commutative geometry*, Rev. Math. Phys. **5** 477.

[28] Kastler D., Schücker T. 1994, *The Standard Model à La Connes-Lott*, Preprint, CPT-94/P.3091.

[29] Koszul J.L. 1960, *Lectures on Fibre Bundles and Differential Geometry*, Tata Institute of Fundamental Research, Bombay.

[30] Macfarlane A.J., Sudbery A., Weisz P.H. 1968, *On Gell-Mann's λ-Matrices, d- and f-Tensors, Octets, and Parametrizations of* $SU(3)$, Commun. Math. Phys. **11** 77.

[31] Madore J. 1991, *Algebraic Structure and Particle Spectra*, Int. J. Mod. Phys. **A6** 1287.

[32] Madore J. 1993, *On a Lepton-Quark Duality*, Phys. Lett. **B305** 84.

[33] Madore J., Mourad J. 1993, *Algebraic-Kaluza-Klein Cosmology*, Class. Quant. Grav. **10** 2157.

[34] Madore J., Masson T., Mourad J. 1995, *Linear Connections on Matrix Geometries*, Class. Quant. Grav. **12** 1429.

[35] Manton N.S. 1979, *A New Six-Dimensional Approach to the Weinberg-Salam Model*, Nucl. Phys. **B158** 141.

[36] Mourad. J. 1995, *Linear Connections in Non-Commutative Geometry*, Class. Quant. Grav. **12** 965.

[37] Várilly, J.C., Gracia-Bondía J.M. 1993, *Connes' noncommutative differential geometry and the Standard Model*, Jour. of Geom. and Phys. **12** 223.

Centre de Physique Théorique CNRS, Luminy, Case 907, F-13288 Marseille 9.
E-mail address: `kastler@@cpt.univ-mrs.fr`

Laboratoire de Physique Théorique et Hautes Energies, Bât. 211, Université de Paris Sud, F-91405 Orsay
E-mail address: `madore@@qcd.th.u-psud.fr`

Laboratoire de Physique Théorique et Hautes Energies, Bât. 211, Université de Paris Sud, F-91405 Orsay
E-mail address: `masson@@qcd.th.u-psud.fr`

Contemporary Mathematics
Volume **203**, 1997

Some Aspects of Noncommutative Differential Geometry

M. Dubois-Violette

ABSTRACT. We discuss in some generality aspects of noncommutative differential geometry associated with reality conditions and with differential calculi. We then describe the differential calculus based on derivations as generalization of vector fields, and we show its relations with quantum mechanics. Finally we formulate a general theory of connections in this framework.

1. Introduction

In [**23**], J.L. Koszul described a powerful algebraic version of differential geometry in terms of a commutative associative algebra $\mathcal{C}$, $\mathcal{C}$-modules and connections ("derivation laws") on these modules. For the applications to differential geometry, $\mathcal{C}$ is the algebra of smooth functions on a manifold and the $\mathcal{C}$-modules are modules of smooth sections of smooth vector bundles over the manifold. The fact that classical differential geometry admits such an algebraic formulation is at the very origin of the idea of noncommutative differential geometry. Historically, the motivation of noncommutative geometry was the development of quantum theory [**12**]. In noncommutative geometry, one replaces the commutative associative algebra $\mathcal{C}$ by an associative algebra $\mathcal{A}$ which is not assumed to be commutative. However this replacement raises several problems which will be discussed in this lecture.

First problem: what should replace the $\mathcal{C}$-modules? The problem arises because there are at least four inequivalent generalizations of the notion of a module over a commutative algebra when the algebra is replaced by a noncommutative algebra $\mathcal{A}$. There is the notion of right $\mathcal{A}$-module and the dual notion of left $\mathcal{A}$-module. If one recalls that a module over a commutative algebra is canonically a bimodule (of a specific kind), there is a notion of bimodule over $\mathcal{A}$. Finally, since a commutative algebra coincides with its center, there is the notion of module over the center $Z(\mathcal{A})$ of $\mathcal{A}$. As will be explained later, there is also a duality between $Z(\mathcal{A})$-modules and bimodules over $\mathcal{A}$.

Second problem: what should be the generalization of the classical notions of reality? For classical differential geometry one can use for $\mathcal{C}$ either the real

1991 *Mathematics Subject Classification.* Primary 47C10; Secondary 14A22 .
* Laboratoire associé au C.N.R.S., URA DOO63.

commutative algebra of smooth real-valued functions or the complex commutative $*$-algebra of smooth complex-valued functions. More generally, if $\mathcal{C}$ is a complex commutative $*$-algebra then the set $\mathcal{C}^h$ of its hermitian elements is a real commutative algebra and $\mathcal{C}$ is the complexification of $\mathcal{C}^h$. Conversely if $\mathcal{C}_\mathbb{R}$ is a real commutative algebra, then its complexification $\mathcal{C}$ is canonically a complex commutative $*$-algebra and one has $\mathcal{C}_\mathbb{R} = \mathcal{C}^h$. In fact $\mathcal{C} \mapsto \mathcal{C}^h$ is an equivalence of the category of commutative associative $*$-algebras over $\mathbb{C}$ and $*$-homomorphisms onto the category of commutative associative algebras over $\mathbb{R}$ and homomorphisms of real algebras. The situation is quite different for noncommutative algebras. If $\mathcal{A}$ is a complex associative $*$-algebra, the set $\mathcal{A}^h$ of its hermitian elements is generally not an associative algebra but a real Jordan algebra. This means that one has two choices for the generalization of the algebra of real-valued functions, either the real Jordan algebra $\mathcal{A}^h$ of all hermitian element of a complex associative $*$-algebra $\mathcal{A}$, which plays the role of the algebra of complex-valued functions, or a real associative algebra. Here we take the first point of view. This choice, which is the standard one, is dictated by quantum theory and, more generally, by spectral theory. This reality problem is not independent of the first problem because if $\mathcal{C}$ is a complex commutative associative $*$-algebra there is again an obvious equivalence between the involutive $\mathcal{C}$-modules and the $\mathcal{C}^h$-modules [**18**].

Third problem: which differential calculus should be used? In other words what should be the generalization of differential forms? Such a generalization is needed for instance to define connections. There is a minimal set of assumptions which must be satisfied and which will be described in the sequel but nevertheless the choice is not straightforward. We can make here the following remarks. In his pioneer work on the subject [**7**], A. Connes defined the cyclic cohomology of an algebra and showed that the correct generalization of the homology of a manifold is the reduced cyclic cohomology. This means that the generalization of the cohomology of a manifold in noncommutative geometry must be the reduced cyclic homology of the algebra $\mathcal{A}$ which replaces the algebra of smooth functions. In classical differential geometry, the de Rham theorem states that the cohomology of a manifold, (a topological invariant), coincides with the cohomology of its differential form s. This does not mean that any cochain complex which has the reduced cyclic homology as cohomology is an acceptable generalization of differential forms, and this for at least two reasons. First, even in the classical situation, there are many ways to compute the cohomology of a manifold and, in particular, there are complexes which are not connected with the differential structure and which have this cohomology. Second, the de Rham theorem is not a tautological result but a deep theorem of differential topology which means that there may well be proper noncommutative generalizations of differential geometry for which the generalization of de Rham theorem fails to be true.

The problems quoted above will be discussed in the first part of this lecture. Then the differential calculus based on the derivations as generalization of vector fields will be introduced [**13**]. This differential calculus is the direct generalization of the one used by J.L. Koszul in [**23**]; it is also connected with the differential calculus used by A. Connes for noncommutative dynamical systems in [**8**]. The noncommutative symplectic structures will be defined in this framework and the relation with quantum mechanics will be described. Finally we shall describe the theory of connections in this framework. Examples of such connections and applications to gauge field theory may be found in [**14**], [**15**], [**16**], [**18**], [**24**].

Let $\mathcal{A}$ be an associative algebra. If M and N are right $\mathcal{A}$-modules, the space of all right $\mathcal{A}$-module homomorphisms of M into N will be denoted by $\mathrm{Hom}^{\mathcal{A}}(M,N)$; if M and N are left $\mathcal{A}$-modules, the space of all left $\mathcal{A}$-module homomorphisms of M into N will be denoted by $\mathrm{Hom}_{\mathcal{A}}(M,N)$. When $\mathcal{A}$ is a commutative algebra $\mathcal{C}$, both notions coincide and the space of $\mathcal{C}$-module homomorphisms of M into N will be denoted by $\mathrm{Hom}_{\mathcal{C}}(M,N)$. If $\mathcal{B}$ is another associative algebra and if M and N are $(\mathcal{A},\mathcal{B})$-bimodules, $\mathrm{Hom}^{\mathcal{B}}_{\mathcal{A}}(M,N)$ will denote the space of all bimodules homomorphisms of M into N. In the sequel, we shall often use the word algebra to mean associative algebra.

This lecture is partly based on joint works with R. Kerner, J. Madore and P.W. Michor [**13**],[**14**],[**15**],[**16**],[**18**],[**19**],[**20**] and the author is grateful to John Madore for discussions and careful reading of the manuscript.

2. Modules, bimodules and reality

In the following $\mathcal{A}$ is a complex unital associative $*$-algebra which is to be considered as a noncommutative generalization of an algebra of complex functions. As a consequence what must replace the algebra of real-valued functions is generally not an associative real algebra but the Jordan algebra $\mathcal{A}^h$ of hermitian elements of $\mathcal{A}$. Although quite familiar in quantum theory, this fact has non trivial consequences for the noncommutative generalization of classical reality conditions. In fact we are here interested in noncommutative differential geometry, which means that $\mathcal{A}$ is to be considered as the generalization of the algebra of complex smooth functions on a manifold. The Jordan algebra $\mathcal{A}^h$ replaces then the algebra of real smooth functions.

Let E be a smooth complex vector bundle of finite rank over a manifold V. Then the set $\Gamma(E)$ of its smooth sections is a finite projective module over the algebra $\mathcal{C}^\infty(V)$ of smooth complex functions on V. Furthermore the correspondence $E \mapsto \Gamma(E)$ is an equivalence of the category of smooth complex vector bundles of finite rank over V onto the category of finite projective modules over $C^\infty(V)$. Let now $E_{\mathbb{R}}$ be a smooth real vector bundle over V, its complexification E is a smooth complex vector bundle over V equipped with a canonical antilinear involution $\xi \mapsto \xi^*$ such that $\xi \in E_{\mathbb{R}}$ if and only if $\xi = \xi^*$. The module $\Gamma(E)$ is then a $*$-module over the $*$-algebra $\mathcal{C}^\infty(V)$ in the sense that it is equipped with an antilinear involution $\psi \mapsto \psi^*$ such that $(f\psi)^* = f^*\psi^*$, $\forall f \in \mathcal{C}^\infty(V)$ and $\forall \psi \in \Gamma(E)$, where $f \mapsto f^*$ is the complex conjugation. A section of $E_{\mathbb{R}}$ is a section $\psi \in \Gamma(E)$ such that $\psi = \psi^*$. Clearly, one can replace $E_{\mathbb{R}}$ by the $*$-module $\Gamma(E)$. With this in mind, let us more generally consider the notion of module and the notion of $*$-module over a commutative $*$-algebra $\mathcal{C}$ and investigate their generalizations when $\mathcal{C}$ is replaced by the noncommutative $*$-algebra $\mathcal{A}$.

As pointed out in the introduction a $\mathcal{C}$-module has several natural generalizations: a right $\mathcal{A}$-module, a left $\mathcal{A}$-module, a module over the center $Z(\mathcal{A})$ of $\mathcal{A}$ and a bimodule over $\mathcal{A}$. Right $\mathcal{A}$-modules and left $\mathcal{A}$-modules are dual in the sense that if M is a right $\mathcal{A}$-module, its dual $M^* = \mathrm{Hom}^{\mathcal{A}}(M,\mathcal{A})$ is a left $\mathcal{A}$-module and if N is a left $\mathcal{A}$-module, its dual $N^* = \mathrm{Hom}_{\mathcal{A}}(N,\mathcal{A})$ is a right $\mathcal{A}$-module; this duality generalizes the duality of $\mathcal{C}$-modules. Similarily, there is a natural duality between bimodules over $\mathcal{A}$ and $Z(\mathcal{A})$-modules [**18**]: if M is a bimodule over $\mathcal{A}$, *its $\mathcal{A}$-dual*

$M^{*_\mathcal{A}} = \mathrm{Hom}^\mathcal{A}_\mathcal{A}(M, \mathcal{A})$ is canonically a $Z(\mathcal{A})$-module and if N is a $Z(\mathcal{A})$-module, *its $\mathcal{A}$-dual* $N^{*_\mathcal{A}} = \mathrm{Hom}_{Z(\mathcal{A})}(N, \mathcal{A})$ is canonically a bimodule over $\mathcal{A}$. This duality ($\mathcal{A}$-duality) also generalizes t he duality of $\mathcal{C}$-modules when the bimodules over $\mathcal{C}$ are the underlying bimodules of $\mathcal{C}$-modules.

Concerning the generalization of $*$-modules over $\mathcal{C}$, (i.e. the generalization of the description of real vector bundles), one notices that one cannot use right or left $\mathcal{A}$-modules because, since the involution of $\mathcal{A}$ reverses the order of the product in $\mathcal{A}$, there cannot be a notion of right or left $*$-module over $\mathcal{A}$. In contrast, since $Z(\mathcal{A})$ is a commutative $*$-algebra, the notion of $*$-module over $Z(\mathcal{A})$ is perfectly defined and one can introduce a dual notion of $*$-bimodule over $\mathcal{A}$: a bimodule M over $\mathcal{A}$ is a *$*$-bimodule* over $\mathcal{A}$ if it is equipped with an antilinear involution $m \mapsto m^*$ such that one has $(xmy)^* = y^* m^* x^*$ $\forall x, y \in \mathcal{A}$ and $\forall m \in M$.

Thus, simple considerations of reality rule out right or left $\mathcal{A}$-modules for the description of a generalization of real vector bundles. This does not mean that one cannot use them for the generalization of complex vector bundles, this simply means that all the above generalizations of the notion of $\mathcal{C}$-module have to be considered when $\mathcal{C}$ is replaced by the noncommutative algebra $\mathcal{A}$. The fact that bimodule structures arise in connection with reality in noncommutative geometry has been also pointed out in [**10**] by A. Connes in the context of his spectral triples approach to noncommutative geometry [**8**],[**9**].

It must be stressed that not every bimodule over a commutative algebra $\mathcal{C}$ is the underlying bimodule of a $\mathcal{C}$-module and therefore not every bimodule over $\mathcal{A}$ can be considered as the generalization of a $\mathcal{C}$-module. One must select an appropriate class of bimodules, for instance the class of central bimodules [**18**],[**19**]. A bimodule M over $\mathcal{A}$ is called a *central bimodule* if one has $zm = mz$, $\forall m \in M$ and $\forall z \in Z(\mathcal{A})$. A central bimodule over a commutative algebra $\mathcal{C}$ is just a $\mathcal{C}$-module for its underlying bimodule structure. In [**19**], the more restrictive notion of diagonal bimodule was introduced. A bimodule M over $\mathcal{A}$ is called a *diagonal bimodule* if it is isomorphic to a subbimodule of $\mathcal{A}^I$ for some set I. A diagonal bimodule is central. A bimodule M over $\mathcal{A}$ is diagonal if and only if the canonical mapping of M into its $\mathcal{A}$-bidual $M^{*_\mathcal{A} *_\mathcal{A}}$ is injective. In particular a diagonal bimodule over a commutative algebra $\mathcal{C}$ is just a $\mathcal{C}$-module such that the canonical mapping in its bidual is injective; projective $\mathcal{C}$-modules are therefore diagonal bimodules. If N is a $Z(\mathcal{A})$-module, its $\mathcal{A}$-dual $N^{*_\mathcal{A}}$ is a diagonal bimodule over $\mathcal{A}$.

3. Differential calculus

In this section we wish to discuss some general features of the noncommutative versions of differential forms.

A *graded differential $*$-algebra* is a complex graded differential algebra $\Omega = \oplus_{n\in\mathbb{N}} \Omega^n$ equipped with an antilinear involution $\omega \mapsto \omega^*$ which preserves the degree and satisfies $(\alpha\beta)^* = (-1)^{ab}\beta^*\alpha^*$ and $(d\omega)^* = d(\omega^*)$ for $\alpha \in \Omega^a, \beta \in \Omega^b$ and $\omega \in \Omega$, where d is the differential of Ω. Notice that then Ω^0 is a $*$-algebra. Given (as before) the complex unital $*$-algebra $\mathcal{A}$, *a differential calculus over $\mathcal{A}$* is a graded differential $*$-algebra Ω with $\Omega^0 = \mathcal{A}$. Among the differential calculi over $\mathcal{A}$, there is a universal one [**21**], $\Omega_u(\mathcal{A})$, which we now review.

Let $\mu : \mathcal{A}\otimes\mathcal{A} \to \mathcal{A}$ be the product $\mu(x\otimes y) = xy$. The mapping μ is a bimodule homomorphism so its kernel $\Omega^1_u(\mathcal{A})$ is a bimodule over $\mathcal{A}$. One defines a derivation d_u of $\mathcal{A}$ into $\Omega^1_u(\mathcal{A})$ by setting $d_u x = \mathbb{1}\otimes x - x\otimes\mathbb{1}$ for $x \in \mathcal{A}$. The pair $(\Omega^1_u(\mathcal{A}), d_u)$ is characterized uniquely (up to an isomorphism) by the following universal property [**4**], [**1**]: given a derivation $\delta : \mathcal{A} \to M$ of $\mathcal{A}$ into a bimodule M over $\mathcal{A}$, there is a unique bimodule homomorphism $j_\delta : \Omega^1_u(\mathcal{A}) \to M$ such that $\delta = j_\delta \circ d_u$. Let $\Omega_u(\mathcal{A})$ be the tensor algebra over $\mathcal{A}$ of the bimodule $\Omega^1_u(\mathcal{A})$ i.e. $\Omega^0_u(\mathcal{A}) = \mathcal{A}$ and $\Omega^n_u(\mathcal{A}) = \otimes^n_{\mathcal{A}} \Omega^1_u(\mathcal{A})$ for $n \geq 1$. The derivation d_u extends uniquely into a differential, again denoted by d_u, of the graded algebra $\Omega_u(\mathcal{A})$. Using the above universal property of $(\Omega^1_u(\mathcal{A}), d_u)$ and the universal property of the tensor product over $\mathcal{A}$, one sees that the graded differential algebra $\Omega_u(\mathcal{A})$ is characterized by the following universal property: given a graded differential algebra $\Omega = \otimes_n \Omega^n$ with $\Omega^0 = \mathcal{A}$, there is a unique homomorphism of graded differential algebra $\varphi : \Omega_u(\mathcal{A}) \to \Omega$ which induces the identity mapping of $\mathcal{A}$ onto itself (i.e. $\varphi \restriction \mathcal{A} = id_{\mathcal{A}}$). Furthermore, there is a unique antilinear involution $\omega \mapsto \omega^*$ on $\Omega_u(\mathcal{A})$ which extends the involution of $\mathcal{A}$ and for which it is a graded differential $*$-algebra [**27**]; this involution is induced on $\Omega^n_u(\mathcal{A})(\subset \otimes^{n+1}\mathcal{A})$ by the involution of $\otimes^{n+1}\mathcal{A}$ defined by $(x_0 \otimes x_1 \otimes \cdots \otimes x_n)^* = (-1)^{\frac{n(n+1)}{2}} x^*_n \otimes \cdots \otimes x^*_1 \otimes x^*_0$. Equipped with this involution, $\Omega_u(\mathcal{A})$ is a differential calculus over $\mathcal{A}$ which is universal in the sense that for any differential calculus Ω over $\mathcal{A}$ there is a unique homomorphism of graded differential $*$-algebra of $\Omega_u(\mathcal{A})$ into Ω which induces the identity mapping of $\mathcal{A}$ onto itself.

One can expect, and it is our point of view here, that a noncommutative generalization of differential forms is a differential calculus over $\mathcal{A}$ when $\mathcal{A}$ replaces the algebra $\mathcal{C}^\infty(V)$ of smooth functions on a manifold V. However not every differential calculus over $\mathcal{A}$ is appropriate. For instance the universal differential calculus is not a proper generalization of the algebra of differential forms. Indeed $\Omega_u(\mathcal{C}^\infty(V))$ does not coincide with the algebra $\Omega(V)$ of differential forms on V although, by the universal property, there is a homomorphism of graded differential algebra of $\Omega_u(\mathcal{C}^\infty(V))$ into $\Omega(V)$. More generally, if $\mathcal{C}$ is a commutative algebra, the bimodule $\Omega^1_u(\mathcal{C})$, for instance, is not the underlying bimodule of a module since left and right multiplications by elements of $\mathcal{C}$ do not coincide. In any case the choice of a differential calculus Ω over $\mathcal{A}$ as generalization of the algebra of complex differential forms is not unique and depends on the applications one has in mind [**7**], [**8**], [**11**], [**13**], [**18**], [**20**], [**22**], [**24**], [**27**]. In the next section we will describe a choice for Ω based on derivations as generalization of vector fields. This choice, which is a direct generalization of [**23**], is natural in the sense that it only depends on the algebra $\mathcal{A}$ (and not on additional structures).

Before leaving this section, two points are worth noticing. First some authors, e.g. G. Maltsiniotis [**25**], consider that a proper generalization of differential geometry is given by a graded differential algebra which then replaces the algebra of differential forms; this point of view is more general than the one, implicit here, where $\mathcal{A}$ replaces the algebra of smooth functions. Second there are generalizations of the space of differential forms which are not differential algebras but merely differential complexes. For instance, it was shown in [**21**] that the subspace $[\Omega_u(\mathcal{A}), \Omega_u(\mathcal{A})]$ of graded commutators in $\Omega_u(\mathcal{A})$ is stable by d_u and that the cohomology of the complex $(\Omega_u(\mathcal{A})/[\Omega_u(\mathcal{A}), \Omega_u(\mathcal{A})], d_u)$, (which is not a differential algebra in general), is the reduced cyclic homology of $\mathcal{A}$ which in many aspects is a good generalization of de Rham cohomology. This is why this complex is a natural generalization of the de Rham complex which is often called *the noncommutative de Rham complex*.

4. Derivations and differential calculus

In this section we explain our approach to the (noncommutative) differential calculus over $\mathcal{A}$, (a complex unital $*$-algebra), based on the derivations of $\mathcal{A}$ as generalization of vector fields **[13]**, **[14]**, **[15]**, **[16]**, **[18]**, **[19]**, **[20]**. This approach is a noncommutative generalization of the one of J.L. Koszul **[23]** which is based on the fact that a vector field on a manifold V, i.e. a smooth section of the tangent bundle over V, is the same thing as a derivation of the algebra $\mathcal{C}^\infty(V)$ of smooth functions on V. More generally, since the derivations are the infinitesimal algebra automorphisms, they are the natural right-hand sides of differential evolution equations. This is why the differential calculus based on derivations is the natural one for commutative and noncommutative dynamical systems i.e. for classsical as well as for quantum mechanics.

Let $\mathrm{Der}(\mathcal{A})$ denote the space of all derivations of $\mathcal{A}$, i.e. the space of all linear mappings X of $\mathcal{A}$ into itself satisfying the Leibniz rule $X(xy) = X(x)y + xX(y)$. The space $\mathrm{Der}(\mathcal{A})$ is in a natural way a module over the center $Z(\mathcal{A})$ of $\mathcal{A}$ and in fact a $*$-module over $Z(A)$ when equipped with the involution $X \mapsto X^*$ defined by $X^*(x) = (X(x^*))^*$. The space $\mathrm{Der}(\mathcal{A})$ is also a Lie algebra with Lie bracket $(X, Y) \mapsto [X, Y] = X \circ Y - Y \circ X$. This bracket satisfies the reality condition $[X, Y]^* = [X^*, Y^*]$. Furthermore, $Z(\mathcal{A})$ is stable under $\mathrm{Der}(\mathcal{A})$ and one has $[X, zY] = X(z)Y + z[X, Y]$, for any $X, Y \in \mathrm{Der}(\mathcal{A})$ and $z \in Z(\mathcal{A})$. This last equality ensures that, in the complex $C(\mathrm{Der}(\mathcal{A}), \mathcal{A})$ of the $\mathcal{A}$-valued Lie-algebra cochains of $\mathrm{Der}(\mathcal{A})$, the subspace $\underline{\Omega}_{\mathrm{Der}}(\mathcal{A})$ of $Z(\mathcal{A})$-multilinear cochains is stable under the differential, i.e. is a subcomplex.

More precisely, let $\underline{\Omega}^n_{\mathrm{Der}}(\mathcal{A})$ be the space of $Z(\mathcal{A})$-multilinear antisymmetric mappings of $(\mathrm{Der}(\mathcal{A}))^n$ into $\mathcal{A}$, (i.e. $\underline{\Omega}^n_{\mathrm{Der}}(\mathcal{A}) = \mathrm{Hom}_{Z(\mathcal{A})}(\Lambda^n_{Z(\mathcal{A})}\mathrm{Der}(\mathcal{A}), \mathcal{A})$). Then the graded space $\underline{\Omega}_{\mathrm{Der}}(\mathcal{A}) = \oplus_n \underline{\Omega}^n_{\mathrm{Der}}(\mathcal{A})$ is in a natural way a graded algebra (the product combining the product of $\mathcal{A}$ with antisymmetrisation in the arguments). One verifies that one defines a differential d of $\underline{\Omega}_{\mathrm{Der}}(\mathcal{A})$, i.e. an antiderivation of degree 1 satisfying $d^2 = 0$, by setting, for $\omega \in \underline{\Omega}^n_{\mathrm{Der}}(\mathcal{A})$ and $X_i \in \mathrm{Der}(\mathcal{A})$,

$$(d\omega)(X_0, \dots, X_n) = \sum_{k=0}^{n} (-1)^k X_k \omega(X_0, \overset{\mathrm{k}}{\check{\dots}}, X_n)$$

$$+ \sum_{0 \leq r < s \leq n} (-1)^{r+s} \omega([X_r, X_s], X_0, \overset{\mathrm{r}}{\check{\dots}} \overset{\mathrm{s}}{\check{\dots}}, X_n)$$

where $\overset{\mathrm{i}}{\check{\vdots}}$ means omission of X_i. Thus, equipped with this differential, $\underline{\Omega}_{\mathrm{Der}}(\mathcal{A})$ is a graded differential algebra and the subalgebra $\underline{\Omega}^0_{\mathrm{Der}}(\mathcal{A})$ coincides with $\mathcal{A}$. If one equips $\underline{\Omega}_{\mathrm{Der}}(\mathcal{A})$ with the involution $\omega \mapsto \omega^*$ defined by $\omega^*(X_1, \dots, X_n) = (\omega(X_1^*, \dots, X_n^*))^*$, it becomes a differential calculus over $\mathcal{A}$.

Let $\Omega_{\mathrm{Der}}(\mathcal{A})$ be the smallest differential subalgebra of $\underline{\Omega}_{\mathrm{Der}}(\mathcal{A})$ which contains $\mathcal{A}$. The differential algebra $\Omega_{\mathrm{Der}}(\mathcal{A})$ is the canonical image of $\Omega_u(\mathcal{A})$ in $\underline{\Omega}_{\mathrm{Der}}(\mathcal{A})$ and is stable by the involution; it consists of finite sums of elements of the form $x_0 dx_1 \dots dx_n$, $x_i \in \mathcal{A}$. The graded differential $*$-algebra $\Omega_{\mathrm{Der}}(\mathcal{A})$ is also a differential calculus over $\mathcal{A}$.

Both $\Omega_{\mathrm{Der}}(\mathcal{A})$ and $\underline{\Omega}_{\mathrm{Der}}(\mathcal{A})$ are generalizations of the algebra of complex differential forms. If V is a finite-dimensional paracompact manifold then $\Omega_{\mathrm{Der}}(C^\infty(V))$

and $\underline{\Omega}_{\mathrm{Der}}(\mathcal{C}^\infty(V))$ both coincide with the graded differential $*$-algebra $\Omega(V)$ of complex differential forms on V. In general the inclusion $\Omega_{\mathrm{Der}}(\mathcal{A}) \subset \underline{\Omega}_{\mathrm{Der}}(\mathcal{A})$ is a strict one: $\Omega_{\mathrm{Der}}(\mathcal{A})$ is the minimal version of noncommutative differential forms based on derivations while $\underline{\Omega}_{\mathrm{Der}}(\mathcal{A})$ is the maximal one. It is worth noticing here that even in the classical situation the above inclusion may be strict, e.g. if V is a manifold which does not admit a partition of unity then the inclusion $\Omega_{\mathrm{Der}}(\mathcal{C}^\infty(V)) \subset \underline{\Omega}_{\mathrm{Der}}(\mathcal{C}^\infty(V))$ is a strict one. There is however a density result of $\Omega_{\mathrm{Der}}(\mathcal{A})$ in $\underline{\Omega}_{\mathrm{Der}}(\mathcal{A})$ which we now describe at the level of one-forms [**18**].

By its very definition, the bimodule $\underline{\Omega}^1_{\mathrm{Der}}(\mathcal{A})$ is the $\mathcal{A}$-dual of the $Z(\mathcal{A})$-module $\mathrm{Der}(\mathcal{A})$,i.e. $\underline{\Omega}^1_{\mathrm{Der}}(\mathcal{A}) = (\mathrm{Der}(\mathcal{A}))^{*_\mathcal{A}} = \mathrm{Hom}_{Z(\mathcal{A})}(\mathrm{Der}(\mathcal{A}), \mathcal{A})$. On the other hand, by the universal property of $(\Omega^1_u(\mathcal{A}), d_u)$, $\mathrm{Der}(\mathcal{A})$ can be identified with $\mathrm{Hom}^\mathcal{A}_\mathcal{A}(\Omega^1_u(\mathcal{A}), \mathcal{A})$ through the canonical mapping $X \mapsto j_X$ (see in last section). However the intersection of the kernels of the bimodule homomorphisms of $\Omega^1_u(\mathcal{A})$ into $\mathcal{A}$, (which is the intersection of the kernels of the j_X when X runs over $\mathrm{Der}(\mathcal{A})$), is just the kernel of the canonical bimodule homomorphism of $\Omega^1_u(\mathcal{A})$ onto $\Omega^1_{\mathrm{Der}}(\mathcal{A})$ [**13**] and therefore one has $\mathrm{Hom}^\mathcal{A}_\mathcal{A}(\Omega^1_u(\mathcal{A}), \mathcal{A}) = \mathrm{Hom}^\mathcal{A}_\mathcal{A}(\Omega^1_{\mathrm{Der}}(\mathcal{A}), \mathcal{A})$. So one has finally $\mathrm{Hom}^\mathcal{A}_\mathcal{A}(\Omega^1_{\mathrm{Der}}(\mathcal{A}), \mathcal{A}) = \mathrm{Der}(\mathcal{A})$ which means that the $Z(\mathcal{A})$-module $\mathrm{Der}(\mathcal{A})$ is the $\mathcal{A}$-dual of the bimodule $\Omega^1_{\mathrm{Der}}(\mathcal{A}) : \mathrm{Der}(\mathcal{A}) = (\Omega^1_{\mathrm{Der}}(\mathcal{A}))^{*_\mathcal{A}}$. Thus $\underline{\Omega}^1_{\mathrm{Der}}(\mathcal{A})$ is the $\mathcal{A}$-bidual bimodule of $\Omega^1_{\mathrm{Der}}(\mathcal{A})$, i.e. one has $\underline{\Omega}^1_{\mathrm{Der}}(\mathcal{A}) = (\Omega^1_{\mathrm{Der}}(\mathcal{A}))^{*_\mathcal{A}*_\mathcal{A}}$. This is an obvious density result which implies in particular that $\Omega^1_{\mathrm{Der}}(\mathcal{A})$ is a diagonal bimodule; however this fact is obvious since $\underline{\Omega}^1_{\mathrm{Der}}(\mathcal{A})$ is diagonal by definition $(\subset \mathcal{A}^{\mathrm{Der}(\mathcal{A})})$.

Using $\mathrm{Hom}^\mathcal{A}_\mathcal{A}(\Omega^1_u(\mathcal{A}), \mathcal{A}) = \mathrm{Hom}^\mathcal{A}_\mathcal{A}(\Omega^1_{\mathrm{Der}}(\mathcal{A}), \mathcal{A})$ one can characterize the pair $(\Omega^1_{\mathrm{Der}}(\mathcal{A}), d)$ consisting of the diagonal bimodule $\Omega^1_{\mathrm{Der}}(\mathcal{A})$ and the derivation d of $\mathcal{A}$ into $\Omega^1_{\mathrm{Der}}(\mathcal{A})$ by the following universal property [**19**]: for any derivation δ of $\mathcal{A}$ into a diagonal bimodule M over $\mathcal{A}$, there is a unique bimodule homomorphism $i_\delta : \Omega^1_{\mathrm{Der}}(\mathcal{A}) \to M$ such that $\delta = i_\delta \circ d$. This means that if δ is a derivation of $\mathcal{A}$ into a diagonal bimodule M, the bimodule homomorphism $j_\delta : \Omega^1_u(\mathcal{A}) \to M$ factorizes through the canonical bimodule homomorphism of $\Omega^1_u(\mathcal{A})$ onto $\Omega^1_{\mathrm{Der}}(\mathcal{A})$. Recall that the underlying bimodule of the module of sections of a vector bundle over a manifold V is diagonal and that $\Omega^1_{\mathrm{Der}}(\mathcal{C}^\infty(V))$ is the space of 1-forms on V, so the above result generalizes a well known result of differential geometry.

Let X be a derivation of $\mathcal{A}$, then one defines an antiderivation i_X of degree -1 of $\underline{\Omega}_{\mathrm{Der}}(\mathcal{A})$ by setting $(i_X\omega)(X_1, \dots, X_{n-1}) = \omega(X, X_1, \dots, X_n)$ for $\omega \in \underline{\Omega}^n_{\mathrm{Der}}(\mathcal{A})$ and $X_i \in \mathrm{Der}(\mathcal{A})$. The mapping $X \mapsto i_X$ is an operation, in the sense of H. Cartan [**3**], of the Lie algebra $\mathrm{Der}(\mathcal{A})$ in the graded differential algebra $\underline{\Omega}_{\mathrm{Der}}(\mathcal{A})$, i.e. one has $i_X i_Y + i_Y i_X = 0$ and, if one sets $L_X = di_X + i_X d$, $L_X i_Y - i_Y L_X = i_{[X,Y]}$ and $L_X L_Y - L_Y L_X = L_{[X,Y]}$. Furthermore $L_X \restriction \mathcal{A} = X$ so $X \mapsto L_X$ is a Lie algebra homomorphism of $\mathrm{Der}(\mathcal{A})$ into the derivations of degree zero of $\underline{\Omega}_{\mathrm{Der}}(\mathcal{A})$ which extends the action of $\mathrm{Der}(\mathcal{A})$ on $\mathcal{A}$. The differential subalgebra $\Omega_{\mathrm{Der}}(\mathcal{A})$ is stable by the $i_X, X \in \mathrm{Der}(\mathcal{A})$, so one has by restriction an operation of $\mathrm{Der}(\mathcal{A})$ in $\Omega_{\mathrm{Der}}(\mathcal{A})$. The operation $X \mapsto i_X$ is of course the generalization of the interior product (or contraction) of forms by vector fields while L_X generalizes the Lie derivative on forms.

5. Noncommutative symplectic structures

It is well known that the structural similarity between classical mechanics and quantum mechanics is the most apparent if one uses the hamiltonian approach for the former and that this is important for the problems of classical and semiclassical limits. In this context the appropriate generalization of the Poisson structures is also well known. *A Poisson bracket on* $\mathcal{A}$ is a Lie algebra structure $(x, y) \mapsto \{x, y\}$ on $\mathcal{A}$ satisfying $\{x, yz\} = \{x, y\}z + y\{x, z\}$ for any elements x, y and z of $\mathcal{A}$. Such a Poisson bracket is *real* if furthermore one has $\{x, y\}^* = \{x^*, y^*\}$ for $x, y \in \mathcal{A}$. For any $\mathcal{A}$, there is the standard real Poisson bracket $\{x, y\} = i[x, y]$ $(= i(xy - yx))$. Although this bracket is trivial for a commutative algebra it is, up to a real factor, the most common Poisson bracket occuring in quantum mechanics. In classical hamiltonian mechanics, the Poisson bracket is associated with the symplectic structure of tho pg hase space. It is the aim of this section to describe the generalization of symplectic structures for $\mathcal{A}$ and to show its relevance for quantum mechanics [**14**], [**15**], [**24**].

The first thing to do is to generalize the notion of a nondegenerate two-form. An element ω of $\underline{\Omega}^2_{\mathrm{Der}}(\mathcal{A})$ will be said to be *nondegenerate* if, for any $x \in \mathcal{A}$, there is a derivation $\mathrm{Ham}(x) \in \mathrm{Der}(\mathcal{A})$ such that one has $\omega(X, \mathrm{Ham}(x)) = X(x)$ for any $X \in \mathrm{Der}(\mathcal{A})$. Notice that if ω is nondegenerate then $X \mapsto i_X\omega$ is an injective linear mapping of $\mathrm{Der}(\mathcal{A})$ into $\underline{\Omega}^1_{\mathrm{Der}}(\mathcal{A})$ but that the converse is not true; the condition for ω to be nondegenerate is stronger than the injectivity of $X \mapsto i_X\omega$. If V is a manifold, an element $\omega \in \underline{\Omega}^2_{\mathrm{Der}}(\mathcal{C}^\infty(V))$ is an ordinary 2-form on V and it is nondegenerate in the above sense if and only if the 2-form ω is nondegenerate in the classical sense (i.e. everywhere nondegenerate).
Let $\omega \in \underline{\Omega}^2_{\mathrm{Der}}(\mathcal{A})$ be nondegenerate, then for a given $x \in \mathcal{A}$ the derivation $\mathrm{Ham}(x)$ is unique and $x \mapsto \mathrm{Ham}(x)$ is a linear mapping of $\mathcal{A}$ into $\mathrm{Der}(\mathcal{A})$. Define then an antisymmetric bilinear bracket on $\mathcal{A}$ by $\{x, y\} = \omega(\mathrm{Ham}(x), \mathrm{Ham}(y))$. One has $\{x, yz\} = \{x, y\}z + y\{x, y\}$ for $x, y, z \in \mathcal{A}$,however the bracket $(x, y) \mapsto \{x, y\}$ is a Lie bracket, (i.e. satisfies the Jacobi identity), if and only if $d\omega = 0$. A closed nondegenerate element ω of $\underline{\Omega}^2_{\mathrm{Der}}(\mathcal{A})$ will be called *a symplectic structure for* $\mathcal{A}$. Let ω be a symplectic structure for $\mathcal{A}$, then the corresponding bracket $(x, y) \mapsto \{x, y\} = \omega(\mathrm{Ham}(x), \mathrm{Ham}(y))$ is a Poisson bracket on $\mathcal{A}$ and one has $[\mathrm{Ham}(x), \mathrm{Ham}(y)] = \mathrm{Ham}(\{x, y\})$, i.e. Ham is a Lie-algebra homomorphism of $(\mathcal{A}, \{,\})$ into $\mathrm{Der}(\mathcal{A})$. If furthermore ω is real, i.e. $\omega = \omega^*$, then this Poisson bracket is real and $\mathrm{Ham}(x^*) = (\mathrm{Ham}(x))^*$ for any $x \in \mathcal{A}$. We shall refer to the above bracket as the Poisson bracket associated to the symplectic structure ω.
If V is a manifold, a symplectic structure for $\mathcal{C}^\infty(V)$ is just a symplectic form on V. Since there are manifolds which do not admit symplectic form, one cannot expect that an arbitrary $\mathcal{A}$ admits a symplectic structure.

Assume that $\mathcal{A}$ has a trivial center $Z(\mathcal{A}) = \mathbb{C}\mathbb{1}$ and that all its derivations are inner (i.e. of the form $ad(x), x \in \mathcal{A}$). Then one defines an element ω of $\underline{\Omega}^2_{\mathrm{Der}}(\mathcal{A})$ by setting $\omega(ad(ix), ad(iy)) = i[x, y]$. It is easily seen that ω is a real symplectic structure for which one has $\mathrm{Ham}(x) = ad(ix)$ and $\{x, y\} = i[x, y]$. Although a little tautological, this construction is relevant for quantum mechanics.
Let $\mathcal{A}$ be, as above, a complex unital $*$-algebra with a trivial center and only inner derivations and assume that there exists a linear form τ on $\mathcal{A}$ which is central, i.e. $\tau(xy) = \tau(yx)$, and normalized by $\tau(\mathbb{1}) = 1$. Then one defines an element $\theta \in \underline{\Omega}^1_{\mathrm{Der}}(\mathcal{A})$ by $\theta(ad(ix)) = x - \tau(x)\mathbb{1}$. One has $(d\theta)(ad(ix), ad(iy)) = i[x, y]$, i.e. $\omega = d\theta$, so in this case the symplectic form ω is exact. As examples of such algebras one can take $\mathcal{A} = M_n(\mathbb{C})$, (a factor of type I_n), with $\tau = \frac{1}{n}$ trace, or $\mathcal{A} = \mathcal{R}$, a von Neumann algebra which is a factor of type II_1 with τ equal to the normalized

trace. The algebra $M_n(\mathbb{C})$ is the algebra of observables of a quantum spin $s = \frac{n-1}{2}$ while $\mathcal{R}$ is the algebra used to describe the observables of an infinite assembly of quantum spin; two typical types of quantum systems with no classical counterpart. Let us now consider the C.C.R. algebra (canonical commutative relations) $\mathcal{A}_{CCR}$ [**14**]. This is the complex unital $*$-algebra generated by two hermitian elements q and p satisfying the relation $[q,p] = i\hbar\mathbb{1}$. This algebra is the algebra of observables of the quantum counterpart of a classical system with one degree of freedom. We keep here the positive constant $\hbar$ (the Planck constant) in the formula for comparison with classical mechanics, although the algebra for $\hbar \neq 0$ is isomorphic to the one with $\hbar = 1$. We restrict here attention to one degree of freedom to simplify the notations but the discussion extends easily to a finite number of degrees of freedom. This algebra has again only inner derivations and a trivial center so $\omega(ad(\frac{i}{\hbar}x), ad(\frac{i}{\hbar}y)) = \frac{i}{\hbar}[x,y]$ defines a symplectic structure for which $\mathrm{Ham}(x) = ad(\frac{i}{\hbar}x)$ and $\{x,y\} = \frac{i}{\hbar}[x,y]$ which is the standard quantum Poisson bracket. In this case one can express ω in terms of the generators q and p and their differentials :

$$\omega = \sum_{n\geq 0} \left(\frac{1}{i\hbar}\right)^n \frac{1}{(n+1)}[\dots \underbrace{[dp,p],\dots,p}_{n}][\dots \underbrace{[dq,q],\dots,q}_{n}]$$

Notice that this formula is meaningful because if one inserts two derivations $ad(ix)$, $ad(iy)$ in it, only a finite number of terms contribute in the sum. For $\hbar = 0$, q and p commute and the algebra reduces to the algebra of complex polynomial functions on the phase space $\mathbb{R}^2$. Furthermore the limit of $\{x,y\} = \frac{i}{\hbar}[x,y]$ at $\hbar = 0$ reduces to the usual classical Poisson bracket as well known and, by using the above formula, one sees that the formal limit of ω at $\hbar = 0$ is $dpdq$.

6. Derivations and Connections

In this section $\mathcal{C}$ is a complex unital commutative $*$-algebra and $\mathcal{A}$ is a complex unital $*$-algebra which is to be considered as the noncommutative generalization of $\mathcal{C}$. Our aim is to discuss the theory of connections on the various objects which generalize the $\mathcal{C}$-modules when $\mathcal{C}$ is replaced by $\mathcal{A}$ in the framework of the differential calculus based on derivations as generalization of vector fields [**18**] (cf. Section 0.4). In most parts of the following the involution is not involved and therefore, in the definitions and results where the reality conditions do not enter, one may assume that $\mathcal{C}$ and its noncommutative counterpart $\mathcal{A}$ are simply algebras (instead of $*$-algebras).

As generalizations of the category of $\mathcal{C}$-modules when $\mathcal{C}$ is replaced by $\mathcal{A}$ we consider the four following categories (cf. Section 0.2), the category $\mathfrak{C}_{(0,0)}$ of $Z(\mathcal{A})$-modules, the category $\mathfrak{C}_{(1,0)}$ of left $\mathcal{A}$-modules, the category $\mathfrak{C}_{(0,1)}$ of right $\mathcal{A}$-modules and the category $\mathfrak{C}_{(1,1)}$ of central bimodules over $\mathcal{A}$ i.e. of left $\mathcal{A}\otimes_{Z(\mathcal{A})}\mathcal{A}^{op}$-modules. In each of these categories, one has a direct sum and if M is an object of any of these categories, it has a canonical underlying structure of $Z(\mathcal{A})$-module. The labelling of these categories by elements $\alpha = (i,j)$ of $\mathbb{Z}_2 \times \mathbb{Z}_2$ will be very convenient to deal with the duality and tensor products. In $\mathbb{Z}_2 \times \mathbb{Z}_2$ one defines an involutive mapping $\alpha \mapsto \alpha'$ by $(i,j)' = (1-i, 1-j)$, i.e. $\alpha' = \alpha + (1,1)$. Correspondingly one has a duality $M \mapsto M'$ of $\mathfrak{C}_\alpha$ into $\mathfrak{C}_{\alpha'}$, where $M' = M^*$ if M is a left or right $\mathcal{A}$-module and $M' = M^{*_\mathcal{A}}$ if M is a $Z(\mathcal{A})$-module or a central bimodule over $\mathcal{A}$. Another bit of notation will be convenient; we set $A_0 = Z(\mathcal{A})$ and

$A_1 = \mathcal{A}$. Using this notation, an object of $\mathfrak{C}_{(i,j)}$ is a (A_i, A_j)-bimodule (of a specific kind) and we can define tensor products $\mathfrak{C}_{(i,j)} \times \mathfrak{C}_{(j,k)} \to \mathfrak{C}_{(i,j)} \tilde{\otimes} \mathfrak{C}_{(j,k)} \subset \mathfrak{C}_{(i,k)}$ by $M \tilde{\otimes} N = M \otimes_{A_j} N$ if M is an object of $\mathfrak{C}_{(i,j)}$ and N an object of $\mathfrak{C}_{(j,k)}$ (one verifies that $M \tilde{\otimes} N$ is then an object of $\mathfrak{C}_{(i,k)}$).

Let M be an object of $\mathfrak{C}_{(i,j)}$. A *connection on* M is a linear mapping ∇, $X \mapsto \nabla_X$, of $\mathrm{Der}(\mathcal{A})$ into the linear endomorphism of M such that one has for any $m \in M$ and any $X \in \mathrm{Der}(\mathcal{A})$

$$\begin{cases} \nabla_{zX}(m) = z\nabla_X(m), & \forall z \in Z(\mathcal{A}) \\ \nabla_X(a_i m a_j) = X(a_i) m a_j + a_i \nabla_X(m) a_j + a_i m X(a_j), & \forall a_i \in A_i, \forall a_j \in A_j \end{cases}$$

remembering that M is canonically a $Z(\mathcal{A})$-module and that since $Z(\mathcal{A}) = A_0$ is stable by $\mathrm{Der}(\mathcal{A}), \mathrm{Der}(\mathcal{A})$ acts by derivations on $Z(\mathcal{A}) = A_0$ and on $\mathcal{A} = A_1$. It should be stressed that elements of $A_0 = Z(\mathcal{A})$ can be moved to the other side. Given ∇ as above, *the curvature* R *of* ∇ is the bilinear antisymmetric mapping $(X,Y) \mapsto R_{X,Y}$ of $\mathrm{Der}(\mathcal{A}) \times \mathrm{Der}(\mathcal{A})$ into the linear endomorphisms of M defined by $R_{X,Y}(m) = \nabla_X(\nabla_Y(m)) - \nabla_Y(\nabla_X(m)) - \nabla_{[X,Y]}(m)$, $\forall X, Y \in \mathrm{Der}(\mathcal{A})$, $\forall m \in M$. One has $R_{zX,Y}(m) = zR_{X,Y}(m)$ and $R_{X,Y}(a_i m a_j) = a_i R_{X,Y}(m) a_j$, $\forall m \in M$, $\forall X, Y \in \mathrm{Der}(\mathcal{A})$, $\forall z \in Z(\mathcal{A})$, $\forall a_i \in A_i$, $\forall a_j \in A_j$. More precisely, R is an antisymmetry $Z(\mathcal{A})$-bilinear mapping of $\mathrm{Der}(\mathcal{A}) \times \mathrm{Der}(\mathcal{A})$ into the $Z(\mathcal{A})$-module $\mathrm{Hom}_{\mathfrak{C}_{i,j}}(M,M)$, ($\mathrm{Hom}_{\mathfrak{C}_{i,j}}$ being the morphisms in $\mathfrak{C}_{(i,j)}$).

There is an obvious connection $\nabla_1 \oplus \nabla_2$ on the direct sum $M_1 \oplus M_2$ of two objects M_1 and M_2 of $\mathfrak{C}_{(i,j)}$ equipped with connections ∇_1 and ∇_2.

Let M be an object of $\mathfrak{C}_{(i,j)}$ then its dual M' is an element of $\mathfrak{C}_{(i,j)'}$ and we denote by $(m, m') \mapsto < m, m' > \in \mathcal{A}$ the bilinear duality bracket obtained by evaluation, $<,>$: $M \times M' \to \mathcal{A}$. Then, for any connection ∇ on M, there is unique *dual connection* ∇' on M' such that $X(< m, m' >) = < \nabla_X(m), m' > + < m, \nabla'_X(m') >$, $\forall m \in M, \forall m' \in M'$ and $\forall X \in \mathrm{Der}(\mathcal{A})$. Indeed the above equality defines ∇' uniquely and one checks that it is a connection. In general, the mapping $\nabla \mapsto \nabla'$ is not injective nor surjective. However if the canonical mapping of M into its bidual M'', (which is a morphism of $\mathfrak{C}_{(i,j)}$), is injective, then ∇'' is an extension of ∇ and therefore $\nabla \mapsto \nabla'$ is injective and of course bijective whenever $M = M''$. An object M of $\mathfrak{C}_{(ij)}$ will be called *diagonal* if the canonical morphism of M in M'' is injective. This generalizes the notion introduced in Section 0.2 (for $\mathfrak{C}_{(1,1)}$) and the terminology is suggested by the following. The algebra $\mathcal{A}$ itself can be considered as an object of $\mathfrak{C}_{(i,j)}$ when it is equipped with the canonical corresponding underlying structure and the same is true for $\mathcal{A}^I$ where I is an arbitrary set, since $\mathfrak{C}_{(i,j)}$ has arbitrary products and, more generally, arbitrary projective limits. Then M is diagonal if and only if there is an injective $\mathfrak{C}_{(i,j)}$-morphism of M into $\mathcal{A}^I$, for some set I. Finally let us notice that any projective limit of diagonal objects is diagonal and that dual objects are diagonal, i.e. if M is an object of $\mathfrak{C}_{(i,j)}$, then its dual M' is a diagonal object of $\mathfrak{C}_{(i,j)'}$. Let M_1 be an object of $\mathfrak{C}_{(i,j)}$ and M_2 be an object of $\mathfrak{C}_{(j,k)}$ and let ∇^1 be a connection on M_1 and ∇^2 be a connection on M_2. Then, for any $X \in \mathrm{Der}(\mathcal{A})$, $D_X = \nabla^1_X \otimes id_{M_2} + id_{M_1} \otimes \nabla^2_X$ is such that it maps into itself the subspace of $M_1 \otimes M_2$ generated by the elements $m_1 a_j \otimes m_2 - m_1 \otimes a_j m_2$, with $m_1 \in M_1$, $m_2 \in M_2$ and $a_j \in A_j$. It follows that the D_X pass to the quotient and define linear endomorphisms ∇_X of $M_1 \tilde{\otimes} M_2$ and one verifies that ∇ so defined is a connection on the object $M_1 \tilde{\otimes} M_2$ of $\mathfrak{C}_{(i,k)}$. This connection will be referred to as *the tensor product of* ∇_1 *and* ∇_2.Thus we have defined connections on $Z(\mathcal{A})$-modules, on left and right $\mathcal{A}$-modules and on central bimodules over $\mathcal{A}$ and we have also defined dual and tensor product of such connections. Let us now come to the problems of reality for such connections. As pointed out in Section 02, the notion of reality makes sense only for $*$-modules over $Z(\mathcal{A})$ or for $*$-bimodules over $\mathcal{A}$. So let M be either a $*$-module over $Z(\mathcal{A})$ or a $*$-bimodule over $\mathcal{A}$ which is central. If ∇ is a connection on M one can define another one ∇^*, *its conjugate*, by setting $\nabla^*_X(m) = (\nabla_{X^*}(m^*))^*$ and ∇ will be said to be a *real* connection if $\nabla = \nabla^*$. Let M' be the dual of M, i.e. $M' = M^{*_\mathcal{A}}$ in this case, then there is a unique involution $m' \mapsto m'^*$ on M' such that $< m, m' > ^* =< m^*, m'^* >$ and, equipped with this involution, M' is a (central) $*$-bimodule over $\mathcal{A}$ if M is a $*$-module over $Z(\mathcal{A})$ or a $*$-module over $Z(\mathcal{A})$ if M is a central $*$-bimodule over $\mathcal{A}$. Furthermore, one has $(\nabla^*)' = (\nabla')^*$, so the dual connection of a real connection is real.

7. Linear connections

In classical differential geometry, a connection on the tangent bundle, or equivalently on the cotangent bundle, of a manifold is usually called a linear connection. Although this terminology is a little misleading, we shall nevertheless use it for the corresponding noncommutative generalizations. Within the framework of Section 0.4 and Section 0.6, one sees that there are three natural definitions of such generalizations. First a connection on $\Omega^1_{\mathrm{Der}}(\mathcal{A})$, second a connection on $\mathrm{Der}(\mathcal{A})$ and third a connection on $\underline{\Omega}^1_{\mathrm{Der}}(\mathcal{A})$. However, as explained in Section 0.4, $\Omega^1_{\mathrm{Der}}(\mathcal{A})$ is a diagonal bimodule with $\mathrm{Der}(\mathcal{A})$ as $\mathcal{A}$-dual, i.e. $\mathrm{Der}(\mathcal{A}) = (\Omega^1_{\mathrm{Der}}(\mathcal{A}))'$ with the notation of Section 0.6, and $\underline{\Omega}^1_{\mathrm{Der}}(\mathcal{A})$ is the $\mathcal{A}$-dual of $\mathrm{Der}(\mathcal{A})$, i.e. $\underline{\Omega}^1_{\mathrm{Der}}(\mathcal{A}) = (\mathrm{Der}(\mathcal{A}))' = (\Omega^1_{\mathrm{Der}}(\mathcal{A}))''$. Therefore, it follows from the discussion of the previous section that, by duality, there is an injective mapping of the (affine) space of connections on $\Omega^1_{\mathrm{Der}}(\mathcal{A})$ into the space of connections on $\mathrm{Der}(\mathcal{A})$ and that there is also on injective mapping of the space of connections on $\mathrm{Der}(\mathcal{A})$ into the space of connections on $\underline{\Omega}^1_{\mathrm{Der}}(\mathcal{A})$. Thus all these connections may be imbedded into the connections on $\underline{\Omega}^1_{\mathrm{Der}}(\mathcal{A})$. A real connection on $\underline{\Omega}^1_{\mathrm{Der}}(\mathcal{A})$ will be called *a linear connection on* $\mathcal{A}$. The connections on $\mathrm{Der}(\mathcal{A})$ form a subclass of connections on $\underline{\Omega}^1_{\mathrm{Der}}(\mathcal{A})$ and an even smaller subclass consists of connections on $\Omega^1_{\mathrm{Der}}(\mathcal{A})$. Given a connection ∇ on $\underline{\Omega}_{\mathrm{Der}}(\mathcal{A})$, one defines a bimodule homomorphism $T : \underline{\Omega}^1_{\mathrm{Der}}(\mathcal{A}) \to \underline{\Omega}^2_{\mathrm{Der}}(\mathcal{A})$, *its torsion*, by setting $(T\omega)(X,Y) = (d\omega)(X,Y) - \nabla_X(\omega)(Y) + \nabla_Y(\omega)(X)$ for $X, Y \in \mathrm{Der}(\mathcal{A})$ and $\omega \in \underline{\Omega}^1_{\mathrm{Der}}(\mathcal{A})$. If ∇ comes from a connection on $\Omega^1_{\mathrm{Der}}(\mathcal{A})$, (by biduality), T restricted to $\Omega^1_{\mathrm{Der}}(\mathcal{A})$, is a bimodule homomorphism of$\Omega^1_{\mathrm{Der}}(\mathcal{A})$ into $\Omega^2_{\mathrm{Der}}(\mathcal{A})$. If ∇ is the dual of a connection, again denoted by ∇, on $\mathrm{Der}(\mathcal{A})$, its torsion can be identified with the $Z(\mathcal{A})$-bilinear antisymmetric mapping T of $\mathrm{Der}(\mathcal{A}) \times \mathrm{Der}(\mathcal{A})$ into $\mathrm{Der}(\mathcal{A})$ defined by $T(X,Y) = \nabla_X(Y) - \nabla_Y(X) - [X,Y]$, $\forall X, Y \in \mathrm{Der}(\mathcal{A})$. For a more complete discussion as well as for the notion of Levi-Civita connection of a generalization of pseudo-riemannian metric, we refer to [**18**].

8. Conclusion: General differential calculi

The above notions of connections are natural ones when one uses the differential calculus based on derivations as generalization of vector fields. However, for some purposes, (see e.g. in [**8**]), it is useful to use other differential calculi and therefore, it is natural to ask for a definition of connections adapted to such calculi. Let Ω

be a differential calculus over $\mathcal{A}$. There is then a well known useful definition of an Ω-connection on a left (or right) $\mathcal{A}$-module [**6**]. The problem arises when one tries to define an Ω-connection on a bimodule over $\mathcal{A}$ such as Ω^1. This problem is unavoidable if one wishes to generalize linear connections since the natural structure of Ω^1 is that of a bimodule. Some authors, e.g. [**5**], define a connection on Ω^1 to be a left module Ω-connection on Ω^1. Besides the fact that it is unnatural to privilege part of a bimodule structure, this definition has two drawbacks if one thinks of it as a generalization of linear connections. First, one cannot introduce then the notion of reality which generalizes the classical notion of reality of a linear connection in differential geometry because the involution of Ω^1 is linked to its bimodule structure, (so the conjugate of a left $\mathcal{A}$-module connection on Ω^1 is rather a right $\mathcal{A}$-module connection). Second, one cannot, in general, define the tensor product over $\mathcal{A}$ of such a connection with a connection say on a left $\mathcal{A}$-module although it is very desirable to have such a tensor product, e.g. for the description of the generalization of the classical coupling of gravitation with a field coupled to a Yang-Mills field. A definition of linear connections for general differential calculi which takes into account the complete bimodule structure of Ω^1 has been proposed by J. Mourad [**26**] and further generalized to other bimodules [**17**]. This definition involves a generalization of the permutations in tensor products and with it the question of reality can be addressed. Furthermore, in this framework, tensor products of connections are defined straightforwardly and it has been recently shown [**2**], (see in appendix A of [**2**]), that conversely, in order that tensor products of connections exist in a very general sense, one has to use this definition of connections for bimodules.

References

1. N. Bourbaki, *Algèbre I*, Chapitre III. Paris, Hermann 1970.
2. K. Bresser, F. Müller-Hoissen, A. Dimakis, A. Sitarz, *Noncommutative geometry of finite groups*, Preprint GOET-TP 95/95.
3. H. Cartan, *Notion d'algèbre différentielle; application aux groupes de Lie et aux variétés où opère un groupe de Lie* and *La trangression dans un groupe de Lie et dans un espace fibré principal*, Colloque de topologie (Bruxelles 1950), Paris, Masson 1951.
4. H. Cartan, S. Eilenberg, *Homological algebras*, Princeton University Press 1973.
5. A.H. Chamseddine, J. Fröhlich, O. Grandjean, *The gravitational sector in the Connes-Lott formulation of the standard model*, J. Math. Phys. **36** (1995) 6255.
6. A. Connes, *Non-commutative differential geometry*, Publ. IHES **62** (1986), 257.
7. A. Connes, *C* algèbres et géométrie différentielle*, C.R. Acad. Sci. Paris, **290**, Série A (1980), 599-604.
8. A. Connes, *Non-commutative geometry*, Academic Press, 1994.
9. A. Connes, *Geometry from the spectral point of view*, Lett. Math. Phys. **34** (1995) 203-238.
10. A. Connes, *Noncommutative geometry and reality*, Preprint IHES/M/95/52 (1995).
11. R. Coquereaux, *Higgs fields and superconnections*, in Differential Geometric Methods in Theoretical Physics, Rapallo 1990 (C. Bartocci, U. Bruzzo, R. Cianci, eds), Lecture Notes in Physics **375**, Springer Verlag 1991.
R. Coquereaux, R. Häußling, F. Scheck, *Algebraic connections on parallel universes*, Int. J. Mod. Phys. **A10** (1995) 89-98.
12. P.A.M. Dirac, *On quantum algebras*, Proc. Camb. Phil. Soc. **23** (1926) 412.
13. M. Dubois-Violette, *Dérivations et calcul différentiel non commutatif*, C.R. Acad. Sci. Paris, **307**, Série I (1988), 403-408.
14. M. Dubois-Violette, *Non-commutative differential geometry, quantum mechanics and gauge Theory*, in Differential Geometric Methods in Theoretical Physics, Rapallo 1990 (C. Bartocci, U. Bruzzo, R. Cianci, eds), Lecture Notes in Physics **375**, Springer-Verlag 1991.

15. M. Dubois-Violette, R. Kerner, J. Madore, *Non-commutative differential geometry of matrix algebras*, J. Math. Phys. **31** (1990) 316.
16. M. Dubois-Violette, R. Kerner, J. Madore, *Gauge bosons in a non-commutative geometry*, Phys. Lett. **B217** (1989) 485.
M. Dubois-Violette, R. Kerner, J. Madore, *Classical bosons in a non-commutative geometry*, Class. Quantum Grav. **6** (1989) 1709.
M. Dubois-Violette, R. Kerner, J. Madore, *Non-commutative differential geometry and new models of gauge theory*, J. Math. Phys. **31** (1990) 323.
17. M. Dubois-Violette, T. Masson, *On the first order operators in bimodules*, Preprint LPTHE-Orsay 95/56, q-alg 9507028.
18. M. Dubois-Violette, P.W. Michor, *Connections on central bimodules in noncommutative geometry*, Preprint L.P.T.H.E.-Orsay 94/100 and ESI-preprint 210, to appear in Journal of Geometry and Physics.
19. M. Dubois-Violette, P.W. Michor, *Dérivations et calcul différentiel noncommutatif.*II, C.R. Acad. Sci. Paris, **319**, Série I (1994), 927-931.
20. M. Dubois-Violette, P.W. Michor, *More on the Frölicher-Nijenhuis bracket in noncommutative differential geometry*, To appear.
21. M. Karoubi, *Homologie cyclique des groupes et algèbres*, C.R. Acad. Sci. Paris **297**, Série I, (1983) 381-384.
22. D. Kastler, *Lectures on Alain Connes' non commutative geometry and applications to fundamental interactions*, in Infinite dimensional geometry, noncommutative geometry, operator algebras, fundamental interactions; Saint-François, Guadeloupe 1993 (R. Coquereaux, M. Dubois-Violette, P. Flad, eds), World Scientific 1995.
23. J.L. Koszul, *Fibre bundles and differential geometry*, Tata Institute of Fundamental Research, Bombay, 1960.
24. J. Madore, *Noncommutative differential geometry and its physical applications*, Cambridge University Press 1995.
25. G. Maltsiniotis, *Le langage des espaces et des groupes quantiques*, Comm. Math. Phys.**151** (1993) 275.
26. J. Mourad, *Linear connections in noncommutative geometry*, Class. Quantum Grav. **12** (1995) 965-974.
27. S.L.Woronowicz, *Differential calculus on compact matrix pseudogroups (quantum groups)*, Commun. Math. Phys. **122** (1989) 125-170.

Laboratoire de Physique Théorique et Hautes Energies*, Bât 211, Université Paris XI, 91405 ORSAY Cedex

E-mail address: `flad@qcd.th.u-psud.fr`

Contemporary Mathematics
Volume **203**, 1997

CONNECTIONS OF BIMODULES IN NON-COMMUTATIVE GEOMETRY

D. KASTLER, J. MADORE and D. TESTARD

ABSTRACT. The bilateral structure of connections on bimodules over non-commutative algebras equipped with general differential calculi is related to Hochschild cohomology with coefficients in an appropriate bimodule. We discuss the particular case of linear connections with their torsion as generalizing the classical example.

In a former paper [**8**] we revisited the connections in non-commutative geometry introduced by Mourad [**1**] and further investigated in [**2**], [**3**], in the general case of arbitrary bimodules, and within a frame, described in section 1 below, abstracting the features of Alain Connes' non-commutative metric geometry [**4**]. In the present paper, section 2, we first investigate without a priori-assumptions the "right Leibniz structure" of quantum connections w.r.t. the left module structure of our bimodule. The latter is shown to implicate in general Hochschild cocycles of the algebra with coefficients in an appropriate bimodule. These cocycles are a priori more general that the objects considered in the above-quoted papers, the latter thus appearing as "special" – this poses the problem of finding criteria for "speciality". In section 3, we consider the case of the "linear connections" treated in [**1**], [**2**], [**3**] and discuss the notion of torsion in non-commutative geometry.

1. General differential calculi

In this section, we consider an unital algebra $\mathbb{A}$ with unit denoted $\mathbb{1}$. Let $(\Omega\mathbb{A} = \oplus_{n\in\mathbb{N}}\Omega\mathbb{A}^n, \delta)$, equipped with the $\mathbb{Z}/2$-grading $\Omega\mathbb{A}^+ = \oplus_{p\in\mathbb{N}}\Omega\mathbb{A}^{2p}$, $\Omega\mathbb{A}^- = \oplus_{p\in\mathbb{N}}\Omega\mathbb{A}^{2p+1}$.

In order to define our general working frame, we need a preliminary result. Motivating examples will be described below.

***Lemma* 1.** *Let $\mathbb{B} = \mathbb{B}^+ \oplus \mathbb{B}^-$ a $\mathbb{Z}/2$-graded unital algebra. Then the following specifications are equivalent:*

(i): *a unital homomorphism $\phi : \Omega\mathbb{A} \to \mathbb{B}$ of $\mathbb{Z}/2$-graded unital algebras, injective in restriction to $\Omega\mathbb{A}^0 \cong \mathbb{A}$.*

(ii): *a pair (ϕ_0, ϕ_1) where:*

1991 *Mathematics Subject Classification.* Primary 47C10; Secondary 14A22 .

(a) $\phi_0 : \mathbb{A} \to \mathbb{B}^+$ *is an injective unital mapping of unital algebras. Note that this allows us to to consider* $\mathbb{B}^-$ *as a* $\mathbb{A}$*-bimodule with the following left and right actions:*

$$\begin{cases} aX = (\phi_0 a)X \\ Xa = X(\phi_0 a) \end{cases}, \qquad [X \in \mathbb{B}^-,\ a \in \mathbb{A}. \tag{1.1}$$

(b) *a homomorphism* $\phi_1 : \Omega\mathbb{A}^1 \to \mathbb{B}^-$ *of* $\mathbb{A}$*-bimodules*

(iii): *a pair* (ϕ_0, d_0) *where:*

(a) ϕ_0 *is an injective unital homomorphism:* $\mathbb{A} \to \mathbb{B}^+$ *of unital algebras,*

(b) d_0 *is a* $\mathbb{C}$*-linear mapping:* $\mathbb{A} \to \mathbb{B}^-$ *fullfilling:*

$$d_0(ab) = (d_0 a)(\phi_0 b) + (\phi_0 a)(d_0 b)\ , \qquad a, b \in \mathbb{A}. \tag{1.2}$$

The bijection (i) $\leftrightarrow$ (ii) is as follows:

$$\begin{cases} \phi_0 = \phi|_{\Omega\mathbb{A}^0 \cong \mathbb{A}} \\ \phi_1 = \phi|_{\Omega\mathbb{A}^1} \end{cases}, \qquad X \in \mathbb{B}^-,\ a \in \mathbb{A}. \tag{1.3}$$

Conversely:

$$\begin{cases} \phi_n(\lambda_1 \otimes_{\mathbb{A}} \ldots \otimes_{\mathbb{A}} \lambda_n) = (\phi_1\lambda_1)\ldots(\phi_1\lambda_n)\ , & \lambda_1, \ldots, \lambda_n \in \Omega\mathbb{A}^1,\ n \geq 1, \\ \phi_0 a \text{ as in (1.3).} \end{cases} \tag{1.4}$$

The bijection (ii) $\leftrightarrow$ (iii) is as follows:

$$d_0 = \phi_1 \circ \delta \quad \text{on} \quad \mathbb{A}, \tag{1.5}$$

***Definition* 2.** The specification of either of the equivalent data $(\phi, \mathbb{B}) \leftrightarrow (\phi_0, \phi_1, \mathbb{B}) \leftrightarrow (\phi_0, d_0, \mathbb{B})$ is called **a differential calculus over** $\mathbb{A}$.

Note that, given such a calculus, we have:

$$\phi(a_0 \delta a_1 \ldots \delta a_n) = (\phi a_0)(d_0 a_1)\ldots(d_0 a_n)\ , \qquad a_0, a_1 \ldots, a_n \in \mathbb{A}. \tag{1.6}$$

(written $\underline{a}_0 d_0 a_1 \ldots d_0 a_n$ with the shorthand $\phi a_0 = \underline{a}_0$).

Examples of differential calculi

(i): $(\Omega\mathbb{A}, id)$ with $d_0 = \delta|_{\Omega\mathbb{A}^0 \cong \mathbb{A}}$ is a differential calculus over $\mathbb{A}$ called its **universal differential calculus**.

(ii): A metric geometry of $\mathbb{A}$ in the sense of Connes [**2**], obtained by specifiying a K-cycle (H, D, χ) of $\mathbb{A}$, yields a differential calculus $(\phi, \mathbb{B})$ over $\mathbb{A}$, where $\mathbb{B}^{\pm} = B(H)^{\pm}$ and $\phi = \pi_D$, i.e.

$$\phi(a_0 \delta a_1 \ldots \delta a_n) = \underline{a}_0 [D, \underline{a}_1] \ldots [D, \underline{a}_n]\ , \qquad a_0, a_1 \ldots, a_n \in \mathbb{A}. \tag{1.7}$$

with

$$d_0 a = [D, \underline{a}]\ , \qquad a \in \mathbb{A}. \tag{1.8}$$

where we used the shorthand $\underline{a} = \pi_D(a)$, $a \in \mathbb{A}$.

(iii): The usual differential calculus on a compact even-dimensional manifold $\boldsymbol{M}$ is obtained as the the special case of (ii) where $\mathbb{A} = C^\infty(\boldsymbol{M})$, $H = L^2(\mathbb{S}_M)$, with $\mathbb{S}_M$ the spin bundle of $\boldsymbol{M}$, and D the Dirac-Atiyah-Singer-Lichnérowicz operator.

***Proposition* 3.** *Let $(\phi,\mathbb{B}) \leftrightarrow (\phi_0,\phi_1,\mathbb{B}) \leftrightarrow (\phi_0,d_0,\mathbb{B})$ be a differential calculus over the algebra $\mathbb{A}$. We obtain as follows a $\mathbb{N}$-graded differential algebra (Ω_ϕ, d). Setting*

$$(1.9)\qquad \begin{cases} \boldsymbol{K} = \ker\phi \\ \boldsymbol{K}^n = K\cap\Omega\mathbb{A}^n\,, & n\in\mathbb{N}, \end{cases}$$

and

$$(1.10)\qquad \begin{cases} \boldsymbol{J}^n = \boldsymbol{K}^n + \delta\boldsymbol{K}^{n-1}\,, & n\geq 1, \\ \boldsymbol{J}^0 = \boldsymbol{K}^0 = \{0\} \end{cases},$$

further[1]

$$(1.11)\qquad \begin{cases} \Omega^n_\phi = \Omega\mathbb{A}^n/\boldsymbol{J}^n = \dfrac{\phi(\Omega\mathbb{A}^n)}{\phi(\delta\boldsymbol{K}^{n-1})}\,, & n\geq 1, \\ \Omega^0_\phi = \Omega\mathbb{A}^0 = \mathbb{A}, \end{cases}$$

we obtain an ideal $\boldsymbol{K}$ (not a differential ideal!) of the algebra $\Omega\mathbb{A}$, a $\mathbb{N}$-graded differential ideal $\boldsymbol{J} = \oplus_{n\in\mathbb{N}}\boldsymbol{J}^n$ of the $\mathbb{N}$-graded differential algebra $(\Omega\mathbb{A} = \oplus_{n\in\mathbb{N}}\Omega\mathbb{A}^n, \delta)$ and therefore a quotient $\mathbb{N}$-graded differential algebra $(\Omega_\phi = \oplus_{n\in\mathbb{N}}\Omega^n_\phi, d)$ with differential given as follows: one has, for $a^i_0\ldots a^i_n\in\mathbb{A}$, $n\geq 1$, $i=1,\ldots,p$:

$$(1.12)\qquad d\left[\sum_{i=1,\ldots,p} a^i_0 d_0 a^i_1\ldots d_0 a^i_n \bmod \phi(\delta\boldsymbol{K}^{n-1})\right] = \sum_{i=1,\ldots,p} d_0 a^i_0 d_0 a^i_1\ldots d_0 a^i_n \bmod \phi(\delta\boldsymbol{K}^n),$$

with

$$(1.13)\qquad \boldsymbol{K}^n = \left\{\sum_{i=1,\ldots,p} a^i_0\delta a^i_1\ldots\delta a^i_n\ ;\ (\phi a^i_0)(d_0 a^i_1)\ldots(d_0 a^i_n) = 0\right\}.$$

The elements of (Ω_ϕ, d) are called the **quantum forms** of the differential calculus $(\phi,\mathbb{B})$ over $\mathbb{A}$.

2. Connections of bimodules

***Definitions* 4.** Let $\boldsymbol{E}$ be an $\mathbb{A}$-bimodule, $\mathbb{A}$ an algebra equipped with a differential calculus $(\phi,\mathbb{B}) \leftrightarrow (\phi_0,\phi_1,\mathbb{B}) \leftrightarrow (\phi_0,d_0,\mathbb{B})$. And let $(\Omega_\phi = \oplus_{n\in\mathbb{N}}\Omega^n_\phi, d)$ be the corresponding differential algebra of quantum forms, cf. **3**.

(i): We denote by $\Gamma(\boldsymbol{E},\phi,\mathbb{B})$ the set of $\mathbb{C}$-linear maps $\nabla : \boldsymbol{E}\to\Omega^1_\phi\otimes_{\mathbb{A}}\boldsymbol{E}$ fullfilling the Leibniz rule:[2]

$$(2.1)\qquad \nabla(a\eta) = a\nabla\eta + da\otimes_{\mathbb{A}}\eta\,, \qquad a\in\mathbb{A},\ \eta\in E.$$

The elements of $\Gamma(\boldsymbol{E},\phi,\mathbb{B})$ are called the **connections of $\boldsymbol{E}$**.

(ii): Given $\nabla\in\Gamma(\boldsymbol{E},\phi,\mathbb{B})$, we define $\tau_\nabla = \tau : \boldsymbol{E}\times\mathbb{A}\to\boldsymbol{E}$ as follows:

$$(2.2)\qquad \tau(\eta,a) = \nabla(\eta a) - (\nabla\eta)a\,, \qquad a\in\mathbb{A},\ \eta\in E.$$

---

[1] Note that $\Omega^1_\phi = \phi(\Omega\mathbb{A}^1)$ owing to $\boldsymbol{K}^0 = \{0\}$.

[2] From now on we write systematically d for $d_0 = d|_{\Omega\mathbb{A}^0\simeq\mathbb{A}}$. Note that requirement (2.1) is coherent with the associativity property $a(b\eta) = (ab)\eta$, $a,b\in\mathbb{A}$, $\eta\in\boldsymbol{E}$.

***Lemma* 5.** *With the preceding definitions and notation we have that* τ *is a* $\mathbb{C}$*-bilinear map with the properties:*

$$\tau(\eta, \mathbb{1}) = 0 \qquad \eta \in E, \tag{2.3}$$

$$\tau(b\eta, a) = b\tau(\eta, a)\ , \qquad a, b \in \mathbb{A},\ \eta \in E, \tag{2.4}$$

$$\tau(\eta, ab) = \tau(\eta, a)b + \tau(\eta a, b)\ , \qquad a, b \in \mathbb{A},\ \eta \in E. \tag{2.5}$$

The latter results will be interpreted using an $\mathbb{A}$-bimodule which we introduce in the following Lemma.

***Lemma* 6.** *Let* $(\phi, \mathbb{B})$ *be a differential calculus over the algebra* $\mathbb{A}$*, with the definitions and notation in* **4.**. *Consider the set* $\mathcal{M}_{\boldsymbol{E}}$ *of left-module maps:* $\boldsymbol{E} \to \Omega^1_\phi \otimes_{\mathbb{A}} \boldsymbol{E}$ *for* $X \in \mathcal{M}_{\boldsymbol{E}}$ *and* $a \in \mathbb{A}$ *we define as follows the maps* aX, Xa*:* $\boldsymbol{E} \to \Omega^1_\phi \otimes_{\mathbb{A}} \boldsymbol{E}$*:*

$$\begin{cases} (aX)\eta = X(\eta a) \\ (Xa)\eta = (X\eta)a \end{cases}, \qquad X \in \mathcal{M}_{\boldsymbol{E}},\ a \in \mathbb{A},\ \eta \in \boldsymbol{E}. \tag{2.6}$$

Then we have that aX *and* Xa *belong to* $\mathcal{M}_{\boldsymbol{E}}$ *and, thus* $\mathcal{M}_{\boldsymbol{E}}$ *becomes an* $\mathbb{A}$*-bimodule.*

***Remark* 7.** $\Gamma(\boldsymbol{E}, \phi, \mathbb{B})$ is an affine space patterened after the vector space $\mathcal{M}_{\boldsymbol{E}}$: fixing any $\nabla_0 \in \Gamma(\boldsymbol{E}, \phi, \mathbb{B})$, we have:

$$\Gamma(\boldsymbol{E}, \phi, \mathbb{B}) = \{\nabla_0 + X;\ X \in \mathcal{M}\}. \tag{2.7}$$

We are now able to interpret the facts in **5** in terms of Hochschild cohomology.

***Proposition* 8.**

(i): *Let* $\mathbb{A}$, $(\phi, \mathbb{B})$, $\nabla \in \Gamma(\boldsymbol{E}, \phi, \mathbb{B})$, *and* $\tau_\nabla = \tau$ *be as in* **4**. *Reinterpreted as the* $\mathbb{C}$*-linear form:*

$$\mathbb{A} \ni a \to \{\boldsymbol{E} \ni \eta \to \tau(\eta, a) = \tau(a)\eta \in \Omega^1_\phi \otimes_{\mathbb{A}} \boldsymbol{E}\}. \tag{2.8}$$

τ appears as a normalized one-cocycle of the Hochschild cohomology of $\mathbb{A}$ with coefficients in $\mathcal{M}_{\boldsymbol{E}}$.[3]

(ii): *The set*

$$\tau_{\boldsymbol{E}} = \{\tau_\nabla,\ \nabla \in \Gamma(\boldsymbol{E}, \phi, \mathbb{B})\} \tag{2.9}$$

is in fact a cohomology class of this cohomology.

The next remark describes a special kind of one-cocycles met in examples.

***Remark* 9.** With $\sigma : \boldsymbol{E} \otimes_{\mathbb{A}} \Omega^1_\phi \to \boldsymbol{E} \otimes_{\mathbb{A}} \Omega^1_\phi$ a bimodule map, set

$$\underline{\sigma}(a, \eta) = \underline{\sigma}(a)\eta = \sigma(\eta \otimes_{\mathbb{A}} da)\ , \qquad a \in \mathbb{A},\ \eta \in E. \tag{2.10}$$

We thus get a normalized one-cocycle of the $\mathcal{M}_{\boldsymbol{E}}$-valued Hochschild cohomology of $\mathbb{A}$. Cocycles of this type will be referred to as **special cocycles**.

---

[3] whose respective sets of n-cocycles, n-coboundary, and n-cohomology classes are denoted by $HZ^n(\mathbb{A}, \mathcal{M}_{\boldsymbol{E}})$, $HB^n(\mathbb{A}, \mathcal{M}_{\boldsymbol{E}})$, and $HH^n(\mathbb{A}, \mathcal{M}_{\boldsymbol{E}})$. We recall (cf. [**6**], p. 284) that the Hochschild coboundary ∂ in grade one is given as follows

$$\partial\tau(a, b) = a\tau(b) - \tau(ab) + \tau(a)b\ , \qquad a, b \in \mathbb{A},$$

and that the elements β of $HB^1(\mathbb{A}, \mathcal{M}_{\boldsymbol{E}})$ are of the form

$$\beta(a)\eta = \theta(\eta a) - (\theta\eta)a\ , \qquad a \in \mathbb{A},\ \eta \in E,$$

with $\theta \in \mathcal{M}_{\boldsymbol{E}}$.

***Proposition* 10.** *Consider the algebra $\mathbb{A}$ equipped with its universal differential calculus $(\Omega\mathbb{A}, id)$, and let $\boldsymbol{E}$ be an $\mathbb{A}$-bimodule: $\tau_{\boldsymbol{E}}$ then consists of special cocycles.*

3. Linear Connections

We now turn our attention to linear connections, that is, the special case $\boldsymbol{E} = \Omega^1_\phi$, in which case we have the notion of torsion.

***Definition* 11.** Let $(\phi, \mathbb{B})$ be a differential calculus over the algebra $\mathbb{A}$ with the definitions and notation of **4**, and denote by π: $\Omega^1_\phi \otimes_{\mathbb{A}} \Omega^1_\phi \to \Omega^2_\phi$ the multiplication map (restriction to $\Omega^1_\phi \otimes_{\mathbb{A}} \Omega^1_\phi$ of the multiplication map of Ω_ϕ, (cf.**3**). Given a linear connection $\nabla \in \Gamma(\Omega^1_\phi, \phi, \mathbb{B})$, the **torsion** $\mathbb{T}$ of ∇ is defined as

$$\mathbb{T} = d_1 - \pi \circ \nabla, \tag{3.1}$$

with d_1 the restriction to Ω^1_ϕ of the differential d of Ω_ϕ:

$$\begin{array}{ccc} \Omega^1_\phi & \xrightarrow{\nabla} & \Omega^1_\phi \otimes_{\mathbb{A}} \Omega^1_\phi \\ & d_1 \searrow & \downarrow \pi \\ & & \Omega^2_\phi \end{array} \tag{3.2}$$

We call the connection ∇ **torsionless** whenever $\mathbb{T}$ vanishes.
The following version of the classical torsion motivates this definition.

***Proposition* 12.** *Assume the setting of* **2** (iii), *that is: $\mathbb{A} = C^\infty(\boldsymbol{M})$, $\Omega^1_\phi = \Omega(\boldsymbol{M})^1$, $\Omega(\boldsymbol{M})$ the DeRham complex of $\boldsymbol{M}$ with differential the exterior derivative d, and denote by $\chi(\boldsymbol{M})$ the set of smooth vector fields of $\boldsymbol{M}$. Given a linear connection ∇ of $\boldsymbol{M}$ (i.e. a connection of the tangent bundle) with torsion 2-tensor T, we have, denoting also by ∇ the corresponding connection on the cotangent vector bundle:*[4]

$$\langle \lambda, T(\xi, \eta)\rangle = [(\boldsymbol{d} - \pi \circ \nabla)\lambda](\xi, \eta)\ , \quad \lambda \in \Omega(\boldsymbol{M})^1,\ \xi, \eta \in \chi(\boldsymbol{M}). \tag{3.3}$$

***Proposition* 13.** *With ∇ and $\mathbb{T}$ as in* **11**, *we have that:*

(i): *The map $\Omega^1_\phi \to \Omega^2_\phi$ is a homomorphism of left $\mathbb{A}$-modules.*[5]

(ii): *One has, with $\underline{id}$ the one-cocycle obtained from the identity map of $\Omega^1_\phi \otimes_{\mathbb{A}} \Omega^1_\phi$ as in Remark* **9**:

$$\mathbb{T}(\lambda a) - \mathbb{T}(\lambda)a = -\pi[(\tau + \underline{id})(a)]\lambda\ , \qquad \lambda \in \Omega^1_\phi,\ a \in \mathbb{A}, \tag{3.4}$$

in other terms $\pi \circ (\tau + \underline{id})$ is a one-coboundary of the Hochschild cohomology $HH^1(\mathbb{A}, \Omega^2_\phi)$:

$$\pi \circ (\tau + \underline{id}) = \partial^0 \mathbb{T}, \tag{3.4a}$$

∂^0 the Hochschild coboundary in zero grade.

(ii): *Consequently the torsion is a bimodule map iff this one-coboundary vanishes. In the case where τ is special, i.e. obtained as $\tau = \underline{\sigma}$ from a bimodule endomorphism σ of $\Omega^1_\phi \otimes_{\mathbb{A}} \Omega^1_\phi$, we have*

---

[4] defined by asking $\lambda(\nabla \xi \eta) + (\nabla \xi \lambda)\eta = \xi(\lambda(\eta))$, $\lambda \in \Omega(\boldsymbol{M})^1$, $\xi, \eta \in \chi(\boldsymbol{M})$

[5] i.e. $\mathbb{T} \in \mathcal{M}_{\boldsymbol{E}}$, cf. **2.3**.

(3.4b) $$\mathbb{T}(\lambda a) - \mathbb{T}(\lambda)a = -\pi[(\underline{\sigma + id})(a)]\lambda\,, \qquad \lambda \in \Omega^1_\phi,\ a \in \mathbb{A},$$

$\mathbb{T}$ is thus a bimodule map iff one has:

(3.5) $$\pi \circ (\sigma + id) = 0.$$

We now discuss in the present frame the case of classical linear connections, finding there again special cocycles.

***Proposition* 14.** *In the setting of* **2** (iii), *that is: $\mathbb{A} = C^\infty(\boldsymbol{M})$, $\Omega^1_\phi = \Omega(\boldsymbol{M})^1$, $\Omega(\boldsymbol{M})$ the DeRham complex of $\boldsymbol{M}$ with its differential the exterior derivative* **d**, *and $\omega\omega'$, $\omega, \omega' \in \Omega(\boldsymbol{M})$, the exterior product of differential forms, we note the commutation rules:*[6]

(3.6) $$a\lambda = \lambda a, \qquad a \in \mathbb{A},\ \lambda \in \Omega(\boldsymbol{M})^1,$$

(3.7) $$\lambda\mu = -\mu\lambda a\,, \qquad \lambda,\ \mu \in \Omega(\boldsymbol{M})^1,$$

implying the following properties of the $\mathbb{A}$-bimodule $\Omega^1_\phi \otimes_{\mathbb{A}} \Omega^1_\phi$:

(3.8) $$\begin{aligned} a(\lambda \otimes_{\mathbb{A}} \mu) &= (a\lambda) \otimes_{\mathbb{A}} \mu = (\lambda a) \otimes_{\mathbb{A}} \mu = \lambda \otimes_{\mathbb{A}} a\mu \\ &= (\lambda \otimes_{\mathbb{A}} \mu)a,\ a \in \mathbb{A},\ \lambda, \mu \in \Omega(\boldsymbol{M})^1. \end{aligned}$$

Our one-cocycles are then of the form $\underline{\kappa}$, with the "flip" of $\Omega^1_\phi \otimes_{\mathbb{A}} \Omega^1_\phi$:

(3.9) $$\kappa(\lambda \otimes_{\mathbb{A}} \mu) = \mu \otimes_{\mathbb{A}} \lambda\,, \qquad \lambda,\ \mu \in \Omega(\boldsymbol{M})^1.$$

In this case $\tau_{\boldsymbol{E}}$ is thus a singlet (note that (3.8) implies that $HB^1(\mathbb{A}, \mathcal{M}_{\boldsymbol{E}})\{0\}$).

After completion of this work we learned about the work [**7**] of M. Dubois-Violette and T. Masson which looks at essentially the same landscape as the present work from a complementary point of view.

References

[1] J. Mourad, Linear connections in non-commutative geometry, *Class. Quant. Grav.*, **12** (1995) 965.

[2] J. Madore, T. Masson, J. Mourad, Linear connections on matrix geometries, *LPTHE*, Orsay, 94/96, (1994).

[3] M. Dubois-Violette, J. Madore, T. Masson, J. Mourad, Linear connections on the quantum plane, *LPTHE*, Orsay, 94/94, (1994).

[4] A. Connes, Non-commutative Geometry, *Academic Press*, (1994).

[5] D. Kastler, Cyclic comomology within the differential envelope, Travaux en cours, *Herrmann*, Paris, **30** (1988).

[6] S. Mac Lane, Homology, Grundlehren d. math. Wiss., *Springer Verlag*, (1970).

[7] M. Dubois-Violette, T. Masson, On the first order operators in bimodules, *LPTHE*, Orsay, 95/56, (1995).

[8] D. Kastler, J. Madore, D. Testard, Connections on bimodules in non-commutative geometry, Erwin Schrödinger Institut preprint, Wien, Austria, (1995).

Current address: Centre de Physique Théorique, Unité Propre de Recherche 7061, CNRS Luminy, Case 907 and Erwin Schrödinger Institut, Wien, Austria.

Current address: LPTHE, Université de Paris-Sud, Bat 211, F-91405, Orsay.

Current address: Centre de Physique Théorique, Unité Propre de Recherche 7061, CNRS Luminy, Case 907

---

[6]We write $\lambda\mu$ for $\lambda \wedge \mu$. Note that the rules (3.8), (3.9) are implied by the Poincaré-duality relations $[\underline{a}, \underline{b}] = [[D, \underline{a}], \underline{b}] = 0,\ a, b \in \mathbb{A}$.

Contemporary Mathematics
Volume **203**, 1997

Riemaniann and Non-Commutative Geometry in Physics

B. IOCHUM, D. KASTLER AND T. SCHÜCKER

ABSTRACT. We feel that non-commutative geometry is to particle physics what Riemannian geometry is to gravity. We try to explain this feeling.

We feel that non-commutative geometry is as fundamental to physics as Minkowskian and Riemannian geometry. Let us try to explain this by comparing the standard model of particle physics and general relativity. From a chronological point of view, this comparison is difficult, because Riemannian geometry existed well before general relativity. However, the field theoretic approach allows to introduce general relativity in close analogy to classical electrodynamics without use of Riemannian geometry. Therefore this approach is well suited for our comparison.

So let us imagine a world ignoring Riemannian geometry where physicists try to describe gravity. They are inspired by Maxwell who takes a field A of spin 1, a second order differential operator D_{Max} and writes down his field equation

$$D_{Max}\, A = \frac{1}{c^2 \epsilon_0}\, j,$$

where j is the source, charge density and currents, and ϵ_0 is the proportionality constant from Coulomb's law. After many ingenious and expensive experiments and theoretical trials and errors, the physicists agree on the standard model of gravity. It starts from a particular spin 2 field g, and a second order differential operator D_{Ein}. The field equation is

$$D_{Ein}\, g = -\frac{8\pi G}{c^4}\, T,$$

where the source T is energy-momentum density and currents, and G is the proportionality constant from Newton's universal law of gravity. Although in

1991 *Mathematics Subject Classification.* Primary 14A22, Secondary 81E13, 47C10.

perfect agreement with experiment, this standard model has draw backs: who ordered spin 2? Maxwell's differential operator D_{Max} contains 8 summands, the gravitational one D_{Ein} results from brute force and contains roughly 80 000 summands. Some of these summands still are unaccessible to experiment. At this stage, Riemannian geometry is discovered, the spin 2 field is recognized as the metric and the differential operator D_{Ein} is recognized as the curvature if the unknown summands are chosen properly. Most physicists say: so what, just fancy mathematics. Some dream of a geometric unification of all forces. Later, even more expensive experiments will test the predictions of Riemannian geometry coming from the unknown summands.

If, in the real world, we qualify general relativity as revolution, we have several criteria.

- Postdiction: the theory correctly reproduces experimental data, that remain unexplained in the old theories, e.g. the precession of perihelia of Mercury.
- Prediction: the theory can be in contradiction with future experimental data, e.g. deflection of light.
- New concepts, e.g. curved spacetimes, absence of universal time.
- Reticence of the majority.

Our purpose is to explain that for non-commutative geometry the analogue of g in the imaginary world is the Higgs field, the analogue of D_{Ein} is the Lagrangian of the standard model of electro-weak and strong interactions. Postdictions of the theory are that fermions sit in fundamental representations, that weak interactions violate parity, that strong interactions are vector like and

$$\rho := \frac{g_1^2 + g_2^2}{g_2^2}\frac{m_W^2}{m_Z^2} = 1 \quad = 1.0004 \pm 0.0042 \quad \text{experimentally},$$

$$m_e < m_W < m_t/\sqrt{3},$$
$$2g_1^{-2} > g_2^{-2} + g_3^{-2}/3.$$

There is also a prediction, the mass of the Higgs, accessible to experiment in about ten years. New concepts are fuzzy spacetimes — that is spacetimes with an uncertainty relation — and discrete spacetimes.

1. The Establishment

Let us briefly summarize today's established theory of particles and interactions. It is a particular Yang-Mills-Higgs theory. To get started, we view this class of theories as black box or slot machine. The input comes in two parts, bills and coins. The output is a particle phenomenology, that is cross sections, branching ratios, life times ... To decide whether a particular input is a winner, its corresponding output is confronted with millions of experimental numbers that cost billions of Swiss Francs.

1.1. Bills and coins. The Yang-Mills-Higgs machine has four slots for one bill each. In the first of these slots you are supposed to put a finite dimensional, real, compact Lie group G. For the remaining slots choose three unitary representations ρ_L, ρ_R, ρ_S defined on Hilbert spaces $\mathcal{H}_L$, $\mathcal{H}_R$, $\mathcal{H}_S$. These Hilbert spaces will accommodate the left- and right-handed fermions and the Higgs scalars.

After having eaten these four bills, the machine will ask you for coins, real or complex numbers. The number of coins depends on the chosen bills.

- An invariant scalar product on the Lie algebra $\mathfrak{g}$ of G. This choice is parameterized by one positive number g, the 'gauge coupling', for every simple factor in G, e.g.
$$(b, b') := \frac{1}{g_1^2}\bar{b}b', \quad b, b' \in u(1),$$
$$(a, a') := \frac{2}{g_n^2}\,\mathrm{tr}\,(a^*a'), \quad a, a' \in su(n).$$
- An invariant, positive polynomial $V(\varphi)$, $\varphi \in \mathcal{H}_S$ of order 4, the 'Higgs potential'. We want this potential to break G spontaneously. This means that no invariant vector in $\mathcal{H}_S$ minimizes V. For example if $G = SU(2)$ with the fundamental representation $\mathcal{H}_S = \mathbb{C}^2$, the most general Higgs potential is
$$V(\varphi) = \lambda(\varphi^*\varphi)^2 - \frac{\mu^2}{2}\varphi^*\varphi, \qquad \varphi \in \mathcal{H}_S, \quad \lambda, \mu > 0.$$
- One complex number or 'Yukawa coupling' g_Y for every trilinear invariant — i.e. for every one dimensional invariant subspace, 'singlet' — in the decomposition of the representation associated to $(\mathcal{H}_L^* \otimes \mathcal{H}_R \otimes \mathcal{H}_S) \oplus (\mathcal{H}_L^* \otimes \mathcal{H}_R \otimes \mathcal{H}_S^*)$. For example if $G = SU(2)$, $\mathcal{H}_L = \mathbb{C}^2$, $\mathcal{H}_R = \mathbb{C}$, $\mathcal{H}_S = \mathbb{C}^2$ there is one singlet:
$$\sum_{j=1}^{2} \bar{\psi}_{Lj}\psi_R\varphi_j, \quad \begin{pmatrix}\psi_{L1}\\ \psi_{L2}\end{pmatrix} \in \mathcal{H}_L, \quad \psi_R \in \mathcal{H}_R, \quad \begin{pmatrix}\varphi_1\\ \varphi_2\end{pmatrix} \in \mathcal{H}_S.$$

Physicist have been playing on this slot machine for the last thirty years. One winner clearly emerged, the so called standard model. Its bills are
$$G = SU(3) \times SU(2) \times U(1)$$
$$\mathcal{H}_L = \bigoplus_1^3 \left[(1, 2, -\frac{1}{2}) \oplus (3, 2, \frac{1}{6})\right],$$
$$\mathcal{H}_R = \bigoplus_1^3 \left[(1, 1, -1) \oplus (3, 1, \frac{2}{3}) \oplus (3, 1, -\frac{1}{3})\right],$$
$$\mathcal{H}_S = (1, 2, -\frac{1}{2}),$$
where (n_3, n_2, y) denotes the tensor product of an n_3 dimensional representation of $SU(3)$, an n_2 dimensional representation of $SU(2)$ and the one dimensional representation of $U(1)$ with hypercharge y:
$$\rho(e^{i\theta}) = e^{iy\theta}, \qquad y \in \mathbb{Q}, \ \theta \in [0, 2\pi).$$
Some vocabulary: particles are basis elements. The spin 1 particles, the gauge bosons, span the Lie algebra $\mathfrak{g}$ of the group G. The eight basis elements of $su(3)$

are called gluons. They are massless and mediate the strong interactions, e.g. nuclear fusion, fission, α-decay. The remaining $su(2) \oplus u(1)$ is spanned by the photon — Maxwell's old friend and later found responsible for γ-decay — and three massive bosons, the W^+, W^- and Z. They mediate the weak interactions, e.g. β-decay. The spin $\frac{1}{2}$ particles or fermions come in three identical copies, 'generations'. The first generation of $\mathcal{H}_L$ is spanned by the left-handed parts (Weyl spinors) of the electronic neutrino, the electron and the up and down quarks. The first two are called leptons, from the greek word for mild, because sitting in $SU(3)$ singlets they are not subject to strong interactions. The other two left-handed generations are spanned by the muonic neutrino, the muon, the charm and strange quarks, and the tau neutrino, the tau, the top and bottom quarks. $\mathcal{H}$ is spanned by the right-handed parts of the same particles, except that there are no right-handed neutrinos. Consequently the neutrinos are massless. The particle count for the spin 0 particles, scalars, is a little bit more complicated. Not all basis elements of $\mathcal{H}_S$ correspond to physical scalars. There is only one in the standard model. It is called Higgs scalar and is still being searched for.

Because of the high degree of reducibility in the bills, there are many coins, among them 27 Yukawa couplings. Not all of them have a physical meaning. They can be converted into 18 physically significant, positive numbers [**1**], three gauge couplings,

$$g_1 = 0.3575 \pm 0.0001, \quad g_2 = 0.6507 \pm 0.0007, \quad g_3 = 1.207 \pm 0.026,$$

eleven particle masses,

$$\begin{aligned} m_W &= 80.22 \pm 0.26 \text{ GeV}, & m_e &= 0.51099906 \pm 0.00000015 \text{ MeV}, \\ m_H &> 58.4 \text{ GeV}, & m_\mu &= 0.105658389 \pm 0.000000034 \text{ GeV}, \\ & & m_\tau &= 1.7771 \pm 0.0005 \text{ GeV}, \end{aligned}$$

$$\begin{aligned} m_u &= 5 \pm 3 \text{ MeV}, & m_d &= 10 \pm 5 \text{ MeV}, \\ m_c &= 1.3 \pm 0.3 \text{ GeV}, & m_s &= 0.2 \pm 0.1 \text{ GeV}, \\ m_t &= 176 \pm 18 \text{ GeV}, & m_b &= 4.3 \pm 0.2 \text{ GeV}, \end{aligned}$$

and quark mixings. These mixings are given in form of a unitary matrix, the Cabbibo-Kobayashi-Maskawa matrix

$$C_{KM} := \begin{pmatrix} V_{ud} & V_{us} & V_{ub} \\ V_{cd} & V_{cs} & V_{cb} \\ V_{td} & V_{ts} & V_{tb} \end{pmatrix}.$$

For physical purposes it can be parameterized by three angles θ_{12}, θ_{23}, θ_{13} and one CP violating phase δ:

$$C_{KM} = \begin{pmatrix} c_{12}c_{13} & s_{12}c_{13} & s_{13}e^{-i\delta} \\ -s_{12}c_{23} - c_{12}s_{23}s_{13}e^{i\delta} & c_{12}c_{23} - s_{12}s_{23}s_{13}e^{i\delta} & s_{23}c_{13} \\ s_{12}s_{23} - c_{12}c_{23}s_{13}e^{i\delta} & -c_{12}s_{23} - s_{12}c_{23}s_{13}e^{i\delta} & c_{23}c_{13} \end{pmatrix},$$

with $c_{kl} := \cos\theta_{kl}$, $s_{kl} := \sin\theta_{kl}$. The absolute values of the matrix elements are:

$$\begin{pmatrix} 0.9753 \pm 0.0006 & 0.221 \pm 0.003 & 0.004 \pm 0.002 \\ 0.221 \pm 0.003 & 0.9745 \pm 0.0007 & 0.040 \pm 0.008 \\ 0.010 \pm 0.006 & 0.039 \pm 0.009 & 0.9991 \pm 0.0004 \end{pmatrix}.$$

Every body agrees that the standard model is ugly, too ugly to be a fundamental theory.

1.2. The general rules. Let us now have a closer look at the inside of the Yang-Mills-Higgs machine. It produces a Lagrangian that consists of five separate pieces. The first is the *Yang-Mills Lagrangian*, well motivated physically as non-abelian generalization of the famous Maxwell Lagrangian, $G = U(1)$. Also on the mathematical side, this Lagrangian needs no further introduction. Its fundamental field are the gauge bosons or connection, $A \in \Omega^1(M, \mathfrak{g})$, a 1-form on the spacetime manifold M with values in the Lie algebra $\mathfrak{g}$:

$$\mathcal{L}_{YM}[A] = \frac{1}{4}(F, *F),$$

where $F := \mathrm{d}A + \frac{1}{2}[A, A]$ denotes the field strength or curvature of A, $*$ is the Hodge star, and $(\cdot, \cdot)$ is the chosen invariant scalar product on $\mathfrak{g}$.

The second piece is the *Dirac Lagrangian*. It is geometricly as noble as the Yang-Mills Lagrangian.

$$\mathcal{L}_D[A, \psi_L, \psi_R] = \psi_L^* \not{D}\psi_L + \psi_R^* \not{D}\psi_R,$$

where ψ_L is a left-handed spinor with values in $\mathcal{H}_L$, ψ_L^* is its dual with respect to the scalar product in $\mathcal{H}_L$, $\not{D}$ is the covariant Dirac operator, $\mathrm{D}\psi_L := \mathrm{d}\psi_L + \tilde{\rho}_L(A)\psi_L$. We denote by $\tilde{\rho}$ the Lie algebra representation belonging to the group representation ρ. For $G = U(1)$ these two Lagrangians yield the very successful quantum electrodynamics and for $G = SU(3)$, $\mathcal{H}_L = \mathcal{H}_R = \mathbb{C}^3$ we get the present day theory for strong interactions, quantum chromodynamics.

In order to incorporate weak interactions and to give masses to gauge bosons and fermions, one is forced to break the symmetry G spontaneously. This is where the patchwork starts. One has to postulate the existence of scalars, til now unobserved. They are 0-forms with values in $\mathcal{H}_S$,

$$\varphi \in \Omega^0(M, \mathcal{H}_S).$$

One also has to add three more Lagrangian pieces involving the scalars, the *Klein-Gordon Lagrangian*

$$\mathcal{L}_{KG}[A, \varphi] = \frac{1}{2}\mathrm{D}\varphi^* * \mathrm{D}\varphi,$$

the *Higgs potential* and the *Yukawa terms*

$$\mathcal{L}_{Yu}[\psi_L, \psi_R, \varphi] = \sum_{j=1}^{n} g_{Yj}\,(\psi_L^*, \psi_R, \varphi)_j + \sum_{j=n+1}^{m} g_{Yj}\,(\psi_L^*, \psi_R, \varphi^*)_j\ + \text{compl. conj..}$$

To summarize, the standard model has two major shortcomings, the general rules of Yang-Mills-Higgs look artificial, as well as the input, bills *and* coins, singled out by nature. Nevertheless the standard model has resisted to an extremely detailed experimental analysis where all concurrent models have failed.

2. The Revolution

The non-commutative formulation improves the situation on all three levels, general rules, bills and coins.

2.1. General rules. The Yang-Mills and Dirac Lagrangians have a geometric origin and Alain Connes found a natural generalization of some of them to non-commutative geometry [**2**]. Connes and Lott have considered these Lagrangians in the particular case of a product geometry of an ordinary four dimensional spacetime geometry by a zero dimensional non-commutative geometry. There a miracle happens [**2,3**]. When decomposing the non-commutative versions of the Yang-Mills and Dirac Lagrangians in terms of ordinary fields they retrieve of course the ordinary Yang-Mills and Dirac Lagrangians. Simultaneously and free of charge, they also get the other three pieces, the Klein-Gordon Lagrangian, the symmetry breaking Higgs potential, and some Yukawa terms. Every such Connes-Lott model yields a particular Yang-Mills-Higgs model. The contrary is far from being true, how far will be discussed in terms of bills and coins in the following subsections.

2.2. Bills. Since the introduction of quantum mechanics, we are used to the description of non-commutative spaces in terms of involution algebras. A zero dimensional non-commutative space is given by a finite dimensional, real involution algebra $\mathcal{A}$. The group G of the ensuing gauge model will be the group of unitaries of $\mathcal{A}$

$$\{a \in \mathcal{A}|\ a^*a = a^*a = 1\}$$

or possibly a subgroup thereof. In order to construct a differential calculus on the non-commutative space, Connes introduces two algebra representations ρ_L and ρ_R on Hilbert spaces $\mathcal{H}_L$ and $\mathcal{H}_R$ such that $\rho_L \oplus \rho_R$ is faithful. In the finite dimensional case, this implies that $\mathcal{A} = M_n(\mathbb{R})$, $M_n(\mathbb{C})$ or $M_n(\mathbb{H})$, $\mathbb{H}$ denoting the quaternions, and that the algebra representations are copies of the defining representation. For $M_n(\mathbb{C})$, there is— in addition to the defining representation — its conjugate. In terms of bills of the resulting Yang-Mills-Higgs model we have the following irreducible possibilities:

$$\begin{aligned} G = O(n,\mathbb{R}), &\mathcal{H}_{L,R} = \mathbb{C}^n, \\ G = U(n) \text{ or } SU(n), &\mathcal{H}_{L,R} = \mathbb{C}^n, \\ G = USp(n), &\mathcal{H}_{L,R} = \mathbb{C}^{2n}. \end{aligned}$$

The restriction on the group bill is mild, only the exceptional groups are excluded. The restrictions on the two fermionic bills is appreciable, e.g. $U(1)$ only admits hypercharge -1, or 1, $SU(2)$ only has one irreducible representation, the two dimensional one, while in the general setting there is an infinite number to choose from.

The restriction on the scalar bill is spectacular. It comes out to be a *group* representation, a unitary representation of the group of unitaries, and is restricted by the fermionic bills: its Hilbert space is an invariant subspace,

$$\mathcal{H}_S\colon\ \subset\colon\ (\mathcal{H}_L^*\otimes\mathcal{H}_R)\oplus(\mathcal{H}_R^*\otimes\mathcal{H}_L)\,.$$

This invariant subspace is entirely determined by the coins.

One is of course tempted to build models with a simple algebra and/or irreducible fermion representations. Besides phenomenological shortcomings, all such models have a degenerate vacuum, an invariant vector in $\mathcal{H}_S$, that minimizes the Higgs potential. All popular Grand Unifies Theories are excluded in Connes and Lott's approach. Similarly, all left-right symmetric models are excluded, because the constraint (666) forbids spontaneous parity violation. The minimal non-commutative model without degeneracy turns out to be the $SU(2)\times U(1)$ model of weak interactions with *two* generations of leptons:

$$\mathcal{H}_L=\bigoplus_1^2(2,0),$$
$$\mathcal{H}_R=\bigoplus_1^2(1,-1).$$

Comparing with (666), we see that the hypercharges are wrong. They are corrected by the inclusion of strong interactions.

This inclusion requires a new ingredient [**4**], a real structure or — in physical terms — a generalization of charge conjugation to non-commutative geometry. The existence of a real structure implies additional constraints on the fermion representations. The representations (666) of the standard model have four features.

- The weak interaction $SU(2)$ violates parity maximally, it acts only on left-handed fermions.
- The strong interaction $SU(3)$ is vectorial, it acts in the same way on left- and right-handed fermions.
- The scalars transform under $SU(2)$, implying spontaneous breaking of $SU(2)$ that renders its gauge bosons, W^+, W^- and Z, massive.
- The scalars do not transform under $SU(3)$. It remains unbroken and its gauge bosons, the gluons, massless.

In a Yang-Mills-Higgs theory these four features are independent, not so in the non-commutative approach. We already stated that the scalar representation is not chosen and the two last features follow from the two first. On top, the existence of a real structure implies that the first feature implies the second [**5**].

The existence of a real structure is intimately related to another mathematical property, a non-commutative version of Poincaré duality which puts still another constraint on the fermion representations. It turns out that this constraint is fulfilled in the standard model (666). However, slightly modifying $\mathcal{H}_R$ by adding right-handed neutrinos — a modification compatible with all constraints so far [**6**] — violates this additional constraint [**4,5**].

2.3. Coins. In a Yang-Mills-Higgs model, that comes from a Connes-Lott model, the coins cannot be chosen independently. In an arbitrary Yang-Mills-Higgs model the choice of coins is a point in the space of direct products of intervals. In a Connes-Lott model this point must lie in a subspace. This subspace is a submanifold with interesting structure. Depending on the choice of bills, this submanifold may be of the same dimension as its surrounding hypercube or not. Due to the high degree of reducibility of its fermionic Hilbert space, the standard model is in the first case. Its Connes-Lott submanifold is an open subset of its Yang-Mills-Higgs hypercube given by the inequalities

$$m_e < m_W < m_t/\sqrt{3},$$
$$2g_1^{-2} > g_2^{-2} + g_3^{-2}/3$$
$$m_{H\,min} < m_H < m_{H\,max}.$$

The bounds on the Higgs mass are complicated functions of the other coins. The fact that the non-commutative constraints on the parameters of the standard model are inequalities rather than equations may be important to insure their stability under renormalization flow. On the other hand, for the experimental values of the parameters

$$\frac{m_{H\,max} - m_{H\,min}}{(m_{H\,max} + m_{H\,min})/2} \simeq \frac{m_\tau^2 - m_e^2}{m_t^2} \simeq 10^{-4}$$

and for all practical purpose the Higgs mass is fixed. To our knowledge, this is the first mass relation, that comes with a (small) conceptual uncertainty and we call it a fuzzy relation. We stress that the fuzziness of the Higgs mass requires the existence of at least two generations.

3. Conclusion

The first miracle of non-commutative geometry applied to particle physics concerns the general rules. Here, this geometry answers the question: Who ordered the Higgs.

The two subspaces of bills and coins accessible to a Connes-Lott model have interesting structure and are tiny compared to the two Yang-Mills-Higgs hypercubes, fig. 1 and 2. To us, it is a second miracle that the two points defining the standard model fall into these tiny subspaces, at least as long as the Higgs mass is unknown.

4. Figure captions

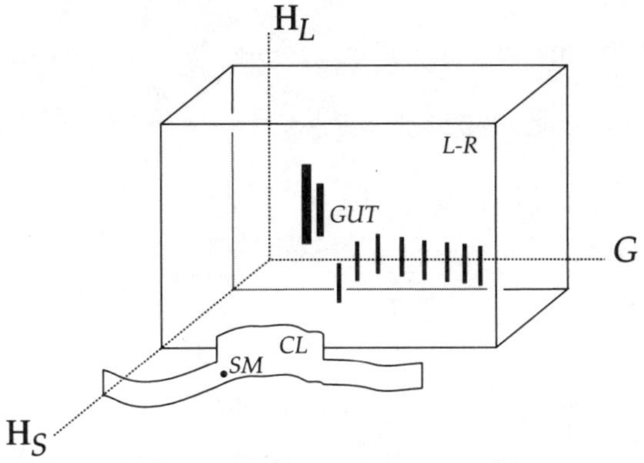

Fig.1: An artist's partial view of the space of bills of all Yang-Mills-Higgs models and some of its subspaces. GUT stands for 'Grand Unified Theories', i.e. G simple. $L-R$ stands for left-right symmetric models, i.e. $\mathcal{H}_L = \mathcal{H}_R$. SM stands for standard model and CL for Connes-Lott models.

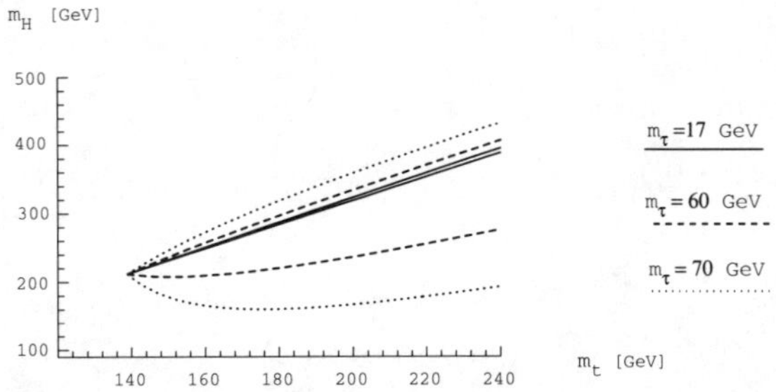

Fig.2: Partial view of the space of coins of the standard model, lower and upper bounds of the Higgs mass as a function of the top and τ masses, all other coins are set to their experimental values. For the experimental value, $m_\tau = 1.8$ GeV, the two bounds differ by 10^{-2} GeV in the indicated range of m_t.

REFERENCES

1. L. Montanet et al., *Review of Particle Properties*, Phys. Rev. D50 1173, 1994.
2. A. Connes, *Noncommutative Geometry*, Academic Press, 1994.
3. A. Connes and J. Lott, *The metric aspect of noncommutative geometry*, in the proceedings of the 1991 Cargèse Summer Conference, eds.: J. Fröhlich et al. (1992), Plenum Press.
4. A. Connes, *Noncommutative geometry and reality*, IHES/M/95/52 (1995).
5. R. Asquith, DTP-95-49, CPT-95/P.3239, hep-th/9509163 (1995).
6. J. M. Gracia-Bondía, *Connes' interpretation of the Standard model and massive neutrinos*, Phys. Lett. B (to appear).

7. D. Testard, *Non-commutative Poincaré duality and right-handed neutrinos* (1995).

CENTRE DE PHYSIQUE THEORIQUE , CNRS - LUMINY, CASE 907, F-13288 MARSEILLE CEDEX

E-mail address: iochum@cpt.univ-mrs.fr schucker@cpt.univ-mrs.fr

ALSO AT UNIVERSITÉ DE PROVENCE AND UNIVERSITÉ DE LA MÉDITERRANNÉE

Contemporary Mathematics
Volume **203**, 1997

Spectral Model and Fuzzy Mass Relations

BRUNO IOCHUM, DANIEL KASTLER AND THOMAS SCHÜCKER

ABSTRACT. We give an introduction to the spectral standard model recently proposed by Alain Connes as an improvement of the Connes-Lott model. We show that the computation of the action, and thus the ensuing fuzzy mass relations obtained in tree-approximation, are essentially unchanged. In particular we find $m_H = 298 \pm 22Gev$.

[A] INTRODUCTION. *K*-CYCLES AND METRIC DUAL PAIRS.

Alain Connes recently replaced the Connes-Lott model by the more harmonious and richer "spectral standard model" [**1**], a move which initiates important developments of the theory (towards quantum and "twisted supersymmetric" invariance). The purpose of the present review is twofold: we first give an introduction to this new development, we then show that our "fuzzy mass relations" yielding a first (tree-level) indication on the mass of the expected Higgs particle [**4**] [**4a**] survive this change of landscape. In fact Connes' new "spectral standard model" is, interestingly, at the same time conceptually innovative and computationally conservative: in particular the derivation of the Yang-Mills and fermionic actions is barely modified and can easily be read off the former computation [**5**] [**5a**], cf. sections [E] [F]; and analogously the novel "adjoint action" of the gauge group practically restitutes the former action leading to the familiar hypercharges, cf. [C.3].

Our paper is organized as follows: in section [B] we describe the "real spectral triples" (d^+-summable K-cycles which feature the last version of "non commutative Poincaré duality" combining the Tomita-Takesaki idea with "charge conjugation" and leading to a "Poincaré self-duality" as one encounters in classical manifolds and electrodynamics). In [B.2] we show that each Poincaré-dual arrangement in the previous sense ("metric dual pair" of [**9**]) can be turned into a (Poincaré self-dual) real spectral triple: thus the algorithm by which Alain

1991 *Mathematics Subject Classification.* Primary 14A22, Secondary 81E13, 47C10.

Connes passes from the Connes-Lott model to the spectral standard model is in fact generic. In sections [C] and [D] we describe this process in detail for the standard model, to the point where it is clear that the computation of the Yang-Mills action, merges into the former computation for the Connes-Lott model. Section [E] states the thus obtained Yang-Mills action. Section [F] describes the tree-level computations and the ensuing constraints on coupling constants and masses maintaining the permanence of our "fuzzy mass relations" [**4**] [**4a**]: under the present knowledge of the top mass this leads us to the tree-approximation value $m_H = 280 \pm 33Gev$ for the Higgs mass.

We conclude this introduction with a short reminder of Alain Connes "metric non-commutative geometry" which underlines both the previous Connes-Lott model and the novel spectral standard model. The basic principle is the fundamental role of the Dirac operator as the carrier of differential geometry. With $\mathbf{M}$ a d-dimensional compact spin manifold, $d = 2m$, $H = L^2(\mathbb{S}_\mathbf{M})$ the Hilbert space of square-integrable sections of its spin bundle $\mathbb{S}_\mathbf{M}$, acted upon by the Dirac operator D, and by the multiplicative action of the functions $a \in \mathbb{A} = C^\infty(\mathbf{M})$, we have the following features:

(i): H is a $\mathbb{Z}/2$-graded Hilbert space (with grading involution χ) provided with a faithful *-representation $a \to \underline{a}$ of the *-algebra $\mathbb{A}$ by bounded even operators;

(ii): D is an odd selfadjoint operator;

(iii): all commutators $[D, \underline{a}]$, $a \in \mathbb{A}$, are bounded;

(iv): the resolvent of D (or equivalently D^{-1} if D has no zero modes) is compact, such that the sum of the N lowest eigenvalues of D^{-d} is $O(\log N)$. Furthermore one has for $p, q \in \mathbf{M}$:

$$\sup\{|a(p) - a(q)|;\ a \in \mathbb{A}, \|[D, \underline{a}]\| \leq 1\} = \text{geodesic distance } \delta(p, q), \tag{A.1}$$

showing that the Dirac operator encodes the whole information of the metric structure.

This striking fact motivates the definition of the differential geometry of (cohomologically d-dimensional) *-algebras $\mathbb{A}$ by means of suitable generalizations of the Dirac operator, the so-called **even d^+-dimensional K-cycles** defined as triples (H, χ, D) possessing all the properties (i) through (iv) above, *which make sense for general *-algebras* $\mathbb{A}$, *commutative or not!* This method turns out to be universal: whilst the known (classical) structures can be reconstructed in this way, a wealth of new structures arises! We sketch how the classical DeRham complex $(\Omega(M), d)$ appears in this landscape. Consider first (for any unital algebra $\mathbb{A}$) the **differential envelope** $\Omega\mathbb{A}$ constructed by means of symbols $a, da, a \in \mathbb{A}$, and natural relations (asking for linearity of da, merging of the product of $\mathbb{A}$ in that of $\Omega\mathbb{A}$, and the Leibniz rule for d, plus $d1 = 0$): $\Omega\mathbb{A}$ is then an algebra generated by the products $\omega = a_0 da_1 \dots da_n$, $\mathbb{N}$-graded by n, and endowed with the differential d coherently defined as $d\omega = da_0 da_1 \dots da_n$. Consider now $\Omega\mathbb{A}$ for $\mathbb{A} = C^\infty(\mathbf{M})$: it appears a priori as a rather loose (free) object carrying little information (in particular acyclic: $d\omega = 0$ entails that $\omega = d\Psi$ for some Ψ). Now the specification of the Dirac operator entails the existence of a bounded representation π_D of $\Omega\mathbb{A}$ on H, namely:

$$\text{(A.2)} \qquad \pi_D(a_0 da_1 \dots da_n) = i^n \underline{a}_0 [D, \underline{a}_1] \dots\dots [D, \underline{a}_n].$$

In this way we have a representation of the algebra $\Omega\mathbb{A}$ on H which is neither graded nor representating as such the differential algebra (since $D^2 \neq 1$!). The kernel $\mathbf{K}$ of π_D is therefore an ideal which is neither graded nor differential. But $\mathbf{K}$ can easily be replaced by a graded differential ideal by taking $\mathbf{K}^* + d\mathbf{K}^*$, with $\mathbf{K}^* = \oplus_{n\in\mathbb{N}} \mathbf{K}^n$, $\mathbf{K}^n = \mathbf{K} \cap \Omega\mathbb{A}^n$. Now the quotient $\Omega_D\mathbb{A} = \Omega\mathbb{A}/\mathbf{K}^* + d\mathbf{K}^*$ is a differential algebra $(\Omega_D\mathbb{A}, d)$ *which turns out as such to be isomorphic to the DeRham complex* $(\Omega(M), d)$: the latter thus appears to be constructible using exclusively the triple algebra-representation-Dirac operator! Moreover, since this construction applies to any algebra equipped with a K-cycle, we have discovered in this context the **non-commutative DeRham complex**, namely $\Omega_D\mathbb{A}$! For the practical handling of $\Omega_D\mathbb{A}$, it is useful to note that

$$\text{(A.3)} \qquad \Omega_D\mathbb{A}^n = \frac{\pi_D(\Omega\mathbb{A}^n)}{d\pi_D(\mathbf{K}^{n-1})} \cdot$$

A further step will now provide us with the equipment for performing Yang-Mills. Indeed, the differential geometry induced by an even d^+-dimensional K-cycle is **metric** in the following sense: we have a "volume form" (a positive trace τ_D of $\Omega\mathbb{A}$) obtained as follows:

$$\text{(A.4)} \qquad \tau_D(\omega) = Tr_\omega\big[D^{-d}\pi_D(\omega)\big], \qquad \omega = a_0 da_1 \dots da_n \in \Omega\mathbb{A},$$

allowing us to perform "integration". The Yang-Mills algorithm consists in "integrating the square of the curvature". The curvature in our non-commutative setting is obtained through an easy paraphrase. Consider the simple case (realized in electrodynamics and in the standard model) where the bundle of interest (or rather the corresponding module) is $\mathbb{A}$ as a right module over itself. A **compatible connection over** $\mathbb{A}$ is then a antihermitean grade-one graded derivation ∇ of $\Omega_D\mathbb{A}$ (concretely realized as $\delta + \rho$, $\rho \in \Omega_D\mathbb{A}^1$ the **connection-one-form**)[1], the corresponding **curvature** being ∇^2 (concretely realized as (multiplication from the left by) $\theta = \delta\rho + \rho^2 \in \Omega_D\mathbb{A}^2 \in \frac{\pi_D(\Omega\mathbb{A}^2)}{d\pi_D(\mathbf{K}^1)}$). If instead of this θ was an element of $\pi_D(\Omega\mathbb{A}^2)$, we would "integrate the square of the curvature" by plugging $\omega = \theta^2$ in (A.3). But we have in addition to effect the division by "the junk" $d\pi_D(\mathbf{K}^1)$: this is the most involved step in the computation, which can be effected either by minimization or by projection, see for details, e.g. [1] or [8], [8a].

---

[1] generalizing the known fact that a connection of $\mathbb{A} = C^\infty(\mathbf{M})$ (= exterior covariant derivative) is a grade-one graded derivation ∇ of the DeRham complex $\Omega(M)$.

[B] REAL SPECTRAL TRIPLES

[B.1]. Alain Connes defines a ***d*-dimensional real spectral triple** $(\mathbf{A}, \mathbf{H}, J)$ as a couple $(\mathbf{A}, \mathbf{H})$ of a *-algebra $\mathbf{A}$ and an even K-cycle $\mathbf{H} = (H, D, \chi)$ with D d^+-summable, $d = 2m$, moreover equipped with a **real structure modulo 8**, i.e. a antilinear operator J of H such that

(i): $JD = DJ$
(ii): $J^*J = 1\!\!1$
(iii): $J^2 = \varepsilon 1\!\!1$ (in view of (i) replaceable by $J^* = \varepsilon J$)
(iv): $J\chi = \varepsilon' \chi J$
(v): $J\underline{b}J^{-1}$, $b \in \mathbf{A}$, commutes with $\underline{a}$ and $[D, \underline{a}]$, $a \in \mathbf{A}$ (hence with $\pi_D(\Omega\mathbf{A})$).

Here ε and ε' are given as follows in terms of d:

(B.1)

d	0	2	4	6
ε	1	-1	-1	1
ε'	1	-1	-1	-1.

The archetype of real spectral triples is the "classical case" obtained by endowing the metric quantum space $(C^\infty, \mathbf{H})$, $C^\infty = C^\infty(\mathbf{M})$, $\mathbf{H} = \big(L^2(\mathbf{S_M}), D, \gamma^{2d+1}\big)$, $\mathbf{S_M}$ the spin bundle of a d-dimensional compact spin manifold $\mathbf{M}$, $d = 2m$, with D the Dirac operator, and the real structure $J = C$ provided by the charge conjugation (of euclidean electrodynamics).

Note that, with $(\mathbf{A}, \mathbf{H}, J)$ any real spectral triple, $\mathbf{H}$ becomes a $\mathbf{A}$-bimodule, by setting:

$$a\Psi b = \underline{a}J\underline{b}^*J^{-1}\Psi = J\underline{b}^*J^{-1}\underline{a}\Psi, \qquad a, b \in \mathbf{A},\ \Psi \in \mathbf{H}, \tag{B.2}$$

[B.2]. Here is a general procedure, basic for the sequel[2], for constructing real spectral triples: each metric dual pair $(\mathbf{B}' \otimes \mathbf{B}'', (H, D, \chi))$ in the sense of [A.2] with $\mathbf{B}' = \mathbf{A}_0 \oplus \mathbf{A}'$, $\mathbf{B}'' = \mathbf{A}_0 \oplus \mathbf{A}''$, is turned through the following construction into a real spectral triple $(\mathbf{A}, (\mathbf{H}, \mathbf{D}, \chi), J)$ with $\varepsilon = \varepsilon' = 1$: the Hilbert space is the direct sum $\mathbb{H} = H \oplus \overline{H}$ of H and its conjugate Hilbert space $\overline{H}$, J is defined as is the antilinear involution $J(\xi, \overline{\eta}) = (\eta, \overline{\xi})$, $\xi, \eta \in H$, the algebra is chosen as the direct sum $\mathbf{A} = \mathbf{A}_0 \oplus \mathbf{A}' \oplus \mathbf{A}''$ (with only *one* summand $\mathbf{A}_0$!), and we set

$$\begin{cases} \chi = (\chi \overline{\oplus} \chi) \\ \pi(a_0, a', a'') = \pi'(a_0, a') \overline{\oplus} \pi''(a_0, a'') \\ \mathbf{D} = (D \overline{\oplus} D) \end{cases}, \tag{B.3}$$

using the shorthand $(S \overline{\oplus} T)$, $S, T \in EndH$ for the element $S \oplus 0 + J(T \oplus 0)J$ of $End\mathbb{H}$.

---

[2]This procedure will be applied in section [C] to the Connes-Lott inner metric dual pair $\mathbf{B}' = \mathbb{C} \oplus \mathbb{H}$, $\mathbf{B}'' = \mathbb{C} \oplus M_3(\mathbb{C})$, $\mathbf{A}_0 = \mathbb{C}$, for turning the latter into the inner space of the "spectral standard model" based on the single algebra $\mathbf{A} = \mathbb{C} \oplus \mathbb{H} \oplus M_3(\mathbb{C})$.

[B.3]. The **gauge group** $\mathcal{G}$ of the real spectral triple $(\mathbf{A}, \mathbf{H}, J)$ is defined as the group of unitaries of $\mathbf{A}$:

$$\mathcal{G} = \{u \in \mathbf{A}; u^* u = uu^* = \mathbb{1}\}, \tag{B.4}$$

with its **adjoint action**[3] $u \to \mathrm{ad}u$ on H given by:

$$(\mathrm{ad}u)\Psi = {}^u\Psi = \underline{u}J\underline{u}J^*\Psi (= J\underline{u}J^*\underline{u}\Psi), \qquad u \in \mathcal{G},\ \Psi \in \mathbf{H}. \tag{B.5}$$

The subspace $\mathbf{H}^J$ of **Majorana spinors** is by definition that of fixpoints under J. $\mathcal{G}$ acts as follows on the connection forms A:

$${}^uA = \underline{u}A\underline{u}^* + \underline{u}[D, \underline{u}^*], \qquad u \in \mathcal{G}. \tag{B.6}$$

Defining the **fermionic action** of real spectral triples as the following functional of $\Psi \in \mathbf{H}$ and the connection one-form A:

$$I_F(\Psi, A) = \left(\Psi, (D + A + JAJ^*)\Psi\right), \tag{B.7}$$

we then have gauge invariance of I_F: $I_F(\Psi, A) = I_F({}^u\Psi, {}^uA)$.

[B.4]. Given real spectral triples $(\mathbf{A}_i, \mathbf{H}_i, J_i)$, $\mathbf{H}_i = (H_i, D_i, \chi_i)$, $i = 1, 2$, $d_i = 2m_i$, with $\varepsilon_1 = \varepsilon_1' = 1$, one obtains a $(d_1 + d_2)$-dimensional real spectral triple $(\mathbf{A}, \mathbf{H}, J)$, $\mathbf{H} = (H, D, \chi)$, called their **tensor product** by specifying:

$$\begin{cases} \mathbf{A} = \mathbf{A}_1 \otimes \mathbf{A}_2 \\ H = H_1 \otimes H_2 \\ \chi = \chi_1 \otimes \chi_2 \\ \underline{a_1 a_2} = \underline{a}_1 \otimes \underline{a}_2, \qquad a_1 \in \mathbf{A}_1,\ a_2 \in \mathbf{A}_2 \\ D = D_1 \otimes id_2 + \chi_1 \otimes D_2 \\ J = J_1 \otimes J_2 \\ \varepsilon = \varepsilon_2, \quad \varepsilon' = \varepsilon_2'. \end{cases} \tag{B.8}$$

[C] SPECTRAL INNER SPACE

We now apply the procedure [B.2] to the Connes-Lott inner metric dual pair $\mathbf{B}' = \mathbb{C} \oplus \mathbb{H}$, $\mathbf{B}'' = \mathbb{C} \oplus M_3(\mathbb{C})$, $\mathbf{A}_0 = \mathbb{C}$, turning the latter into the inner space of the spectral standard model, a real spectral triple $(\mathbf{A}, \mathbf{H}, J)$ based on the algebra $\mathbf{A} = \mathbb{C} \oplus \mathbb{H} \oplus M_3(\mathbb{C})$. We first recall the inner space of the Connes-Lott theory [3], [6] based on the Poincaré-dual pairs of algebras $\mathbf{A}_{ew} = \mathbb{C} \oplus \mathbb{H}$ and $\mathbf{B}_{chrom} = \mathbb{C} \oplus M_3(\mathbb{C})$.[4]

[3] Terminology justified by (B.2). Note that in this "spectral formalism" the fermions transform according to the adjoint rather than to the fundamental representation of the gauge group. One checks that ${}^v({}^u\Psi) = {}^{vu}\Psi$, $({}^u\Psi, {}^u\Psi') = (\Psi, \Psi')$, and ${}^u(J\Psi) = J({}^u\Psi)$, $\Psi, \Psi' \in H$, $u \in \mathcal{G}$, thus leaves stable the subset of Majorana spinors.

[4] ew stands for "electroweak", chr for "chromodynamic".

[C.0] The leptonic and quarkonic $\mathbf{A}_{ew}$-*K*-cycles. We shall throughout identify as follows the real algebra $\mathbb{C}$ of complex numbers with the real algebra $\mathbb{H}_{diag}$ of diagonal quaternions (the context indicating the relevant object):

$$\mathbb{C} \ni p = \begin{pmatrix} \overline{p} & 0 \\ 0 & p \end{pmatrix} \in \mathbb{H}_{diag}. \tag{C.0}$$

Quarkonic K-cycle (H_q, D_q, χ_q). The Hilbert space is:[5]

$$H_q = (\underset{u_R\ d_R}{\mathbb{C}^2_R} \oplus \underset{u_L\ d_L}{\mathbb{C}^2_L}) \otimes \mathbb{C}^N \tag{C.1}$$

with its endomorphisms specified as 2×2 matrices corresponding to the decomposition $R - L$, with entries in $M_2(\mathbb{C}) \otimes M_N(\mathbb{C})$. One then takes respectively for the grading involution (parity), the representation of $\mathbf{A}_{ew}$, and the generalized Dirac operator:

$$\chi_q = \begin{array}{c} \begin{matrix} R & \qquad L \end{matrix} \\ \begin{pmatrix} \mathbb{1} \otimes \mathbb{1}_N & 0 \\ 0 & -\mathbb{1} \otimes \mathbb{1}_N \end{pmatrix} \end{array} \begin{matrix} R \\ L \end{matrix}, \tag{C.2}$$

$$\pi_q((p,q)) = \begin{array}{c} \begin{matrix} R & \qquad L \end{matrix} \\ \begin{pmatrix} p \otimes \mathbb{1}_N & 0 \\ 0 & q \otimes \mathbb{1}_N \end{pmatrix} \end{array} \begin{matrix} R \\ L \end{matrix}, \tag{C.3}$$

$$D_q = \begin{array}{c} \begin{matrix} R & L \end{matrix} \\ \begin{pmatrix} 0 & \mathbb{M}^* \\ \mathbb{M} & 0 \end{pmatrix} \end{array} \begin{matrix} R \\ L \end{matrix}, \tag{C.4}$$

where $\mathbb{M} = \begin{pmatrix} M_u & 0 \\ 0 & M_d \end{pmatrix}$, M_u a diagonal strictly positive matrix with eigenvalues the masses of the upper, charmed, and top quark; and $M_d = C|M_d|$, $|M_d|$ a diagonal strictly positive matrix with eigenvalues the masses of the lower, strange, and bottom quark, C the unitary **Kobayashi-Maskawa matrix**.[6]

---

[5]type of particle indicated under direct summands, tensorial factor $\mathbb{C}^N$ for the N families of fermions: u and d stand for up, resp. down, R and L stand for right, resp. left.

[6]We comply to the common usage of choosing our fermion mass-matrices such that *M_e and M_u are diagonal positive matrices, whilst $M_d = C|M_d|$, with C* (the **Kobayashi-Maskawa matrix**) *unitary and $|M_d|$ strictly positive.* Furthermore we assume that *all fermion masses are different* (the eigenvalues of M_e, M_u and $|M_d|$ consists of positive numbers (the masses of leptons and quarks) all different from one another - experiment!). We further assume that *no eigenstate of $|M_d|$ is an eigenstate of C* .

Leptonic K-cycle (H_l, D_l, χ_l). The Hilbert space is:[7]

(C.5)
$$H_l = (\underset{e_R}{\mathbb{C}^1_R} \oplus \underset{\nu_L\, e_L}{\mathbb{C}^2_L}) \otimes \mathbb{C}^N$$

with endomorphisms now 2×2 matrices

$$\begin{pmatrix} M_1(\mathbb{C}) \otimes M_N(\mathbb{C}) & M(\mathbb{C}^2, \mathbb{C}) \otimes M_N(\mathbb{C}) \\ M(\mathbb{C}, \mathbb{C}^2) \otimes M_N(\mathbb{C}) & M_2(\mathbb{C}) \otimes M_N(\mathbb{C}) \end{pmatrix},$$

and:

(C.6)
$$\chi_l = \begin{array}{c} \begin{matrix} R & L \end{matrix} \\ \begin{pmatrix} 1\!\!1 \otimes 1 & 0 \\ 0 & -1\!\!1 \otimes 1\!\!1_N \end{pmatrix} \end{array} \begin{matrix} R \\ L \end{matrix}.$$

(C.7)
$$\pi_l((p,q)) = \begin{array}{c} \begin{matrix} R & L \end{matrix} \\ \begin{pmatrix} \overline{p} \otimes 1\!\!1_N & 0 \\ 0 & q \otimes 1\!\!1_N \end{pmatrix} \end{array} \begin{matrix} R \\ L \end{matrix},$$

(C.8)
$$D_l = \begin{array}{c} \begin{matrix} R & L \end{matrix} \\ \begin{pmatrix} 0 & (0 M_e^*) \\ \begin{pmatrix} 0 \\ M_e \end{pmatrix} & 0 \end{pmatrix} \end{array} \begin{matrix} R \\ L \end{matrix},$$

where M_e is a diagonal strictly positive matrix with eigenvalues the masses of the electron, muon, and tau.

We described so far the uncoloured K-cycles (H_l, D_l, χ_l) and (H_q, D_q, χ_q). The **coloured *K*-cycles** $(\underline{H}_l, \underline{D}_l, \underline{\chi}_l)$ and $(\underline{H}_q, \underline{D}_q, \underline{\chi}_q)$ are specified as follows:

(C.9)
$$\begin{cases} \underline{H}_l = H_l \otimes \mathbb{C} \\ \underline{\chi}_l = \chi_l \otimes 1\!\!1 \\ \underline{D}_l = D_l \otimes 1\!\!1 \\ \underline{\pi}_l(p,q) = \pi_l(p,q) \otimes 1\!\!1 \\ \underline{\pi}_l(p', m) = 1\!\!1 \otimes p' = p' \end{cases}, \quad \begin{cases} \underline{H}_q = H_q \otimes \mathbb{C}^3, & \underline{\chi}_q = \chi_q \otimes 1\!\!1 \\ \underline{D}_q = D_q \otimes 1\!\!1 & \\ \underline{\pi}_q(p,q) = \pi_q(p,q) \otimes 1\!\!1, & (p,q) \in \mathbb{C} \oplus \mathbb{H} \\ \underline{\pi}_q(p', m) = 1\!\!1 \otimes m, & (p', m) \in \mathbb{C} \oplus M_3(\mathbb{C}) \end{cases}$$

and we shall henceforth be concerned with the direct-sum K-cycle $(\underline{H}, \underline{D}, \underline{\chi}) = (\underline{H}_l, \underline{D}_l, \underline{\chi}_l) \oplus (\underline{H}_q, \underline{D}_q, \underline{\chi}_q)$ with data

(C.10)
$$\begin{cases} \underline{H} = \underline{H}_l \oplus \underline{H}_q \text{ with grading involution } \underline{\chi} = \underline{\chi}_l \oplus \underline{\chi}_q \\ \underline{\pi} = \underline{\pi}_l \oplus \underline{\pi}_q \\ \underline{D} = \underline{D}_l \oplus \underline{D}_q \end{cases},$$

---

[7]type of particle indicated under direct summands, tensorial factor $\mathbb{C}^N$ for the N families of fermions: u and d stand for up, resp. down, R and L stand for right, resp. left.

Remark. One passes from the matrices $(C, 2, 3, 4)$ to the matrices $(C, 6, 7, 8)$ through the changes $M_u \to 0$, $M_d \to M_e$ followed by restriction to the right-lower corner 3×3 matrix. This procedure applied to a 3×3 depending upon M_u and M_d is called **leptonic reduction**.

We now specify the inner space of the spectral standard model, according to [B.2] a real spectral triple.

[C.1] The inner-space real spectral triple. The **algebra** identified with its faithful representation is now the direct sum:

$$\mathbf{A} = \mathbb{C} \oplus \mathbb{H} \oplus M_3(\mathbb{C}) = \{(p, q, m);\ p \in \mathbb{C},\ q \in \mathbb{H},\ m \in M_3(\mathbb{C})\} \tag{C.11}$$

(thus with gauge group $U(1) \times SU(2) \times U(3)$:

$$\mathcal{G} = \{\boldsymbol{u} = (u, v, w) \in \mathbf{A}; u^*u = uu^* = 1\!\!1, v^*v = vv^* = 1\!\!1, w^*w = ww^* = 1\!\!1\} \tag{C.12}$$

The **inner-space real spectral triple** is then $(\mathbf{A}, (\mathcal{H}, \chi, \mathcal{D}), J)$, where the Hilbert space $\mathcal{H}$ is the direct sum of $\underline{H}$ and its conjugate Hilbert space $\underline{\overline{H}}$:

$$\mathcal{H} = \underline{H} \oplus \underline{\overline{H}}, \tag{C.13}$$

and the K-cycle $(\mathcal{H}, \chi, \mathcal{D})$ is specified as follows: one has with $\Psi = (\xi, \overline{\eta}) \in \mathcal{H}$:[8]

$$\chi = (\underline{\chi} \overline{\oplus} \underline{\chi}) \qquad \text{i.e.} \qquad \chi(\xi, \overline{\eta}) = (\underline{\chi}\xi, \overline{\underline{\chi}\eta}) \tag{C.14}$$

$$\begin{gathered}(p, q, m) = ((p, q) \overline{\oplus} (p, m)) \qquad \text{i.e.} \\ (p, q, m)(\xi, \overline{\eta}) = ((p, q)\xi, \overline{(p, m)\eta}),\ (p, q, m) \in \mathbf{A},\end{gathered} \tag{C.15}$$

$$\mathcal{D} = (\underline{D} \overline{\oplus} \underline{D}) \qquad \text{i.e.} \qquad \mathcal{D}(\xi, \overline{\eta}) = (\underline{D}\xi, \overline{\underline{D}\eta}), \tag{C.16}$$

the real structure being:

$$J(\xi, \overline{\eta}) = (\eta, \overline{\xi}). \tag{C.17}$$

(note that we have:

$$JaJ(\xi, \overline{\eta}) = ((p, m)\xi, \overline{(p, q)\eta}), \qquad a = (p, q, m) \in \mathbf{A})). \tag{C.18}$$

[C.3] Action of the gauge group, weak isotopic spin and hypercharge. The ensuing adjoint action of the gauge group on H is as follows: one has for $(u, v, w) \in \mathcal{G}$:

$$\operatorname{ad}(u, v, w)(\xi, \overline{\eta}) = ((u, v)(u, w)\xi, \overline{(u, v)(u, w)\eta}), \qquad (\xi, \overline{\eta}) \in \mathcal{H}, \tag{C.19}$$

---

[8] We use the notation $(S \overline{\oplus} T)$ introduced earlier (cf. (B.3)), and denote throughout by the same symbol (p, q, m) elements of $\mathbf{A}$ and their action on $\mathcal{H}$.

(corresponding to the matrix:

(C.19a)
$$\begin{pmatrix} \begin{pmatrix} \overline{u} & 0 \\ 0 & u \end{pmatrix} \otimes \mathbb{1}_N & 0 \\ 0 & v \otimes \mathbb{1}_N \end{pmatrix} \otimes w \quad \oplus \quad \begin{pmatrix} u \otimes \mathbb{1}_N & 0 \\ 0 & v \otimes \mathbb{1}_N \end{pmatrix}$$
(columns labelled $u_R\ d_R\ u_L\ d_L$ and $e_R\ \nu_L\ e_L$)

acting on particles – for antiparticles one has the complex-conjugate matrix). The infinitesimal generators of the one-parameter subgroups $t \to (\mathbb{1}, e^{iVt}, \mathbb{1})$, $\to (e^{iUt}, \mathbb{1}, \mathbb{1})$, $t \to (\mathbb{1}, \mathbb{1}, e^{iWt})$ are thus as follows:

(C.20)

	u_R	d_R	u_L	d_L	e_R	ν_L	e_L
ad$V$	0	0	1	-1	0	1	-1
ad$U$	-1	1	0	0	2	1	1
ad$W$	1	1	1	1	0	0	0.

Compared this with the following values of T_3 (**weak isotopic spin**) and Y (**hypercharge**) (we also plotted the **charge** $Q = T_3 + \frac{1}{2}Y$):

(C.21)

	u_R	d_R	u_L	d_L	e_R	ν_L	e_L
T_3	0	0	$\frac{1}{2}$	$-\frac{1}{2}$	0	$\frac{1}{2}$	$-\frac{1}{2}$
Y	$\frac{4}{3}$	$-\frac{2}{3}$	$\frac{1}{3}$	$\frac{1}{3}$	-2	-1	-1
Q	$\frac{2}{3}$	$-\frac{1}{3}$	$\frac{2}{3}$	$-\frac{1}{3}$	-1	0	-1

yields $T_3 = \frac{1}{2}\mathrm{ad}V$, $Y = \mathrm{ad}U - \frac{1}{3}\mathrm{ad}W$. Thus the "coalescence subgroup" $\{(u, \mathbb{1}, u^{1/3}); u \in \mathbb{C}\}$ of $\mathcal{G}$ yields the wanted gauge-group $U(1) \times SU(2) \times U(3)$.

[C.4]. Here is an operator $\boldsymbol{\Gamma}$ later to be used as a generalized coupling constant, cf. (E.1), depending upon positive scalars $\alpha_l, \alpha_q, \alpha_l', \alpha_q'$, and positive functions Γ_N, Γ_N' of $\mu_e = M_e^2$ such that $Tr_N \Gamma_N = tr\Gamma_N' = N$. The matrix-form of the general element $\boldsymbol{\Gamma}$ of $\{\pi_{\mathcal{D}}(\mathbf{A}), \mathcal{D}\}' \cap \{J\pi_{\mathcal{D}}(\mathbf{A})J, \mathcal{D}\}'$ is:

(C.22)
$$\boldsymbol{\Gamma} = \alpha_q \begin{pmatrix} \mathbb{1}_2 \otimes \mathbb{1}_N & 0 \\ 0 & \mathbb{1}_2 \otimes \mathbb{1}_N \end{pmatrix} \otimes \mathbb{1}_3 \quad \oplus \quad \alpha_l \begin{pmatrix} 1 \otimes \Gamma_N & 0 \\ 0 & \mathbb{1}_2 \otimes \Gamma_N \end{pmatrix}$$
(columns labelled $u_R\ d_R\ u_L\ d_L$ and $e_R\ \nu_L\ e_L$)

$$\oplus \quad \alpha_q' \begin{pmatrix} \mathbb{1}_2 \otimes \mathbb{1}_N & 0 \\ 0 & \mathbb{1}_2 \otimes \mathbb{1}_N \end{pmatrix} \otimes \mathbb{1}_3 \quad \oplus \quad \alpha_l' \begin{pmatrix} 1 \otimes \Gamma_N' & 0 \\ 0 & \mathbb{1}_2 \otimes \Gamma_N' \end{pmatrix}.$$
(columns labelled $\overline{u}_R\ \overline{d}_R\ \overline{u}_L\ \overline{d}_L$ and $\overline{e}_R\ \overline{\nu}_L\ \overline{e}_L$)

[D] THE FULL SPECTRAL TRIPLE. QUANTUM POTENTIALS AND THEIR CURVATURES

We now tensor the zero-dimensional real spectral triple $(\mathbf{A}, (\mathcal{H}, \chi, \mathcal{D}), J)$ described in the preceding section [C] by the space-time (classical) real spectral triple $(C^\infty(\mathbf{M}),(L^2(\mathbf{S_M}),\gamma^5,D),C)$ where space-time is a compact riemannian spin 4-manifold $\mathbf{M}$.

The **full spectral algebra** $\mathcal{A}$ is the tensor product of the space-time algebra $C^\infty(\mathbf{M})$ by the inner-space algebra $\mathbf{A}=\mathbb{C}\oplus\mathbb{H}\oplus M_3(\mathbb{C})$:

$$\begin{aligned}\mathcal{A}&=C^\infty(\mathbf{M})\otimes\mathbf{A}\\&=\big\{(p,q,m);\ p\in C^\infty(\mathbf{M},\mathbb{C}),q\in C^\infty(\mathbf{M},\mathbb{H}),m\in C^\infty(\mathbf{M},M_3(\mathbb{C}))\big\}\end{aligned}\tag{D.1}$$

with gauge group:

$$\begin{aligned}U(\mathcal{A})&=C^\infty(\mathbf{M},U(\mathbf{A}))\\&=\big\{(u,v,w);\ u\in C^\infty(\mathbf{M},U(1)),\\&\qquad v\in C^\infty(\mathbf{M},SU(2)),w\in C^\infty(\mathbf{M},U(3))\big\}.\end{aligned}\tag{D.2}$$

The **full real spectral triple** is the tensor product of the zero-dimensional inner real spectral triple $(\mathbf{A},\mathbf{H},J)$ (cf. [B4]) by the classical real spectral triple where $d=4$. As such it is a 4-dimensional real spectral triple.

These definitions lead to the following quantum potentials and curvatures, in matrix form:

$V=i\rho(\mathbf{a},\mathbf{b},\mathbf{H},\mathbf{c}_0,\mathbf{c}^{\cdot})\in i\Omega_{\mathbb{D}}\mathcal{A}^1$ has the matrix:

(D.3)

$$i\rho=(\mathbf{a},\mathbf{b}^{\cdot}_{\cdot},\mathbf{H},\mathbf{c}_0,\mathbf{c}^{\cdot})$$

$$=\begin{pmatrix}\underline{\gamma}(\mathbf{a}^{\cdot}_{\cdot})\otimes 1\!\!1_N & \mathbb{M}^*(\gamma^5\underline{\mathbf{H}}^*\otimes 1\!\!1_N)\\ (\gamma^5\underline{\mathbf{H}}\otimes 1\!\!1_N)\mathbb{M} & \underline{\gamma}(\mathbf{b}^{\cdot}_{\cdot})\otimes 1\!\!1_N\end{pmatrix}_{u_R\ d_R\ |\ u_L\ d_L}\oplus\ \underset{e_R\quad\nu_L\quad e_L}{\text{Leptonic reduction of latter}}$$

$$\oplus\ \underset{\overline{u}_R\ \overline{d}_R\quad \overline{u}_L\ \overline{d}_L}{\tfrac{i}{2}\gamma(\mathbf{c}_0)1\!\!1_2\otimes 1\!\!1_N\otimes 1\!\!1_3+\tfrac{i}{2}\gamma(\mathbf{c}^a)1\!\!1_2\otimes 1\!\!1_N\otimes\tfrac{\lambda_a}{2}}\ \oplus\ \underset{\overline{e}_R\quad\overline{\nu}_L\ \overline{e}_L}{\tfrac{i}{2}\gamma(\mathbf{a})1\!\!1_3\otimes 1\!\!1_N}$$

(D.4)

$$\boldsymbol{\theta}=\begin{pmatrix}\Big[\tfrac{i}{2}\gamma(\mathbf{f})1\!\!1-(\alpha_l+6\alpha_q)^{-1}N^{-1}Lv_\Phi 1\!\!1\Big]\otimes 1\!\!1_N+v_\Phi\mathbb{M}^*\mathbb{M} & \mathbb{M}^*[\gamma^5\underline{\gamma}(i\mathbf{D}\Phi^*)\otimes 1\!\!1_N]\\ [\gamma^5\underline{\gamma}(i\mathbf{D}\Phi)\otimes 1\!\!1_N] & \mathbb{M}\Big[\tfrac{i}{2}\gamma(\mathbf{h}^{\cdot}_{\cdot})-(\alpha_l+3\alpha_q)^{-1}(2N)^{-1}Lv_\Phi 1\!\!1\Big]\otimes 1\!\!1_N+v_\Phi 1\!\!1\otimes\Sigma\end{pmatrix}_{u_R\ d_R\ |\ u_L\ d_L}$$

$$\oplus\ \underset{e_R\quad\nu_L\quad e_L}{\text{Leptonic reduction of latter}}$$

$$\oplus\ \underset{\overline{u}_R\ \overline{d}_R\quad \overline{u}_L\ \overline{d}_L}{\tfrac{i}{2}\gamma(\mathbf{g}_0)1\!\!1_2\otimes 1\!\!1_N\otimes 1\!\!1_3+\tfrac{i}{2}\gamma(\mathbf{g}^a)1\!\!1_2\otimes 1\!\!1_N\otimes\tfrac{\lambda_a}{2}}\ \oplus\ \underset{\overline{e}_R\quad\overline{\nu}_L\ \overline{e}_L}{\tfrac{i}{2}\gamma(\mathbf{f})1\!\!1_3\otimes 1\!\!1_N}$$

where

$$L = Tr\big[3\alpha_q(\mu_u + \mu_d) + \alpha_l \mathbb{1}_N \mu_e\big]. \tag{D.5}$$

Here:

- $\mathbf{a}$ and $\mathbf{c}_0$ are classical hermitean $U(1)$-connection: $\mathbf{a} \in \Omega(\mathbf{M}, \mathbb{C})^1$,
- $\mathbf{b}^{\cdot}_{\cdot}$ is a classical hermitean $SU(2)$-connection: $\mathbf{b}^{\cdot}_{\cdot} = \begin{pmatrix} \mathbf{b}^1_1 \mathbf{b}^1_2 = & \overline{\mathbf{b}}^2_1 \\ \mathbf{b}^2_1 \mathbf{b}^2_2 = & -\overline{\mathbf{b}}^1_1 \end{pmatrix} \in \Omega(\mathbf{M}, i\mathbb{H})^1$,
- $\mathbf{c}^{\cdot} = (\mathbf{c}^a)_{a=1,\dots,8}$ is a hermitean $SU(3)$-connection (the λ_a are the eight Gell-Man matrices).
- $\mathbf{H} \in C^\infty(\mathbf{M}, \mathbb{H})$ is identified with a doublet-field:

$$\mathbf{H} = \begin{pmatrix} \overline{\mathbf{H}}^2 & \mathbf{H}^1 \\ -\overline{\mathbf{H}}^1 & \mathbf{H}^2 \end{pmatrix} \leftrightarrow \mathbf{H}^{\cdot} = (\mathbf{H}^1, \mathbf{H}^2) \leftrightarrow \mathbf{H}_{\cdot} = (\overline{\mathbf{H}}^1, \overline{\mathbf{H}}^2). \tag{D.6}$$

We used the shorthands

$$\gamma(\mathbf{a}^{\cdot}_{\cdot}) = \begin{pmatrix} -\underline{\gamma}(\overline{\mathbf{a}}) & 0 \\ 0 & \underline{\gamma}(\mathbf{a}) \end{pmatrix} \; (\mathbf{a}^{\cdot}_{\cdot} = \begin{pmatrix} -\overline{\mathbf{a}} & 0 \\ 0 & \mathbf{a} \end{pmatrix} \in \Omega(\mathbf{M}, i\mathbb{H}_{diag})^1),$$

$$\underline{\gamma}(\mathbf{b}^{\cdot}_{\cdot}) = \begin{pmatrix} \underline{\gamma}(\mathbf{b}^1_1) & \underline{\gamma}(\mathbf{b}^1_2) \\ \underline{\gamma}(\mathbf{b}^2_1) & \underline{\gamma}(\mathbf{b}^2_2) \end{pmatrix}, \text{ and}$$

$$\begin{cases} \underline{\gamma}(\mathbf{f}^{\cdot}_{\cdot}) = \begin{pmatrix} -\underline{\gamma}(\mathbf{f}) & 0 \\ 0 & \underline{\gamma}(\mathbf{f}) \end{pmatrix}, & \mathbf{f} = \mathbf{da} \in C^\infty(\mathbf{M}, \mathbb{C}), \\ \underline{\gamma}(\mathbf{h}^{\cdot}_{\cdot}) = \begin{pmatrix} \underline{\gamma}(\mathbf{h}^1_1) & \underline{\gamma}(\mathbf{h}^1_2) \\ \underline{\gamma}(\mathbf{h}^2_1) & \underline{\gamma}(\mathbf{h}^2_2) \end{pmatrix}, & \mathbf{h}^j_k = \mathbf{db}^j_k - i\mathbf{b}^j_m \wedge \mathbf{b}^m_k, ^9 \\ \Phi = \mathbf{H} + \mathbb{1}, \; v_\Phi = \Phi_i \Phi^i - 1, & \mathbf{H} \in C^\infty(\mathbf{M}, \mathbb{H}), \\ \mathbf{D}\Phi = \mathbf{d}\Phi + i(\mathbf{a}\Phi - \mathbf{b}\Phi), \text{ i.e. } \mathbf{D}\Phi^j = \mathbf{d}\Phi^j + i(\mathbf{a}\Phi^j - \mathbf{b}^j_k \Phi^k) & \\ \mathbf{D}\Phi_j = \mathbf{d}\Phi_j + i(\mathbf{a}\Phi_j - \mathbf{b}^k_j \Phi_k) = \overline{\mathbf{D}\Phi^j}. & \end{cases} \tag{D.7}$$

These formulae, where the "electroweak" items act on H and the "chromodynamics" items on $\overline{H}$, make it clear that the computation of the Yang-Mills action merges into that of the Connes-Lott model.

[E] YANG-MILLS ACTION

[E.1] Proposition. *The bosonic action of the spectral standard model, defined as*

$$YM = Tr_\omega(\boldsymbol{\Gamma} \mathbb{D}^{-4} \boldsymbol{\theta}^2) \tag{E.1}$$

[9] i.e. $\mathbf{h}_{\mu\nu} = \partial_\mu \mathbf{b}_\nu - \partial_\nu \mathbf{b}_\mu - i[\mathbf{b}_\mu, \mathbf{b}_\nu]$.

with $\boldsymbol{\Gamma}$ *the generalized coupling constant in* [C.4], *is computed to be*

$$\text{(E.2)}\quad YM = -A\mathbf{g}^a_{\mu\nu}\mathbf{g}^{\mu\nu}_a - B\mathbf{f}_{\mu\nu}\mathbf{f}^{\mu\nu} - \frac{1}{4}C\mathbf{h}^s_{\mu\nu}\mathbf{h}^{\mu\nu}_s + 2L\mathbf{D}_\mu\Phi_j\mathbf{D}^\mu\Phi^j + K(\Phi_i\Phi^i - 1)^2,$$

where:

$$\text{(E.3)}\quad \begin{cases} B = \frac{1}{2}(1+C) + \frac{2}{3}A + y' \\ C = 1 + Try \\ L = Tr(y\mu_e) + \frac{1}{N}Tr(\mu_u + \mu_d) \\ K = \frac{3}{2}Tr(y\mu_e^2) + \frac{1}{N}Tr\left[\frac{3}{2}(\mu_u^2 + \mu_d^2) + \mu_u\mu_d\right] - \frac{L^2}{2}\left(\frac{1}{C} + \frac{1}{B - 2A/3}\right) \end{cases}$$

Here $\mu_e = M_e^2$, $\mu_u = M_u^2$, $\mu_d = M_d M_d^*$, A *and* y' *are positive constants, and* y *is a positive function of* μ_e.[10] Tr *denotes the trace of* $N \times N$ *matrices.*

[F] TREE-LEVEL COMPUTATIONS. FUZZY MASS RELATIONS

Identifying the Yang-Mills action (E.1) of the spectral model with the bosonic part of the traditional full standard model:

$$\text{(F.1)}\quad \begin{aligned} \mathcal{L}_{\text{gauge}} + \mathcal{L}_{\text{Higgs}} = & -\frac{1}{4}G^a_{\mu\nu}G^{\mu\nu}_a - \frac{1}{4}B_{\mu\nu}B^{\mu\nu} - \frac{1}{4}W^s_{\mu\nu}W^{\mu\nu}_s \\ & + (D_\mu\phi)^*(D^\mu\phi) + \frac{\mu^2}{v^2}\left(\phi^*\phi - \frac{v^2}{2}\right)^2, \end{aligned}$$

with identification of the corresponding covariant derivatives yields the relations $\mathbf{c}^a_\mu = g_3 G^a_\mu$, $\mathbf{a}_\mu = -\frac{1}{2}g_1 B_\mu$, $\mathbf{b}^s_\mu = g_2 W^s_\mu$ (implying $\mathbf{g}^a_{\mu\nu} = g_3 G^a_{\mu\nu}$, $\mathbf{f}_{\mu\nu} = -\frac{1}{2}g_1 B_{\mu\nu} = -\frac{1}{2}g_2\cos\theta_W B_{\mu\nu}$, $\mathbf{h}^s_{\mu\nu} = g_2 W^s_{\mu\nu}$) and $\frac{v}{\sqrt{2}}\Phi = \phi$, this leading to the following tree-level evaluations

$$\text{(F.2)}\quad g_3 = \frac{1}{2}(C/A)^{1/2}g_2,$$

$$\text{(F.3)}\quad \frac{g_2^2}{g_1^2} = \text{tang}^{-2}\theta_W = \frac{B}{C} \quad \text{whence} \quad \sin^2\theta_W = \frac{C}{B+C},$$

$$\text{(F.4)}\quad m_W^2 = L/C$$

$$\text{(F.5)}\quad m_H^2 = 2K/L.$$

From this we deduce the following constraints on coupling constants and fuzzy mass relations:

---

[10]the relationship of the notation in (E.2) and the generalized coupling constant $\boldsymbol{\Gamma}$ in [C.4] is $y = \alpha_l\Gamma_N/3N\alpha_q$ (whence $Try = \alpha_l/3\alpha_q$) and $y' = \alpha_l'/2\alpha_q$.

[F.2] Proposition. *From now on we assume $N = 3$.*

(i): *With θ_W the Weinberg angle, we have the constraint:*

$$\sin^2\theta_W < \frac{2}{3}\left[1 + \frac{1}{9}\left(\frac{g_2}{g_3}\right)^2\right]^{-1}. \tag{F.6}$$

(ii): *Assume the inequality $m_\tau^2 < \frac{1}{3}(m_d^2 + m_u^2 + m_s^2 + m_c^2 + m_b^2 + m_t^2)$: one has then:*[11]

$$m_e^2 < m_W^2 < \frac{1}{3}(m_d^2 + m_u^2 + m_s^2 + m_c^2 + m_b^2 + m_t^2) \tag{F.7}$$

(iii): *Assume in addition the inequality $m_\tau < m_W$, and assume that the quark masses belong to the following domain D_0:*

$$D_0 : \begin{cases} u + d < \min\{c, s\} \\ c + s + \min\{c, s\} < \min\{t, b\} \end{cases}, \quad \begin{cases} \min\{t, b\} < \dfrac{1}{1+\varepsilon}\max\{t, b\} \\ \min\{c, s\} < \dfrac{1}{1+\varepsilon}\max\{c, s\} \end{cases}, \tag{F.8}$$

where $\varepsilon := 1 - \min\{|C_{tb}|^2, |C_{cs}|^2, |C_{ud}|^2\}$ measures the deviation of the Cabbibo-Kobayashi-Maskawa matrix from the identity. Then one has the fuzzy mass relation:

$$m_{H\,\min}^2 < m_H^2 < m_{H\,\max}^2 \tag{F.9}$$

with bounds given in (F.19) *below. The "fuzzyness" $m_{H\,\max} - m_{H\,\min}$ factorizes through $\sqrt{(m_\tau^2 - m_e^2)}$ and $\sqrt{(m_d^2 + m_u^2 + m_s^2 + m_c^2 + m_b^2 + m_t^2 - 3m_W^2)}$, and is given as follows if one neglects all fermion masses but m_τ and m_t:*

$$m_{H\,\max} - m_{H\,\min} = \left[k\left(\frac{m_\tau}{m_t}\right)^2 + O\left(\frac{m_\tau}{m_t}\right)^4\right] m_t\,. \tag{F.10}$$

$k = 2.2$ for experimental values of m_τ and m_t and we have

$$m_H^2 \sim 3t - W - \frac{g_2^{-2}}{g_1^{-2} - g_3^{-2}/6} W.$$

<u>Proof:</u>

(i): stems from (F.3) yields by the first line (F.1) and by (F.13):

$$\sin^2\theta_W = \frac{3/2C}{\frac{1}{2}(3 + C) + C\left[1 + \frac{1}{9}\left(\frac{g_2}{g_3}\right)^2\right] + y'}, \tag{F.11}$$

bounded by (F.6) for $y' = 0$.

We pause to introduce some notation, and rewrite the mass relations accordingly. We shall use the symbol of the particles to denote the square of their masses: $e = m_e^2$, $\mu = m_\mu^2$, $\tau = m_\tau^2$, $d = m_d^2$, $u = m_u^2$, $s = m_s^2$,

[11] For the left inequality it suffices to assume $m_e^2 < \frac{1}{3}(m_d^2 + m_u^2 + m_s^2 + m_c^2 + m_b^2 + m_t^2)$.

$c = m_c^2$, $b = m_b^2$, $t = m_t^2$, $W = m_W^2$, $H = m_H^2$: with this notation (F.4) and (F.5) read:

$$W = L/C, \tag{F.4a}$$

$$H = 2K/L, \tag{F.5a}$$

whence

$$\frac{H}{W} = \frac{2CK}{L^2} = \frac{2K}{CW^2} \,. \tag{F.12}$$

Using the shorthands:

$$\begin{cases} q = d + u + s + c + b + t = Tr(\mu_u + \mu_d) \\ Q = d^2 + u^2 + s^2 + c^2 + b^2 + t^2 = Tr(\mu_u^2 + \mu_d^2) \\ \alpha_0 = \frac{q}{3W},\ \alpha_1 = \frac{e}{W},\ \alpha_2 = \frac{\mu}{W},\ \alpha_3 = \frac{\tau}{W} \\ Y = \Sigma_{i=0,1,2,3}\ \alpha_i^2 y_i \\ E = \frac{Q}{W^2} + \frac{2Tr(\mu_u \mu_d)}{3W^2} - \frac{1q^2}{3W^2} \end{cases} , \tag{F.13}$$

where the y_i, $i = 1, 2, 3$, are the eigenvalues of the matrix y, and we set $y_0 = 1$, and $0 < \alpha_1 < \alpha_2 < \alpha_3 < 1 < \alpha_0$ owing to $m_\tau < m_W$, the mass relations (F.4) and (F.5), equivalent to (F.4a) and (F.12), then read:

$$C(= 1 + y_1 + y_2 + y_3) = \Sigma_{i=0,1,2,3}\ \alpha_i y_i \quad (\Leftrightarrow \Sigma_{i=0,1,2,3}(1 - \alpha_i) y_i = 0), \tag{F.14}$$

$$\frac{H}{W} = \frac{E}{C} + \frac{3Y}{C} - \Big(1 + \frac{C}{B - 2A/3}\Big) - 1 = f(C, Y), \tag{F.15}$$

as follows from the two last lines of (E.2) rewritten as

$$\begin{cases} L = \frac{1}{3} q + e y_1 + \mu y_2 + \tau y_3 \\ 2K = Q + 3(e^2 y_1 + \mu^2 y_2 + \tau^2 y_3) + \frac{2}{3} Tr(\mu_u \mu_d) - L^2 \Big(\frac{1}{C} + \frac{1}{B - 2A/3}\Big) \end{cases} . \tag{F.16}$$

(ii): the right inequality follows from $1 - \alpha_0 = 1 - \frac{q}{3W} < 0$, the left inequality from the first line (F.16) and $\frac{1}{3} q > e$.

(iii): With

$$\begin{cases} a_i = \frac{\alpha_i - \alpha_3}{1 - \alpha_3}, & b_i = a_i c_i, & c_i = \alpha_i + \alpha_3 - \alpha_i \alpha_3, & i = 0, 1, 2, \\ a'_j = \frac{\alpha_i - \alpha_1}{1 - \alpha_1}, & b'_j = a'_j c'_j, & c'_j = \alpha_j + \alpha_1 - \alpha_j \alpha_1, & j = 0, 2, 3, \end{cases} \tag{F.17}$$

(giving rise to the inequalities $a_0, b_0, c_0, a'_0, b'_0, c'_0 > 1$; $c_1, c_2 > 0$, $a_1, a_2, b_1, b_2 < 0$; $a'_1 a'_2, b'_1, b'_2, c'_1, c'_2 > 0$; $b_0 > b'_0 > a'_0$) the functions C and Y of the y_i, expressible as

$$\text{(F.18)} \qquad \begin{cases} C = a_0 + a_1 y_1 + a_2 y_2 = a'_0 + a'_1 y_1 + a'_3 y_3 \\ Y = b_0 + b_1 y_1 + b_2 y_2 = b'_0 + b'_2 y_2 + b'_3 y_3 \end{cases},$$

uncorrelated owing to the non-vanishing of the functional determinant $\frac{D(C,Y)}{D(y_1,y_2)}$, range independently in the intervals $a'_0 < C < a_0$, $b'_0 < Y < b_0$. Since the function f in (F.15) increases in X and decreases in C, we have the fuzzy mass relation:

$$\text{(F.19)} \qquad f(a_0, b'_0) < \frac{H}{W} < f(a'_0, b_0).$$

References

1. A. CONNES, *Noncommutative geometry and reality*, IHES/95/52 (1995).
2. A. CONNES, *Noncommutative geometry*, Academic Press (1994).
3. A. CONNES, J. LOTT, *The metric aspect of noncommutative geometry (1992)*, Proceedings of the 1991 Cargèse Summer Conference, eds.: J. Fröhlich et al., Plenum Press (1992).
4. B. IOCHUM, D. KASTLER, T. SCHÜCKER, *Fuzzy mass relations in the standard model*, CPT-95/P.3235, hep-th/9507150 (1995).

4a. B. IOCHUM, D. KASTLER, T. SCHÜCKER, *Fuzzy mass relations for the Higgs*, CPT-95/P.3197, hep-th/9506044 (1995) (to appear in J. of Math. Phys).

5. D. KASTLER, T. SCHÜCKER, *A detailed account of Alain Connes' version of the standard model in non-commutative differential geometry* IV, CPT-94/P.3092, hep-th/9501077 (to appear in Rev. Math. Phys).

5a. D. KASTLER, T. SCHÜCKER, *The standard model à la Connes-Lott*, CPT-94/P.3091, hep-th/9412185 (1994).

6. D. KASTLER, *A detailed account of Alain Connes' version of the standard model in non-commutative differential geometry* III, CPT-92/P.2824, in print Rev. Math. Phys.
7. D. KASTLER, D. TESTARD, *Quantum forms of tensor products*, Comm. Math. Phys **155** (1993), 135.
8. D. KASTLER, M. MEBKHOUT, *Lectures on non-commutative geometry and applications to fundamental interactions.*

8a. D. KASTLER, *Lectures on Alain Connes' non-commutative geometry and applications to fundamental interactions. Infinite dimensional geometry. Non-commutative geometry. Operator algebras. Fundamental interactions. First carribbean Spring School of Mathematics and Theoretical Physics*, R. Coquereaux, M. Dubois-Violette, P. Flad eds., World Scientific (1995).

9. D. KASTLER, *Dual pairs of quantum spaces*, CPT-94/P.3056 (to appear in Letters in theoretical physics).
10. T. SCHÜCKER, J.M. ZYLINSKI, *Connes'model building kit*, Journ. of Geom. Phys **16** (1994), 1.
11. B. IOCHUM, T. SCHÜCKER, *Yang-Mills-Higgs versus Connes-Lott*, CPT-94/P.3090, hep-th/9501142 (to appear in Comm. Math. Phys).

CENTRE DE PHYSIQUE THÉORIQUE, UNITÉ PROPRE DE RECHERCHE 7061, CNRS LUMINY, CASE 907

E-mail address: iochum@cpt.univ-mrs.fr, schucker@cpt.univ-mrs.fr

Part V

COSMOLOGY AND GENERAL RELATIVITY

Contemporary Mathematics
Volume **203**, 1997

Not So Simple Universe

M. DEMIAŃSKI

ABSTRACT. We review recent developments in cosmology.

"We may conceive our space to have everywhere a nearly uniform curvature, but that slight variations of the curvature may occur from point to point, and themselves vary with time. These variations of the curvature with the time may produce effects which we not unnaturally attribute to physical causes independent of the geometry of our space"
W. K. Clifford "*The common sense of the exact sciences*".

Cosmology as a respectable branch of science has been established some 70 years ago about 10 years after Albert Einstein formulated his general theory of relativity. At the beginning of 1920's astronomers were not sure about the scale of the Milky Way galaxy. The extent of the Milky Way galaxy and the nature of nebulae were thoroughly discussed at the Great Debate [**1**] before the US National Academy of Sciences in 1920. Particularly intriguing were the high velocities of radial motions of the spiral nebulae discovered in 1913 by Vesto M. Slipher [**2**]. In 1919 new 100 inch telescope went into operation at Mount Wilson Observatory in California. Early in 1920's Edwin Hubble made several crucial discoveries [**3**]. He noticed that many nebulae were rather round and spherical and could hardly be classified as spirals. Using the 100 inch telescope Hubble was able to resolve the nebula in Andromeda into stars finally proving that at least some of the nebulae are large conglomerates of stars. Hubble discovered the realm of galaxies. Soon Hubble noticed several variable stars in the Andromeda galaxy. Using the empirically established period-luminosity relation for Cepheid stars Hubble was able to derive the distance of about 0.9 million light years to the Andromeda galaxy (the true distance to the Andromeda galaxy is 2.28 million light years). This was the final and definite proof that the Andromeda

1991 *Mathematics Subject Classification.* Primary 83F05; Secondary 85A40.

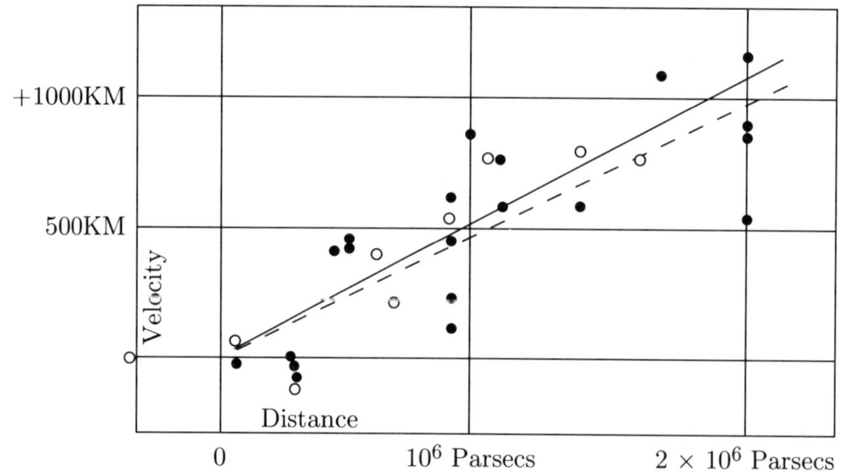

FIGURE 1. The original Hubble diagram of the distance-velocity relation.

galaxy lies far outside the boundaries of our Galaxy. In the next few years using Cepheid stars as distance indicators Hubble was able to determine distances to several galaxies. In 1929 Hubble [**4**] plotted then known distances to galaxies and velocities of recession measured by Slipher [**2**] and Humason [**5**] and established his famous velocity-distance relation.

Hubble insisted that for galaxies their velocity of recession is proportional to their distance as measured relative to our Galaxy. This relation, now known as the Hubble law, can be expressed as

$$v = Hd, \tag{1}$$

where d stands for distance measured in Mpc (1Mpc=10^6pc=3.26 million light years), v is the velocity measured in km/s, and H is the proportionality coefficient known as the Hubble constant. Hubble initially estimated that $H \approx$ 500km/s Mpc, more recent measurements restrict the value of H to 50km/s Mpc $\leq H \leq$ 100km/s Mpc. Because the Hubble constant is so poorly known we will use the following notation $H =$ 100h km/sMpc where h is a coefficient $0.4 \leq \mathrm{h} \leq 1$.

It is worth mentioning that Hubble never fully appreciated his discovery and never accepted the concept of an expending universe.

Parallelly with astronomical discoveries there were theoretical developments. A year after finishing the final paper containing exposition of the general theory of relativity Einstein published a paper "Kosmologische Betrachtungen zür allgemeinen Relativitätstheorie" [**6**]. This was the first paper applying the new general theory of relativity to the problem of structure of the Universe. In this paper Einstein has spelled out his philosophical views concerning the structure of the universe. Let us remind that at that time the Milky Way was the only one known large system of stars. The Universe as a whole was assumed to be static and populated only by stars. There were hypotheses not substantiated by observations that many structures like our Milky Way exist and they all form so called island Universes.

To solve the set of fundamental equations describing the local geometry of the Universe Einstein had to make several assumptions. Einstein correctly realized that the Universe at a large scale should be simple and he assumed that the Universe is spatially homogeneous. 4 dimensional physical spacetime is spatially homogeneous if it admits 3 dimensional isometry group acting transitively on spacelike sections. Actually to simplify considerations further Einstein made implicitly even stronger assumption demanding that the spacetime should be not only spatially homogeneous but also spatially isotropic. Spacetime is spatially isotropic if it admits 6 dimensional isometry group acting on spacelike sections. In such a space not only all points but also all directions are equivalent.

In his paper Einstein makes another strong assumption that the Universe is spatially closed and its volume is finite. Einstein also inherited another philosophical idea namely he viewed the whole Universe as a static system. Local evolution, for example within the Milky Way, was allowed but the Universe as a whole was assumed to be static and eternal. Einstein believed so strongly that the Universe is static that in order to obtain static cosmological model he changed the basic field equations by adding so called cosmological constant. In the Einstein model the cosmological constant plays a role of a constant negative effective pressure counter balancing the effect of gravitational attraction.

There was also another reason of introducing the cosmological term. Einstein was strongly influenced by Mach and he thought that spacetime can exist only when matter is present and there should be no solutions to his field equations when the density of matter is zero.

Almost simultaneously with the first Einstein cosmological paper Willem de Sitter published his solution of the Einstein equations [**7**]. De Sitter also assumed that the spacetime is spatially isotropic and static. De Sitter kept the cosmological term but he considered empty universe with $p = \varrho = 0$. De Sitter model is a counterexample to the Einstein hypothesis that the general relativistic field equations with the cosmological term possess non trivial solutions only when the density of matter is different from zero. Einstein immediately realized that de Sitter was right and admitted that the introduction of the cosmological constant was a mistake.

Early in 1920's Russian mathematician and meteorologist Alexander Friedman showed that Einstein field equations have a simple solution describing non stationary universe [**8**]. Constructing his cosmological model Friedman assumed that general theory of relativity is the correct relativistic theory of gravitation and that matter in the universe is distributed homogeneously and isotropically. In his first paper published in 1921 Friedman considered only closed universe, three years later he discussed also the open universe.

In 1932 Einstein and de Sitter [**9**] completed the catalogue of possible isotropic universes and showed that evolving Universe can possess flat Euclidian (zero curvature) space.

In 1927 after the expansion of the universe was observationally established George Lemaître using Friedman results constructed the first cosmological model of an evolving universe [**10**]. He identified galaxies as the most important constituents of the Universe and derived the linear relation between the red shift (velocity) of a galaxy and its distance. Lemaître realized that the mean density of matter was larger at earlier epochs and assumed that the Universe began its evolution from a very dense state which he called the primeval atom.

It was later formally proved by Robertson and Walker [**11**] that the general line element of a 4-dimensional spacetime admitting homogeneous and isotropic spatial sections can be written in the form

$$ds^2 = c^2dt^2 - R^2(t)(\frac{dr^2}{1-kr^2} + r^2(d\theta^2 + \sin^2\theta d\varphi^2)), \tag{2}$$

where $R(t)$ is an arbitrary function of time - so called scale factor of the universe, and k is a parameter which can be reduced to $+1$, 0, or -1. When $k = 0$, $t = \text{const}$ sections have Euclidean geometry and the universe is called flat. When $k = +1$, $t = \text{const}$ sections have geometry of S^3, and the universe is closed. When $k = -1$ $t = \text{const}$ sections have geometry of 3 dimensional pseudosphere. $k = 0$ and $k = -1$ cases represent an open, infinite universe. More complicated topologies compatible with the metric (2) are also possible.

The Einstein field equations for the metric (2) reduce to

$$\frac{\dot{R}^2}{R^2} = \frac{8\pi G}{3}\varrho - \frac{kc^2}{R^2} + \frac{\Lambda}{3}, \tag{3}$$

$$2\frac{\ddot{R}}{R} + \frac{\dot{R}^2}{R^2} + \frac{8\pi G}{c^2}p = -\frac{kc^2}{R^2} + \Lambda, \tag{4}$$

where ϱ is the mean density of matter, p its pressure, and Λ is the cosmological constant.

The integrability condition of the system of equations (3) and (4) is

$$\frac{d}{dt}(\varrho R^3) - \frac{p}{c^2}\frac{d}{dt}(R^3) = 0. \tag{5}$$

This equation has very simple physical interpretation - it assures that the universe is evolving adiabatically. To solve the Friedman equations it is necessary to specify k and to choose an equation of state $F(\varrho, p) = 0$. In cosmological considerations only three equations of state are frequently used:

a) $p = 0$ describing a dust like matter

b) $p = \frac{1}{3}\varrho c^2$ describing radiation and relativistic gas, and

c) $p = -\varrho c^2$ describing false vacuum.

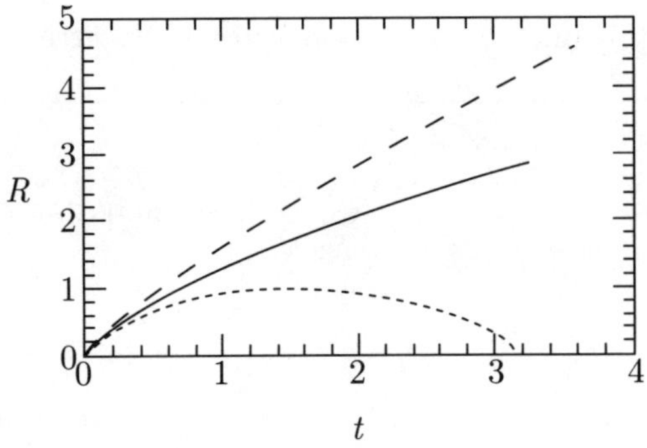

FIGURE 2. Time dependence of the scale factor in the Friedman cosmological models. Solid line represents the flat universe ($k = 0$), short dashed line a closed universe ($k = +1$), and long dashed line an open universe ($k = -1$).

General solutions of the Friedman equations with an equation of state such that $p > -1/3\varrho c^2$ are shown on Fig. 2.

In all cases the universe begins its evolution from a singular state characterized by $R = 0$, and $\varrho = \infty$. When $k = -1$ or 0 the universe expands forever. When $k = +1$ the initial expansion is slowed down by gravitational attraction and after reaching maximal size the universe contracts to a final singularity.

Due to the lack of new observations and theoretical ideas cosmology was dormant for about 20 years. After the Second World War George Gamow initiated an ambitious program of linking the observed abundance of chemical elements with processes occurring close to the initial singularity [**12**]. Gamow realized that early universe was not only very dense but also very hot. Sufficiently early not only atoms but also atomic nuclei could not exist and matter formed a hot plasma consisting of protons, neutrons, electrons, and positrons. The idea that the very early universe was very dense and very hot was popularized by Fred Hoyle who coined the name Big Band model.

Gamow's ambitious program was carried out by Alpher and Herman [**13**]. After making detailed calculations they realized that in the early universe it is practically impossible to synthesize elements heavier than helium. Their calculations allowed however to estimate the present temperature of microwave background radiation. It turned out that to produce the observed abundance of light elements (about 25% of helium by weight) the microwave background radiation should now have temperature of about 5K. It was generally believed that this radiation will be overshadowed by radiation of stars and interstellar gas and hence it will be unobservable. In 1964 A. Doroshkevich and I. Novikov [**14**] analyzed radiation produced by stars and gas in our Galaxy and showed that the microwave background radiation should be visible at centimeters and shorter wavelengths. Almost at the same time R. Dicke, P. J. E. Peebles, P. G. Roll, and D. T. Wilkinson started to construct a microwave antenna to detect the microwave background radiation. Late in 1964 the microwave background radiation was unexpectedly detected by A. Penzias and R. Wilson [**15**] who tested a horn type antenna at Holmdale. Their discovery was soon confirmed by Dicke's

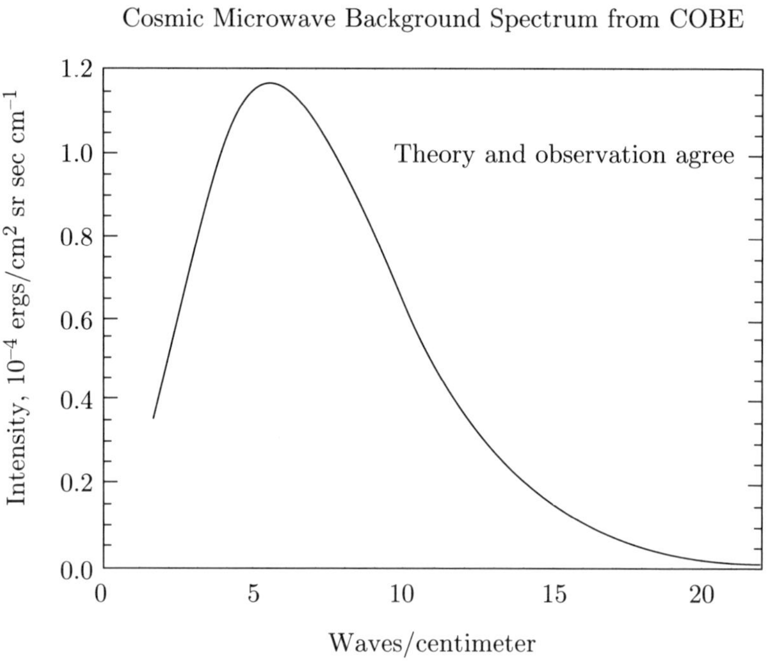

FIGURE 3. Spectrum of the microwave background radiation observed by the COBE satellite.

group [**16**] and by a group of radioastronomers at Cambridge.

The discovery of the microwave background radiation strongly supported the Big Bang model of the universe. Recent observations made by the COBE satellite definitely established thermal character of the microwave background (see Fig. 3) and its temperature $T = 2.726 \pm 0.005\mathrm{K}$ [**17**]. Physicists and cosmologists started to seriously consider physical processes at early stages of evolution of the universe.

Observational measurements of temperature of the microwave background radiation made it possible to return to Gamow's idea of primordial nucleosynthesis and calculate more exactly expected abundances of helium, lithium and isotopes of hydrogen and helium. Such calculations were done by Peebles [**18**] and in a much more elaborate form by Wagoner, Fowler, and Hoyle [**19**]. Observationally established abundances of light elements were in a very good agreement with theoretical predictions. This agreement strongly supported the hot Big Bang model.

To calculate the expected abundance of light elements it is necessary to specify present temperature of the universe, mean density of baryons, and the rate of expansion of the universe during the nucleosynthesis period. The rate of expansion of the universe depends on the number of families of relativistic particles present. Comparing theoretical predictions with observational abundances of light elements it is possible to obtain the present mean density of baryons

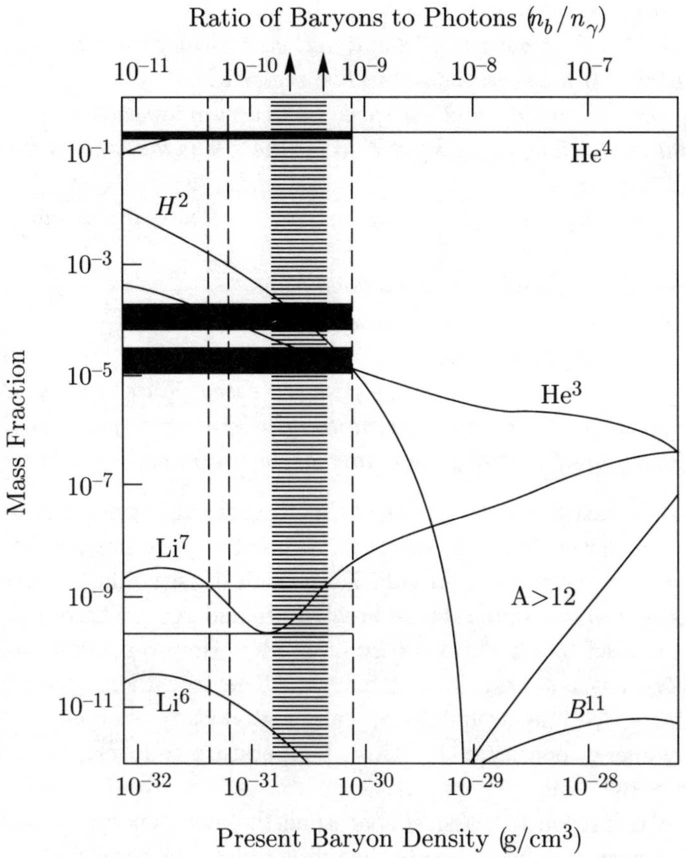

FIGURE 4. Abundance of light elements, theoretical predictions and observational limits.

($\approx 5 \times 10^{-31}$g/cm^3), and the number of families of neutrinos.

It is worth mentioning that cosmological considerations provided the first reliable confirmation that in Nature there are only three families of neutrinos. This result was later substantiated by accelerator experiments.

The discovery of microwave background radiation and successes of primordial nucleosynthesis idea established the Big Bang model as the standard cosmological scenario. On the basis of this scenario it is possible to reconstruct the whole history of the universe.

According to the standard Big Bang model the universe started its evolution from a singular state of infinite density and temperature. This singular state is an unavoidable feature of classical cosmological models based on general theory of relativity. The most general theorems on the existence of singularities were proven by Hawking and Penrose. Just to illustrate the flavour of such theorems let me quote the general one:

THEOREM (HAWKING AND PENROSE [**20**]).
Spacetime $(\mathcal{M}, g)$ *is not timelike and null geodesically complete if:*
(1) $R_{ab}k^a k^b \geq 0$ *for every non-spacelike vector* $\mathbf{k}$.
(2) The generic condition is satisfied, i.e. every non-spacelike geodesic contains a point at which $k_{[a}R_{b]cd[e}k_{f]}k^c k^d \neq 0$, *where* $\mathbf{k}$ *is the tangent vector to the geodesic.*
(3) The chronology condition holds on $\mathcal{M}$ *(i.e. there are no closed timelike curves).*
(4) There exists at least one of the following:
(i) a compact achronal set without edge,
(ii) a closed trapped surface,
(iii) a point p *such that on every past (or every future) null geodesic from* p *the divergence* Θ *of the null geodesic from* p *becomes negative (i.e. the null geodesics from* p *are focused by the matter or curvature and start to reconverge).*

Our universe satisfies assumptions of this theorem and hence it started evolving from a singular state. From the physical point of view singular beginning of the universe is unacceptable. At sufficiently high density and/or curvature the classical concept of spacetime should break down and general theory of relativity should be replaced by quantum theory of gravity. However despite lot of effort such a theory has not been formulated yet. The transition between classical and quantum spacetime occurs at exotically high density of about 10^{94}g/cm^3 or equivalently energy density $\approx 10^{19}\text{GeV}$, temperature $\approx 10^{32}\text{K}$, time $\approx 10^{-43}\text{s}$, and length $\approx 10^{-33}\text{cm}$.

Just after transition to classical spacetime the mean energy of particles and density are so high that according to the present ideas the basic elementary interactions - strong and electroweak - are unified, read indistinguishable. Generally accepted theory unifying elementary interactions does not exist. Model considerations indicate however that at energies above 10^{14}GeV strong and electroweak interactions should be unified. The universe expands and its temperature drops down inversely proportional to the increase of the scale factor R. When temperature of the universe drops to about 10^{14}GeV strong interactions acquire their identity and the matter is then composed of quarks, gluons, leptons, gauge bosons $(W^{\pm}, Z^0)$, photons and gravitons. Relativistic particles dominate the mean energy density and they determine the expansion rate of the universe $(R \sim t^{1/2})$. The universe is radiation dominated.

Let me briefly explain why the early evolution of such a complicated system as the whole universe can be described in such simple terms. There are two main reasons for that. The universe is homogeneous and as we now know from observations of anisotropy of the microwave background radiation made by the COBE satellite it was even more homogeneous earlier. $\Delta T/T|_{\text{at recombination}} \sim 10^{-5}$ implying that density perturbations were then of the same order while now even on the scale of clusters of galaxies $\delta\varrho/\varrho > 1$. Homogeneity guaranties that there are no spatial gradients and all quantities depend only on time. Extrapolating in-

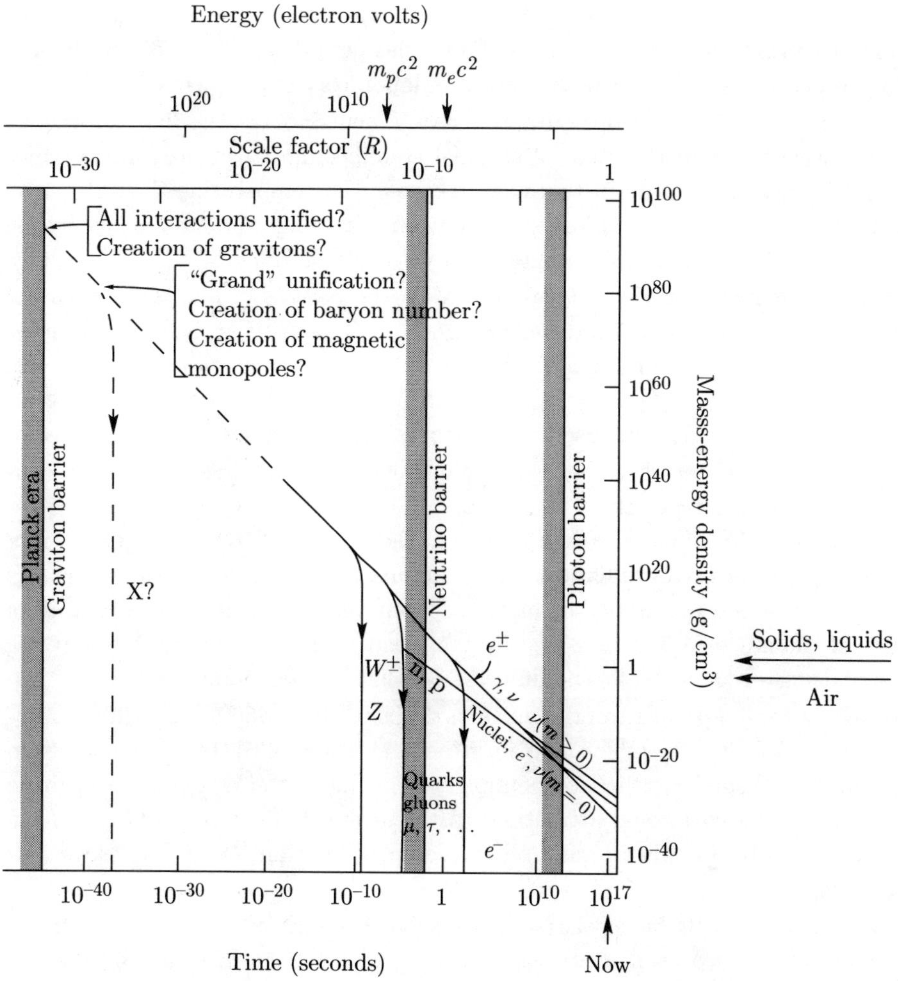

FIGURE 5. Thermal history of the universe.

formation gained at high energy accelerators to even higher energies it turns out that the time rate of elementary processes occurring at early stages of evolution of the universe is shorter than the expansion rate of the universe and though the universe is expanding and constantly changing the elementary processes are fast enough to preserve the state of thermodynamic equilibrium. This property greatly simplifies the description of matter. In the state of thermal equilibrium the mean energy density of matter depends on the number of effective degrees of freedom (spins of particles) and temperature and the composition of matter is determined by masses of particles.

When temperature drops to about several GeV quark gluon plasma recombines to form protons and neutrons. From that period on physics of the elementary processes is well known and tested in laboratory experiments.

When temperature drops to about 3MeV neutrinos as the first know particles decouple from the state of thermal equilibrium. Cosmological neutrinos should now form a uniform thermal background of temperature $T \approx 2$K. It is expected that at the transition epoch between quantum and classical spacetime cosmological background of gravitational waves was formed.

When temperature drops to about 1MeV the process of nucleosynthesis begins. Protons and almost all available neutrons fuse to form helium and other light elements up to lithium. Approximately at the same time electrons and positrons annihilate slightly warming up the universe. The process of nucleosynthesis stops when the universe is only about 3 minutes old.

Long and uneventful era begins. The universe expands causing the mean density of matter and radiation to decrease. The mean energy density of matter declines as $1/R^3$ while the mean energy of radiation as $1/R^4$. At a certain moment energy density of radiation, which dominated up to that moment, becomes equal to the energy density of matter. From that moment on the expansion rate of the universe is determined by the mean density of matter, the universe expands faster ($R \sim t^{2/3}$). The universe becomes matter dominated.

Soon after temperature of the universe drops to about $4500K$ allowing proton-helium plasma to recombine and form neutral atoms. Matter decouples from radiation and small perturbations implanted in the matter distribution begin to grow due to gravitational attraction. After several billions of years they form galaxies, cluster of galaxies, and all the complex distribution of matter we now observe.

The standard Big Bang model of the universe which is based on very simple assumptions that on sufficiently large scale matter is distributed homogeneously and isotropically and the evolution of the universe is governed by the Friedman equations describes surprisingly well the observed structure of the real universe and its evolution.

There are however some properties of the universe which are not accounted by the standard Big Bang model. The observed isotropy of the universe on a large scale so convincingly confirmed by the COBE satellite can be explained in the framework of the standard model only by a very special choice of initial conditions in a region containing many causally disconnected patches. From a physical point of view this is not a satisfactory solution.

In recent years more and more accurate observations allow to make more reliable estimates of the mean density of the universe. It turns out that the parameter $\Omega = \varrho/\varrho_{crit}$ is restricted to the range $0.1 < \Omega \leq 1$. Ω is changing with time according to

$$\frac{1-\Omega}{\Omega} = \frac{1-\Omega_0}{\Omega_0}\frac{1}{(1+z)^n}, \tag{6}$$

where Ω_0 is the present value, z is the redshift $(z+1 = R(t_0)/R(t))$, and $n=1$ for matter dominated universe, and $n=2$ for radiation dominated universe. From this relation it follows that Ω falls in the presently observed range only if at very early epochs ($z \to \infty$) it was very close to 1, for example at Planck time $|\Omega - 1|_{t=t_{Planck}} \sim 10^{-60}$. The standard cosmological model does not provide any explanation why the initial value of Ω was so close to 1 and how it was set so precisely.

More detailed investigation of symmetry breaking transition at energy $\approx 10^{14} GeV$ showed that topological defects like domain walls, cosmic strings, and monopoles could have been formed. It turns out that it is very difficult to dissipate monopoles and they once formed could survive up to the present epoch. The estimated mean mass density of monopoles is however several orders of magnitude larger than the observed mean mass density of the universe. This apparent inconsistency cannot be explained in the framework of the standard Big Bang model.

To solve these conundrums of the Big Bang model Alan Guth [**21**] proposed in 1981 a slightly modified scenario of the very early evolution of the universe called the inflationary model. The main feature of this scenario is a short period when the universe expands exponentially increasing its scale by about 30 orders of magnitude. A few different models of inflation have been suggested. Let me briefly describe stochastic inflationary model invented by Andrei Linde [**22**].

Let us consider a homogeneous massive scalar field ϕ. If the scalar field determines the expansion rate of the universe its time evolution is governed by

$$\ddot{\phi} + 3H\dot{\phi} + m^2\phi = 0, \tag{7}$$

$$H^2 = \frac{4\pi}{3M_{Pl}^2}(\dot{\phi}^2 + m^2\phi^2) - \frac{k}{R^2}, \tag{8}$$

where $H = \dot{R}/R$ is the Hubble parameter, and M_{Pl} denotes the Planck mass ($M_{Pl} = 10^{19} GeV$). If at some sufficiently early moment $t_0 \sim t_{Planck}$ $\phi(t_0) = \phi_0$ then equations (7) and (8) have the following solution

$$\phi(t) = \phi_0 - \frac{nM_{Pl}}{2\sqrt{3\pi}}t, \tag{9}$$

$$R(t) = R_0 \exp\left[\frac{2\pi}{M_{Pl}^2}(\phi_0^2 - \phi^2(t))\right]. \tag{10}$$

For $t < \phi/mM_{Pl}$ we have

$$R(t) = R_0 \exp(H(\phi_0)t), \tag{11}$$

where $H(\phi_0) = \sqrt{\frac{4\pi}{3}} \frac{m}{M_{Pl}} \phi_0$ and the universe expands exponentially. If the scalar field is initially highly excited and $\phi_0 \gg M_{Pl}$ then $H(\phi_0) \gg 1$. The initial highly excited state of the scalar field linearly declines with time. If the period of time when $\phi(t) \gg M_{Pl}$ is sufficiently long so that $H(\phi_0)t \sim 70$ then during that period the scale factor of the universe increases to $R(t) \sim R_0 10^{30}$.

Such rapid exponential growth of the scale factor will dramatically decrease the number density of monopoles and it will reduce the curvature term in equation (8) negligibly small or in other words will make the parameter Ω very close to one even if initially, before the inflation sets in, Ω was different from 1.

The chaotic inflation is generated by a highly excited scalar field. In almost all considered so far models of unification of elementary interactions at least one massive self interacting scalar field appears as a necessary ingredient. In this sense the chaotic inflation makes minimal requirements.

Inflationary models provide a natural mechanism of generation of primordial density perturbations and background gravitational waves. Inflationary models make two strong predictions - the means density of matter should be almost exactly equal to the critical density and the spectrum of created primordial density perturbations should be scale free. These predictions of inflationary models stimulated interest in theories of formation of the large scale distribution of matter and in observational attempts to measure more precisely the basic cosmological parameters.

Theories of formation of large scale structure in the universe use theory of gravitational instability formulated early in forties by E. M. Lifschitz [**23**]. This theory has been later refined by J. Silk [**24**] who included the microwave background radiation. Interactions between radiation and plasma smooth small scale density perturbations and only sufficiently large perturbations of a scale larger than clusters of galaxies survive till the recombination period. After recombination perturbation grow undisturbed.

When astronomers started to use more refined techniques to estimate the mean density of the universe it turned out that galaxies and clusters of galaxies contain large fraction of dark non luminous matter. Already in mid thirties F. Zwicky noticed that clusters of galaxies could not have survived for so long if they contain only visible light emitting matter. The dark matter shows up most dramatically in the mass to luminosity ratio for galaxies and clusters of galaxies when mass is estimated by dynamical methods. For galaxies the mass to luminosity ratio $M/L \sim 10-40 M_{\odot}/L_{\odot}$ and for clusters of galaxies $M/L \sim 100-400 M_{\odot}/L_{\odot}$. Using this information it is possible to estimate the Ω parameter, we obtain $\Omega_{\text{observed}} \sim 0.2$. From the theory of primordial nucleosynthesis it follows that $\Omega_{\text{baryons}} h^2 \leq 0.015$. This leads to very important and unexpected conclusion that most probably the universe is dominated by some exotic non baryonic dark matter.

Physicist have been already hunting for the dark matter for several years without success. In the theory of large scale structure formation three different families of dark matter particles are considered. For a few years it seemed that neutrinos do have a non zero mass of the order of several eV. Such neutrinos perfectly fitted for the role of dark matter particles but unfortunately it was too good to be true. Neutrinos decouple from the rest of matter when they were relativistic and dark matter of this type is called hot. When dark matter particles decouple from the state of thermal equilibrium when they are non relativistic or when they are created with small velocities they form cold dark matter. If dark matter particles are unstable but survive past the recombination period and decay later they are called warm.

For several years cold dark matter scenario of large scale structure formation was very popular. This model was very attractive because in conjunction with the inflationary scenario it contained only few arbitrary parameters [**25**]. Now the cold dark matter model is facing great difficulties. First of all it cannot simultaneously reproduce the observed correlations in the distribution of galaxies and perturbations in the microwave background radiation. This model gives also systematically too large values of expected peculiar velocities of galaxies.

Large scale structure formation in the hot dark matter model gives also results not compatible with observations. The warm dark matter model can probably account for all observations [**26**] but it contains too many free parameters and lacks acceptable candidates for warm dark matter particles.

As I already mentioned estimated mean density of matter seems to contradict predictions of the inflationary scenario. Observations favour low values of $\Omega \sim 02. - 0.3$ while inflationary models predict that $\Omega = 1$. Only observations of large scale flows of galaxies indicate that $\Omega \approx 1$ but these measurements are uncertain. Theorists confronted with crisis of the inflationary model invented new inflationary scenarios which produce low Ω universes overcoming at the same time the horizon problem of the standard Big Bang cosmology. It is too early to say if the new inflationary models would survive thorough comparison with observational data. There is recently a trend to revive the cosmological constant but this is rather an ad hoc approach.

Fortunately new observations are planned and soon, before the end of this century, we should know more about distribution of galaxies in space. More detailed observations of temperature anisotropy of the microwave background radiation at small scales of 1^0 and smaller [**27**] will make it easier to discriminate between different theoretical models.

The universe turned out to be more complicated but fortunately more interesting.

References

1. H. Shapley, *The Scale of the Universe. Part I*, Bull. Nat. Research Council **2** (1921), 171.
H. D. Curtis, *The Scale of the Universe. Part II*, Bull. Nat. Research Council **2** (1921), 194.
2. V. M. Slipher, *Spectrographic observations of nebulae*, Paper presented at the 17th meeting of the AAS, August, 1914.
3. E. Hubble, *Extragalactic nebulae*, Ap. J. **64** (1926), 321.
4. E. Hubble, *A relation between distance and radial velocity among extragalactic nebulae*, Proc. Nat. Acad. Sci. (USA) **15** (1929), 168.
5. M. L. Humason, *The Apparent Radial Velocities of 100 Extra-Galactic Nebulae*, Publ. Astr. Soc. Pacific **83** (136), 10.
6. A. Einstein, *Kosmologische betrachtungen zür allgemeinen Relativtätstheorie*, Preuss. Akad. Wiss. Berlin, Sitzber, (1917), 142.

7. W. de Sitter, *On the relativity of inertia: remarks concerning Einstein's latest hypothesis*, roc. Akad. Weteusch. Amsterdam **19** (1917), 1217.
8. A. Friedman, *Über die krummung des reumes*, Z. Phys. **10** (1922), 377.
______, Z. Phys. **21** (1924), 326.
9. A. Einstein and W. de Sitter, *On the relation between the expansion and the mean density of the universe*, Proc. Nat. Acad. Sci. (USA) **18** (1932), 213.
10. G. Lemaître, *Un Universe Homogéne de Masse Constante et de Rayon Croissant, Rendant Compte de la Vitese Radiale des Nébuleuses Extra-Galactiques*, Ann. Soc. Sci. Brux. **A47** (1927), 49 - 59.
11. H. P. Robertson, *Kinematics and world structure*, Ap.J. **82** (1935), 248.
A. G. Walker, *On Milne's theory of world-structure*, Proc. Lond. Math. Soc. **42** (1936), 90.
12. G. Gamow, *Expanding universe and the origin of elements*, Phys. Rev. **70** (1946), 572.
13. R. A. Alpher and R. C. Herman, Phys. Rev. **75** (1949), 1089.
14. A. G. Doroshkevich and I. D. Novikov, Dokl. Akad. Nauk USSR **154** (1964), 809.
15. A. A. Penzias and R. W. Wilson, *Measurement of excess antenna temperature at 4080 Mc/s*, Ap. J. **142** (1965), 419.
16. R. H. Dicke, P. J. E. Peebles, P. G. Roll, and D. T. Wilkinson, *Cosmic Black-body Radiation*, Ap. J. **142** (1965), 414.
17. J. C. Mather et. al., *Measurement of the Cosmic Background Spectrum by the COBE FIRAS Instrument*, Ap. J. **420** (1994), 439 - 444.
18. P. J. E. Peebles, *Primordial helium abundance and the primordial fireball*, Ap. J. **146** (1966), 542.
19. R. V. Wagoner, W. A. Fowler and F. Hoyle, *On the synthesis of elements at very high temperature*, Ap. J. **148** (1967), 3.
20. S. W. Hawking, and R. Penrose, *The singularities of gravitational collapse and cosmology*, Proc. Roy. Soc. London **A 300** (1970), 529 - 548.
21. A. H. Guth, *Inflationary Universe: A Possible Solution to the Horizon and Flatness Problem*, Phys. Rev. **D23** (1981), 347 - 356.
22. A. Linde, *Chaotic Inflation*, Phys. Lett. **129B** (1983), 177 - 181.
23. E. Lifshitz, *On the gravitational instability of the expanding universe*, J. Phys. (USSR) **10** (1946), 116.
24. J. Silk, *Fluctuations in the Primeval Fireball*, Nature **151** (1967), 1155 - 1156.
25. M. Davis, G. Efstathiou, C. S. Frank and S. D. M. White, *The Evolution of Large-scale Structure in the Universe Dominated by Cold Dark Matter*, Ap. J. **292** (1985), 371 - 394.
26. A. Klypin, J. A. Holtzman, J. Primack, and E. Regos, *Structure Formation with Cold Plus Hot Dark Matter*, Ap. J. **416** (1993), 1 - 16.
27. M. White, D. Scott, and J. Silk, *Anisotropies of the Cosmic Microwave Background*, Ann. Rev. Astr. Astrophys. **32** (1994), 319 - 370.

INSTITUTE FOR THEORETICAL PHYSICS, UNIVERSITY OF WARSAW, HOZA 69, 00-681 WARSAW, POLAND AND N. COPERNICUS ASTRONOMICAL CENTER, WARSAW, POLAND

E-mail address: mde@fuw.edu.pl

Contemporary Mathematics
Volume **203**, 1997

Extended Tensorial Curvature Analysis for Embeddings and Foliations

B. CARTER

ABSTRACT. It is shown how a pure background tensor formalism provides a concise but explicit and highly flexible machinery for the generalised curvature analysis of individual embedded surfaces and foliations such as arise in the theory of topological defects in cosmological and other physical contexts. Obtaining processes of such networks. The learning rules are discussed on various examples.

1. Introduction

The recent rise of interest in the theory of topological defect structures such as cosmic strings and higher dimensional cosmic membranes, as well as the related phenomenon of vortex foliations in neutron stars, has led to the investigation of a wide range of new problems of equilibrium and dynamical evolution in a special or general relativistic framework. Various aspects of these problems[1-8] and in recent years most particularly the requirements of general purpose perturbation analysis[9-16], have shown, and in some cases helped to satisfy, the need to adapt and develop pre-existing mathematical machinery for describing the relevant geometry and particularly the various kinds of curvature that are involved.

As usual, new physical problems have given rise to new mathematical problems, while even when what was needed has in principle been already available in the mathematical literature [17-21] it has often been in a form that is inaccessible or inconvenient for the purposes of physicists, many of whom have remained dependent on obsolete sources such as Eisenhart's still very influential textbook[22] (written in ignorance of the modern concept of generalised curvature, which was at that time under development by Cartan, and which was made familiar to physicists much later by the theory of Yang and Mills). Whereas some

1991 *Mathematics Subject Classification.* Primary 57R30, 53B21 ; Secondary 83C5,70G05.

treatments have obtained a neatly concise abstract formulation at the expense of flexibility, others have obtained general purpose adaptability at the price of using complicated and potentially confusing reference systems involving specially adapted coordinates and frames that require the simultaneous use of many different kinds of indices.

For the purpose of obtaining an optimal compromise between these two undesirable extremes, an advantageous tactic [4,7,8,10,16] is to rely as much as possible on ordinary tensors as defined with respect to the relevant n-dimensional background space-time with local coordinates x^μ say, (μ=0, ..., n-1). The advantage of avoiding explicit dependence on a specialised internal coordinate system becomes particularly clear in cases[4,8] where one is concerned with mutual contractions of tensors constructed on distinct but mutually intersecting embedded surfaces whose internal coordinates could not in general be made to be mutually compatible.

The present treatment will employ the same notation as was used in the most mathematically detailed presentation[7] of this approach, in which, rather that describing the induced curvature of an embedded spacelike or timelike p-surface, with internal coordinates σ^i say, in terms of the intrinsic version, with components $R_{ijk\ell}$ say, of its Riemann tensor, one prefers to describe it in terms of the corresponding background spacetime tensor $R_{\lambda\mu\nu\rho}$. The latter is definable (using the abbreviation $x^\mu_{,i} = \partial x^\mu/\partial\sigma^i$) as the index lowered verson – obtained by contraction with the background spacetime metric $g_{\mu\nu}$ – of the projection $R^{\lambda\mu\nu\rho} = R^{ijkl}\, x^\lambda_{,i} x^\mu_{,j} x^\nu_{,k} x^\rho_{,l}$ of the contravariant version of the intrinsic curvature that is obtained by internal index raising, using the inverse, with contravariant components η^{ij} (whose existence depends on the postulate that the surface is spacelike or timelike, but not null) of the induced metric with components $\eta_{ij} = g_{\mu\nu} x^\mu_{,i} x^\nu_{,j}$. This background spacetime representation $R_{\lambda\mu\nu\rho}$ of the internal curvature of the embedded p-surface has of course to be distinguished from the ordinary n-dimensional Riemann curvature of the background spacetime itself, whose components will be denoted simply by $B_{\lambda\mu\nu\rho}$. In the same way, rather than working with the intrinsic components η_{ij} and the contravariant inverse components η^{ij} of the induced metric, one prefers to use the corresponding background coordinate components $\eta_{\mu\nu}$ and the contravariant components of the corresponding projected tensor specified by $\eta^{\mu\nu} = \eta^{ij} x^\mu_{,i} x^\nu_{,j}$, which is what is referred to as the (first) *fundamental tensor* of the embedding.

To set up a systematic analysis of curvature in exclusively background tensorial terms, the natural starting point is obviously the fundamental tensor that has just been defined, whose mixed (contra/covariant) version with components $\eta^\mu{}_\nu$ is interpretable as a rank-p *tangential projection* operator that sends a vector into the tangent subspace of the embedding. The notation $\gamma^\mu{}_\nu$ will be used to denote the complementary rank-(n-p) operator of *lateral projection* orthogonal to the surface, whose components will evidently be given in terms of those of the fundamental (tangential projection) tensor by the defining relation

$$\gamma^\mu{}_\nu + \eta^\mu{}_\nu = g^\mu{}_\nu \ , \tag{1.1}$$

since the mixed version $g^\mu{}_\nu$ of the metric tensor is of course interpretable as representing the identity operator. As well as having the separate operator properties

$$\eta^\mu{}_\rho \eta^\rho{}_\nu = \eta^\mu{}_\nu \ , \qquad \gamma^\mu{}_\rho \gamma^\rho{}_\nu = \gamma^\mu{}_\nu \tag{1.2}$$

the tensors thus defined will evidently be related by the conditions

$$\eta^\mu{}_\rho \gamma^\rho{}_\nu = 0 = \gamma^\mu{}_\rho \eta^\rho{}_\nu \ . \tag{1.3}$$

In the familiar way, the specification of the background metric $g_{\mu\nu}$ characterises a corresponding orthonormal frame bundle, and automatically endows it with a natural Riemannian or pseudo Riemnnian connection (with components $\Gamma_\mu{}^\nu{}_\rho$ say) and its associated curvature tensor, (with components $B_{\lambda\mu}{}^\nu{}_\rho$) with respect to the action of the relevant n-dimensional orthogonal or pseudo-orthogonal (usually Lorentz type) rotation group. The decomposition expressed by (1.1) of the local tangent space to the manifold into a p-dimensional subspace tangential to the embedded surface, and an (n-p)-dimensional subspace orthogonal to the surface, evidently determines a reduced bundle of orthonormal frames that are correspondingly *adapted* in the sense of being restricted to consist exclusively of vectors that are either tangential or orthogonal to the embedded surface. Moreover this adapted frame bundle will be automatically endowed with a naturally induced *foliation connection* and hence with a corresponding *foliation curvature* (which will however be no longer of the long familiar Riemannian type) with respect to the action of a subgroup the full n-dimensional rotation group that will evidently have the form of a direct product of a p-dimensional surface tangential rotation group with an (n-p)-dimensional rotation group acting on the orthogonal subspace.

Since each adapted frame in the reduced bundle is constructed as the combination of a surface tangential subframe with a surface orthogonal subframe, it is evident that the full adapted frame bundle – whose group is the direct product of p-dimensional and (n-p)-dimensional rotation subgroups – can itself be analysed in terms of a distinct tangential subframe bundle with just a p-dimensional rotation group and with a naturally associated "inner" connection and curvature, together with a separate orthogonal subframe bundle with just an (n-p)-dimensional subgroup and a naturally associated "outer" connection and curvature. The tangential subframe bundle is of course interpretable as the ordinary Riemannian frame bundle of the the embedded p-dimensional submanifold, whose curvature is representable in background tensorial form by the components $R_{\lambda\mu}{}^\nu{}_\rho$. The orthogonal subframe bundle is not of the traditional Riemannian type: it can easily be shown[7] that (unlike its "inner" analogue $R_{\lambda\mu}{}^\nu{}_\rho$) the corresponding "outer" curvature tensor with components $\Omega_{\lambda\mu}{}^\nu{}_\rho$ is automatically both conformally invariant and trace free, so that it does not have any analogue of the Ricci type contraction $R_{\mu\nu} = R_{\lambda\mu}{}^{\lambda}{}_{\nu}$ and the associated scalar $R = R^{\nu}{}_{\nu}$, though in the important 4-dimensional case it can give a non vanishing pseudoscalar Ω that is given in terms of the 4-dimensional measure tensor by $2\Omega = \varepsilon_{\lambda\mu\nu\rho}\Omega^{\lambda\mu\nu\rho}$. (One of the reasons why the "outer" curvature is so much less widely known than its "inner" analogue is that in a three dimensional background it always vanishes as a trivial identity, and that even for n=4 the only non-trivial case is that for which p=2, which applies to the case of a string, but not to the more familiar cases of a point particle or a membrane.)

Although it is less detailed, the present article extends the results of the preceeding development[7] of the background tensorial analysis of embedding curvature by taking account of situations in which one is concerned not just with an individual embedded p-surface but with a smooth foliation by a space filling congruence of such surfaces. It also innovates in using a for many purposes more powerful approach, by working not just with the separate tangential ("inner") and surface-orthogonal ("outer") subframe bundles but with the full ("amalgamated") adapted frame bundle whose group is the product of the p-dimensional ("inner") rotation group with the (n-p)-dimensional "outer" rotation group, thus giving rise to the concept of the *full* ("amalgamated") embedding curvature tensor $F_{\lambda\mu}{}^{\nu}{}_{\rho}$, which is obtainable as the surface tangential part of the extended *foliation curvature* tensor $\mathcal{F}_{\lambda\mu}{}^{\nu}{}_{\rho}$, and of which the distinct "inner" and "outer" curvature tensors $R_{\lambda\mu}{}^{\nu}{}_{\rho}$ and $\Omega_{\lambda\mu}{}^{\nu}{}_{\rho}$ are just decomposed parts.

2. The first deformation tensor and the second fundamental tensor

The strategy of the present approach will be to treat the case of an individual embedded *p*-surface within the more extended context of a smooth foliation of such surfaces, for which there will be corresponding background tensor fields of tensors $\eta^{\mu}{}_{\nu}$ and $\gamma^{\mu}{}_{\nu}$ satisfying the relations (1.1), (1.2), and (1.3), so that they will have well defined (Riemannian or pseudo-Riemannian) covariant derivatives. These derivatives given will be fully determined by the specification of a certain *(first) deformation tensor*, $\mathcal{K}_{\mu}{}^{\nu}{}_{\rho}$ say, via an expression of the form

$$\nabla_{\mu}\,\eta^{\nu}{}_{\rho} = -\nabla_{\mu}\,\gamma^{\nu}{}_{\rho} = \mathcal{K}_{\mu}{}^{\nu}{}_{\rho} + \mathcal{K}_{\mu\nu}{}^{\rho}\ . \tag{2.1}$$

It can easily be seen from (1.2) that the required deformation tensor will be given simply by

$$\mathcal{K}_{\mu}{}^{\rho}{}_{\nu} = \eta^{\rho}{}_{\sigma}\,\nabla_{\mu}\,\eta^{\sigma}{}_{\nu} = -\,\gamma^{\sigma}{}_{\nu}\nabla_{\mu}\gamma^{\rho}{}_{\sigma}\ . \tag{2.2}$$

The middle and last indices of this tensor will evidently have the respective properties of tangentiality and orthogonality that are expressible as

$$\gamma^{\sigma}{}_{\nu}\mathcal{K}_{\mu}{}^{\rho}{}_{\sigma} = 0\ , \qquad \mathcal{K}_{\mu}{}^{\rho}{}_{\sigma}\,\eta^{\sigma}{}_{\nu} = 0\ . \tag{2.3}$$

There is no automatic tangentiality or orthogonality property for the first index of the deformation tensor (2.2), which is thus reducible with respect to the tangential and orthogonally lateral projections (1.1) to a sum

$$\mathcal{K}_{\mu}{}^{\rho}{}_{\nu} = K_{\mu}{}^{\rho}{}_{\nu} - L_{\mu\nu}{}^{\rho} \tag{2.4}$$

in which such a property is obtained for each of the parts

$$K_{\mu}{}^{\rho}{}_{\nu} = \eta^{\sigma}{}_{\mu}\mathcal{K}_{\sigma}{}^{\rho}{}_{\nu} \,, \qquad L_{\mu\nu}{}^{\rho} = -\gamma^{\sigma}{}_{\mu}\mathcal{K}_{\sigma}{}^{\rho}{}_{\nu} \,, \tag{2.5}$$

which satisfy the conditions

$$\gamma^{\sigma}{}_{\mu}K_{\sigma}{}^{\rho}{}_{\nu} = \gamma^{\rho}{}_{\sigma}K_{\mu}{}^{\sigma}{}_{\nu} = 0 = K_{\mu}{}^{\rho}{}_{\sigma}\eta^{\sigma}{}_{\nu} \,. \tag{2.6}$$

and

$$\eta^{\sigma}{}_{\mu}L_{\sigma}{}^{\rho}{}_{\nu} = \eta^{\rho}{}_{\sigma}L_{\mu}{}^{\sigma}{}_{\nu} = 0 = L_{\mu}{}^{\rho}{}_{\sigma}\gamma^{\sigma}{}_{\nu} \,. \tag{2.7}$$

It is evident that the first of these decomposed parts is appropriately describable as the *tangential turning* tensor, since by (2.2) and (2.5) it is given by the expression

$$K_{\mu}{}^{\rho}{}_{\nu} = \eta^{\rho}{}_{\sigma}\eta^{\tau}{}_{\mu}\nabla_{\tau}\eta^{\sigma}{}_{\nu} \,, \tag{2.8}$$

in which the only differentiation involved is contained in the tangential gradient operator $\eta^{\tau}{}_{\mu}\nabla_{\tau}$ – which is well defined even for fields whose support is restricted to a single embedded surface – so that, unlike the full deformation tensor $\mathcal{K}_{\mu}{}^{\rho}{}_{\nu}$ of the foliation, the tangential turning tensor $K_{\mu}{}^{\rho}{}_{\nu}$ is well defined just for an individual embedded p-surface. As such, this turning tensor $K_{\mu}{}^{\rho}{}_{\nu}$ is identifiable as what has been defined[4,7] as the ordinary *second fundamental tensor* of the particular p-surface passing through the point under consideration.

Up to this point, none of the relations formulated in this section depends on the condition that the (first) fundamental tensor field $\eta^{\mu}{}_{\nu}$ is actually tangential to well behaved p-surfaces rather than just being an arbitrary field of rank-p projection tensors as characterised by the purely algebraic conditions (1.2). As pointed out in the previous analysis [7], the Frobenius type integrability condition that is necessary and sufficient for the local existence of well behaved p-surfaces tangential to $\eta^{\mu}{}_{\nu}$ is that the second fundamental tensor defined by (2.6) should have the generalised Weingarten property

$$K_{[\mu\nu]}{}^{\rho} = 0 \,, \tag{2.9}$$

(using square brackets to denote index antisymmetrisation) which means that it is symmetric with respect to its two surface tangential indices.

It is to be noticed that the second part of the decomposition (2.4), namely the *lateral turning tensor*, $L_{\mu}{}^{\rho}{}_{\nu}$, is expressible by the formula

$$L_{\mu}{}^{\rho}{}_{\nu} = \gamma^{\rho}{}_{\sigma}\gamma^{\tau}{}_{\mu}\nabla_{\tau}\gamma^{\sigma}{}_{\nu} \,, \tag{2.10}$$

which differs from that in (2.8) only by the substitution of $\gamma^{\nu}{}_{\mu}$ for $\eta^{\nu}{}_{\mu}$. It evidently follows that the necessary and sufficient integrability condition for the local existence of an (n-p)-dimensional foliation orthogonal to the p-dimensional foliation whose existence is guaranteed by (2.9) is that this lateral turning tensor should have the analogous symmetry property, i.e. that the *rotation tensor* $L_{[\mu\nu]}{}^{\rho}$ of the foliation should vanish.

3. The second deformation tensor and the third fundamental tensor

Just as the first order derivatives of the fundamental tensor $\eta^{\mu}{}_{\nu}$ of a foliation are specifiable by the first deformation tensor $\mathcal{K}_{\mu}{}^{\nu}{}_{\rho}$ defined in the previous section, so analogously the derivatives of the next order will be fully determined by the further specification of a corresponding *second deformation tensor*, $\mathcal{X}_{\lambda\mu}{}^{\nu}{}_{\rho}$ say, via an expression of the slightly less simple form

$$\nabla_{\lambda}\mathcal{K}_{\mu}{}^{\nu}{}_{\rho} = \mathcal{X}_{\lambda\mu}{}^{\nu}{}_{\rho} + \mathcal{K}_{\lambda}{}^{\nu}{}_{\sigma}\mathcal{K}_{\mu}{}^{\sigma}{}_{\rho} - \mathcal{K}_{\lambda}{}^{\sigma}{}_{\rho}\mathcal{K}_{\mu}{}^{\nu}{}_{\sigma} \,, \tag{3.1}$$

that is easily derivable from the definition

$$\mathcal{X}_{\lambda\mu}{}^{\nu}{}_{\rho} = \eta^{\nu}{}_{\sigma}\gamma^{\tau}{}_{\rho}\nabla_{\lambda}\mathcal{K}_{\mu}{}^{\sigma}{}_{\tau} \,, \tag{3.2}$$

using the projection properties (2.3).

Just as the second fundamental tensor of an individual surface was obtained from the first deformation tensor of the foliation by the simple tangential projection operation (2.4), so analogously the corresponding *third fundamental tensor*, $\Xi_{\lambda\mu}{}^{\nu}{}_{\rho}$ say, will be obtainable from the second deformation tensor by the slightly less simple tangential projection operation

$$\Xi_{\lambda\mu}{}^{\nu}{}_{\rho} = \eta^{\alpha}{}_{\lambda}\eta^{\beta}{}_{\mu}\big(\mathcal{X}_{\alpha\beta}{}^{\nu}{}_{\rho} + \mathcal{K}_{\alpha\beta}{}^{\sigma}\mathcal{K}_{\sigma}{}^{\nu}{}_{\rho}\big) \,. \tag{3.3}$$

It can be seen to follow from the Weingarten integrability property (2.9) that the antisymmetric part of this third fundamental tensor will be given directly by the simple projection operation

$$\Xi_{[\lambda\mu]}{}^{\nu}{}_{\rho} = \eta^{\alpha}{}_{\lambda}\eta^{\beta}{}_{\mu}\mathcal{X}_{[\alpha\beta]}{}^{\nu}{}_{\rho} \,. \tag{3.4}$$

It is easy to use the projection properties (2.3) to verify that the tensor obtained by this procedure does indeed satisfy the original definition[7]

$$\Xi_{\lambda\mu}{}^{\nu}{}_{\rho} = \eta^{\alpha}{}_{\lambda}\eta^{\beta}{}_{\mu}\eta^{\nu}{}_{\sigma}\gamma^{\tau}{}_{\rho}\nabla_{\alpha}K_{\beta}{}^{\sigma}{}_{\tau} \,, \tag{3.5}$$

of the third fundamental tensor, which can thereby be seen to be unambigously determined just by the specification of an individual embedded surface, independently of any extension to a space covering foliation.

The higher order differential information contained in the second deformation tensor and the third fundamental tensor is not needed for the evaluation of the curvature tensors discussed in the following sections, even though they are dependent on the second derivatives of the background metric itself. However a useful biproduct of the curvature analysis that follows is the derivation of a separation identity (interpretable as a generalisation of the historic Codazzi identity) that expresses the antisymmetric part of the second deformation tensor – and hence by (3.4) that of the third fundamental tensor – directly as the corresponding projection of the background spacetime curvature $B_{\lambda\mu}{}^{\nu}{}_{\rho}$.

4. The adapted foliation connection

Due to existence of the decomposition whereby a background spacetime vector, with components ξ^μ say, is split up by the projectors (1.1) as the sum of its surface tangential part $\eta^\mu{}_\nu\,\xi^\nu$ and its surface orthogonal part $\gamma^\mu{}_\nu\xi^\nu$, there will be a corresponding adaptation of the ordinary concept of parallel propagation with respect to the background connection $\Gamma_\mu{}^\nu{}_\rho$. The principle of the adapted propagation concept is to follow up an ordinary operation of infinitesimal parallel propagation by the projection adjustment that is needed to ensure that purely tangential vectors propagate onto purely tangential vectors while purely orthogonal vectors propagate onto purely orthogonal vectors. Thus for purely tangential vectors, the effect of the adapted propagation is equivalent to that of ordinary internal parallel propagation with respect to the induced metric in the embedded surface, while for purely orthogonal vectors it is interpretable as the natural generalisation of the standard concept of Fermi-Walker propagation. For an infinitesimal displacement dx^μ the deviation between the actual component variation $(dx^\nu)\,\partial_\nu\xi^\mu$ and the variation that would be obtained by the corresponding adapted propagation law will be expressible in the form $(dx^\nu)\,D_\nu\xi^\mu$ where D denotes the corresponding *foliation adapted differentiation* operator, whose effect will evidently be given by

$$D_\nu\,\xi^\mu = \eta^\mu{}_\rho\nabla_\nu\big(\eta^\rho{}_\sigma\xi^\sigma\big) + \gamma^\mu{}_\rho\nabla_\nu\big(\gamma^\rho{}_\sigma\xi^\sigma\big)\ . \tag{4.1}$$

It can thus be seen that this operation will be expressible in the form

$$D_\nu\,\xi^\mu = \nabla_\nu\xi^\mu + \alpha_\nu{}^\mu{}_\sigma\xi^\sigma = \partial_\nu\,\xi^\mu + \mathcal{A}_\nu{}^\mu{}_\sigma\xi^\sigma\ , \tag{4.2}$$

where the adapted *foliation connection* components $\mathcal{A}_\mu{}^\nu{}_\rho$ are given by the formula

$$\mathcal{A}_\mu{}^\nu{}_\rho = \Gamma_\mu{}^\nu{}_\rho + \alpha_\mu{}^\nu{}_\rho\ , \tag{4.3}$$

in which the $\alpha_\mu{}^\nu{}_\rho$ are the components of the relevant *adaptation tensor*, whose components can be seen from (4.1) to be given by

$$\alpha_\mu{}^\nu{}_\rho = \big(\eta^\nu{}_\sigma - \gamma^\nu{}_\sigma\big)\nabla_\mu\,\eta^\sigma{}_\rho\ . \tag{4.4}$$

The entirely covariant index lowered version of this adaptation tensor can thus be seen to be expressible in terms of the deformation tensor defined by (2.2) in the form

$$\alpha_{\mu\nu\rho} = 2\mathcal{K}_{\mu[\nu\rho]}\ . \tag{4.5}$$

An important example of the application of this operation is to the case of the first deformation tensor itself, whose foliation adapted derivative will be expressible in terms of the second deformation tensor (3.2) by the relation

$$D_\mu\mathcal{K}_\nu{}^\rho{}_\sigma = \mathcal{X}_{\mu\nu}{}^\rho{}_\sigma + \alpha_{\mu\nu}{}^\tau\mathcal{K}_\tau{}^\rho{}_\sigma\ . \tag{4.6}$$

The fact that the expression (4.5) is manifestly antisymmetric with respect to the last two indices of the adaptation tensor makes it evident that, like the usual Riemannian differentiation operator ∇, the adapted differentiation operator D

will commute with index raising or lowering, since the metric itself remains invariant under adapted propagation:

$$D_\mu \, g_{\nu\rho} = 0 \; . \tag{4.7}$$

However, unlike ∇, the adapted differentiation operator has the very convenient property of also commuting with tangential and orthogonal projection, since it can be seen to follow from (2.1) and (2.3) the corresponding operators also remain invariant under adapted propagation:

$$D_\mu \, \eta^\nu{}_\rho = 0 \; , \qquad D_\mu \gamma^\nu{}_\rho = 0 \; . \tag{4.8}$$

There is of course a price to be paid in order to obtain this considerable advantage of D over ∇, but it is not exorbitant: all that has to be sacrificed is the analogue of the symmetry property

$$\Gamma_{[\mu}{}^\nu{}_{\rho]} = 0 \tag{4.9}$$

expressing the absence of torsion in the Riemannian case. For the adapted foliation connection $\mathcal{A}_\mu{}^\nu{}_\rho$, the torsion tensor defined by

$$\Theta_\mu{}^\nu{}_\rho = 2\mathcal{A}_{[\mu}{}^\nu{}_{\rho]} = 2\alpha_{[\mu}{}^\nu{}_{\rho]} \; , \tag{4.10}$$

will not in general be zero.

When one is dealing not with an extended foliation but just with an individual embedded surface, the general adapted differentiation operator D will not be well defined, but adapted differentiation in tangential directions will still be specified by the corresponding unambiguously defined restricted differentiation operator, $\overline{D}$ say, component representation is given by

$$\overline{D}_\mu = \eta^\nu{}_\mu D_\nu \; . \tag{4.11}$$

Restriction to a single embedded surface will not invalidate the properties of commuting with index raising and tangential or orthogonal projection as expressed in the general case by (4.7) and (4.8), so in abstract symbolical notation we shall have not only $\big[\overline{D}, g\big] = 0$, but also $\big[\overline{D}, \eta\big] = 0$, and $\big[\overline{D}, \gamma\big] = 0$. It is evident from (4.2) and (4.11) that the effect of this tangentially restricted adapted differentiation operator on any surface supported (but not necessarily surface tangential) vector field with components ξ^μ will be given by

$$\overline{D}_\mu \, \xi^\nu = \overline{\nabla}_\mu \xi^\mu + a_\mu{}^\nu{}_\rho \xi^\rho \; , \tag{4.12}$$

using the notation

$$\overline{\nabla}_\mu = \eta^\nu{}_\mu \nabla_\nu \; , \qquad a_\mu{}^\nu{}_\rho = \eta^\sigma{}_\mu \alpha_\sigma{}^\nu{}_\rho \tag{4.13}$$

where $\overline{\nabla}$ indicates tangential covariant differentiation of the usual Riemannian type, and where the relevant tangentially restricted part of the adaptation tensor, which may naturally described as the *acceleration tensor*, is given explicitly as function only of the ordinary (restricted) second fundamental tensor of the embedded surface by the simple formula

$$a_{\mu\nu\rho} = 2K_{\mu[\nu\rho]} \ . \tag{4.14}$$

As a simple illustration, it is to be remarked that in the particular case for which the embedded surface is a particle world-line (with p=1) having a timelike tangent vector u^μ satisfying the unit normalisation condition $u^\nu u_\nu = -1$, the first and second fundamental tensors will be given respectively by $\eta^{\mu\nu} = -u^\mu u^\nu$ and $K_\mu{}^\nu{}_\rho = u_\mu u^\nu \dot{u}_\rho$ where $\dot{u}^\rho = u^\sigma \nabla_\sigma u^\rho$ are the components of the ordinary acceleration vector, so that the corresponding acceleration tensor will be given by $a_{\mu\nu\rho} = 2u_\mu u_{[\nu}\dot{u}_{\rho]}$.

5. The amalgamated foliation curvature tensor

The curvature associated with the adapted connection introduced by (4.2) in the preceding section can be read out from the ensuing commutator formula, which, for an arbitrary vector field with components ξ^μ, will take the standard form

$$2D_{[\mu}D_{\nu]}\xi^\rho = \mathcal{F}_{\mu\nu}{}^\rho{}_\sigma \xi^\sigma - \Theta_\mu{}^\sigma{}_\nu D_\sigma \xi^\rho \ , \tag{5.1}$$

in which the torsion tensor components $\Theta_\mu{}^\sigma{}_\nu$ are as defined by (3.9) while the components $\mathcal{F}_{\mu\nu}{}^\rho{}_\sigma$ are defined by a Yang-Mills type curvature formula of the form

$$\mathcal{F}_{\mu\nu}{}^\rho{}_\sigma = 2\partial_{[\mu}\mathcal{A}_{\nu]}{}^\rho{}_\sigma + 2\mathcal{A}_{[\mu}{}^{\rho\tau}\mathcal{A}_{\nu]\tau\sigma} \ . \tag{5.2}$$

Although the connection components $\mathcal{A}_\mu{}^\nu{}_\rho$ from which it is constructed are not of tensorial type, the resulting curvature components (5.2) are of course strictly tensorial. This is made evident by evaluating the components (5.2) of this *amalgamated foliation curvature* in terms of the background curvature tensor

$$B_{\mu\nu}{}^\rho{}_\sigma = 2\partial_{[\mu}\Gamma_{\nu]}{}^\rho{}_\sigma + \Gamma_\mu{}^\rho{}_\tau\Gamma_\nu{}^\tau{}_\sigma - \Gamma_\nu{}^\rho{}_\tau\Gamma_\mu{}^\tau{}_\sigma \ , \tag{5.3}$$

and the adaptation tensor $\alpha_\mu{}^\nu{}_\rho$ given by (4.4), which gives the manifestly tensorial expression

$$\mathcal{F}_{\mu\nu}{}^\rho{}_\sigma = B_{\mu\nu}{}^\rho{}_\sigma + 2\nabla_{[\mu}\alpha_{\nu]}{}^\rho{}_\sigma + 2\alpha_{[\mu}{}^{\rho\tau}\alpha_{\nu]\tau\sigma} \ . \tag{5.4}$$

Although it does not share the full set of symmetries of the Riemann tensor, the foliation curvature obtained in this way will evidently be antisymmetric in both its first and last pairs of indices:

$$\mathcal{F}_{\mu\nu\rho\sigma} = \mathcal{F}_{[\mu\nu][\rho\sigma]} \ . \tag{5.5}$$

Using the formula (4.5), it can be see from (2.3) and and (3.1), that the difference between this adapted curvature and the ordinary background Riemann curvature will be given by

$$\mathcal{F}_{\mu\nu}{}^{\rho\sigma} - B_{\mu\nu}{}^{\rho\sigma} = 4\mathcal{X}_{[\mu\nu]}{}^{[\rho\sigma]} + 2\mathcal{K}_{[\mu}{}^{\sigma\tau}\mathcal{K}_{\nu]}{}^\rho{}_\tau + 2\mathcal{K}_{[\mu}{}^{\tau\rho}\mathcal{K}_{\nu]}{}^\sigma_\tau \ , \tag{5.6}$$

which superficially appears to depend on the higher order derivatives involved in the second deformation tensor $\mathcal{X}_{\mu\nu}{}^{\rho\sigma}$.

This appearance is however deceptive. The reason for qualifying the foliation curvature tensor (5.2) as "amalgamated" is that because the adapted derivation operator has been constructed in such a way as to map tangential vector fields into purely tangential vector fields, and lateral (surface orthogonal) vector fields into lateral vector fields, it follows that the same applies to the corresponding curvature, which will therefore consist of an additive amalgamation of two separate parts having the form

$$\mathcal{F}_{\mu\nu}{}^{\rho}{}_{\sigma} = \mathcal{P}_{\mu\nu}{}^{\rho}{}_{\sigma} + \mathcal{Q}_{\mu\nu}{}^{\rho}{}_{\sigma} , \tag{5.7}$$

in which the "inner" curvature acting on purely tangential vectors is given by a doubly tangential (surface parallel) projection as

$$\mathcal{P}_{\mu\nu}{}^{\rho}{}_{\sigma} = \mathcal{F}_{\mu\nu}{}^{\kappa}{}_{\lambda}\, \eta^{\rho}{}_{\kappa}\, \eta^{\lambda}{}_{\sigma} , \tag{5.8}$$

while the "outer" curvature acting on purely orthogonal vectors is given by a doubly lateral projection as

$$\mathcal{Q}_{\mu\nu}{}^{\rho}{}_{\sigma} = \mathcal{F}_{\mu\nu}{}^{\kappa}{}_{\lambda} \gamma^{\rho}{}_{\kappa}\, \gamma^{\lambda}{}_{\sigma} . \tag{5.9}$$

It is implicit in the separation expressed by (5.7) that the mixed tangential and lateral projection of the adapted curvature must vanish:

$$\mathcal{F}_{\mu\nu}{}^{\kappa}{}_{\lambda}\, \eta^{\rho}{}_{\kappa} \gamma^{\lambda}{}_{\sigma} = 0 . \tag{5.10}$$

The non-trivial separation identity (5.10) can be considered as a generalisation to the case of foliations of the relation that is itself interpretable as an extended generalisation to higher dimensions of the historic Codazzi equation that was originally formulated in the restricted context of 3-dimensional flat space. It can be seen from (5.6) that this extended Codazzi identity is expressible as

$$2\mathcal{X}_{[\mu\nu]}{}^{\rho}{}_{\sigma} + B_{\mu\nu}{}^{\kappa}{}_{\lambda}\, \eta^{\rho}{}_{\kappa} \gamma^{\lambda}{}_{\sigma} = 0 , \tag{5.11}$$

which shows that the relevant higher derivatives are all determined entirely by the Riemannian background curvature so that no specific knowledge of the second deformation tensor is needed. The formula (5.6) can also be used to evaluate the "inner" tangential part of the foliation curvature tensor as

$$\mathcal{P}_{\mu\nu}{}^{\rho}{}_{\sigma} = 2\mathcal{K}_{[\nu}{}^{\rho\tau}\mathcal{K}_{\mu]\sigma\tau} + B_{\mu\nu}{}^{\kappa}{}_{\lambda}\, \eta^{\rho}{}_{\kappa}\, \eta^{\lambda}{}_{\sigma} \tag{5.12}$$

and to evaluate the "outer" orthogonal part of the foliation curvature tensor as

$$\mathcal{Q}_{\mu\nu}{}^{\rho}{}_{\sigma} = 2\mathcal{K}_{[\mu}{}^{\tau\rho}\mathcal{K}_{\nu]\tau\sigma} + B_{\mu\nu}{}^{\kappa}{}_{\lambda} \gamma^{\rho}{}_{\kappa} \gamma^{\lambda}{}_{\sigma} . \tag{5.13}$$

The formula (5.12) for the "inner" foliation curvature is evidently classifiable as an extended generalisation of the historic Gauss equation, while the formula (5.13) for the "outer" foliation curvature is an extension of what has sometimes been referred to as the "Ricci equation" but what would seem more appropriately describable as the *Schouten equation*, with reference to the earliest relevant source with which I am familiar[17], since long after the time of Ricci it was not yet understood even by such a leading geometer as Eisenhart[22].

It will be shown in the final section how, by doubly tangential projection of the first two indices, the extended generalisations (5.11), (5.12), (5.13) respectively give back the already familiar versions[7] of the generalised Codazzi, Gauss, and Schouten identities for the individual embedded p-surfaces of the foliation. The corresponding doubly lateral projections would give the analogous results for the orthogonal foliation by (n-p)-surfaces that would exist in the irrotational case for which the lateral turning tensor given by (2.10) is symmetric. Finally the corresponding mixed tangential and lateral projections give results of a less familiar kind, whose discussion will be omitted here due to lack of space and time: in particular, it will be shown elsewhere that the recently derived generalisation[15] of the well known Raychaudhuri equation (that was originally formulated for a foliation by 1-dimensional curves) is interpretable as the special case given by the mixed projection of the identity (5.11), of which complete version presented here might therefore appropriately be described as the extended *Raychaudhuri-Codazzi* identity.

6. The amalgamated embedding curvature tensor

To get back from the extended foliation curvature tensors that have just been introduced, to their antecedent analogues for an individual embedded surface, the first step is to construct the *amalgamated embedding curvature* tensor, $F_{\mu\nu}{}^{\rho}{}_{\sigma}$ say, which will be obtainable from the corresponding amalgamated foliation curvature $\mathcal{F}_{\mu\nu}{}^{\rho}{}_{\sigma}$ by a doubly tangential projection having the form

$$F_{\mu\nu}{}^{\rho}{}_{\sigma} = \eta^{\alpha}{}_{\mu}\,\eta^{\beta}{}_{\nu}\mathcal{F}_{\alpha\beta}{}^{\rho}{}_{\sigma} \,. \tag{6.1}$$

As did the extended foliation curvature, so also this amalgated embedding curvature will separate as the sum of "inner" and "outer" parts in the form

$$F_{\mu\nu}{}^{\rho}{}_{\sigma} = R_{\mu\nu}{}^{\rho}{}_{\sigma} + \Omega_{\mu\nu}{}^{\rho}{}_{\sigma} \,, \tag{6.2}$$

in which the "inner" embedding curvature is given by another doubly tangential projection as

$$R_{\mu\nu}{}^{\rho}{}_{\sigma} = F_{\mu\nu}{}^{\kappa}{}_{\lambda}\,\eta^{\rho}{}_{\kappa}\,\eta^{\lambda}{}_{\sigma} = \eta^{\alpha}{}_{\mu}\eta^{\beta}{}_{\nu}\mathcal{P}_{\alpha\beta}{}^{\rho}{}_{\sigma} \,, \tag{6.3}$$

while the "outer" embedding curvature is given by the corresponding doubly lateral projection as

$$\Omega_{\mu\nu}{}^{\rho}{}_{\sigma} = F_{\mu\nu}{}^{\kappa}{}_{\lambda}\gamma^{\rho}{}_{\kappa}\,\gamma^{\lambda}{}_{\sigma} = \eta^{\alpha}{}_{\mu}\eta^{\beta}{}_{\nu}\mathcal{Q}_{\alpha\beta}{}^{\rho}{}_{\sigma} \,. \tag{6.4}$$

Again, as in the case of an extended foliation, the corresponding mixed tangential and lateral projection of the amalgamated embedding curvature must vanish identically,

$$F_{\mu\nu}{}^{\kappa}{}_{\lambda}\,\eta^{\rho}{}_{\kappa}\gamma^{\lambda}{}_{\sigma} = 0 \,, \tag{6.5}$$

so that by (3.4) one recovers the ordinary tensorial version[7] of the Codazzi identity, according to which the antisymmetric part of the third fundamental tensor is given by

$$2\Xi_{[\mu\nu]}{}^{\rho}{}_{\sigma} + \eta^{\alpha}{}_{\mu}\, g^{\beta}{}_{\nu}\, B_{\alpha\beta}{}^{\kappa}{}_{\lambda}\, \eta^{\rho}{}_{\kappa} \gamma^{\lambda}{}_{\sigma} = 0\ . \tag{6.6}$$

One similarly obtains the familiar generalised Gauss equation for the (Riemann type) "inner" curvature of the embedding in the standard tensorial form

$$R_{\mu\nu}{}^{\rho}{}_{\sigma} = 2K_{[\nu}{}^{\rho\tau} K_{\mu]\sigma\tau} + \eta^{\alpha}{}_{\mu}\eta^{\beta}{}_{\nu} B_{\alpha\beta}{}^{\kappa}{}_{\lambda}\, \eta^{\rho}{}_{\kappa}\, \eta^{\lambda}{}_{\sigma}\ , \tag{6.7}$$

while finally the ordinary albeit not so widely familiar Schouten equation for the "outer" embedding curvature is analogously obtained as

$$\Omega_{\mu\nu}{}^{\rho}{}_{\sigma} = 2K_{[\mu}{}^{\tau\rho} K_{\nu]\tau\sigma} + \eta^{\alpha}{}_{\mu} g^{\beta}{}_{\nu} B_{\alpha\beta}{}^{\kappa}{}_{\lambda} \gamma^{\rho}{}_{\kappa} \gamma^{\lambda}{}_{\sigma}\ . \tag{6.8}$$

As pointed out in the recent detailed analysis of the embedding case, this last expression is replacable by a more informative formula [7] expressing the outer curvature $\Omega_{\mu\nu}{}^{\rho}{}_{\sigma}$ in a form that is independent both of the Ricci trace part of the background curvature (whose vanishing expresses Einstein's vacuum gravitational field equations) defined by

$$B_{\mu\nu} = B_{\lambda\mu}{}^{\lambda}{}_{\nu}\ , \qquad R = B_{\mu}{}^{\mu}\ , \tag{6.9}$$

and also of the trace vector of the second fundamental tensor (whose vanishing expresses the dynamical equations of motion for a Dirac membrane or Goto Nambu string) defined by

$$K^{\mu} = K^{\lambda}{}_{\lambda}{}^{\mu}\ . \tag{6.10}$$

In terms just of the trace free *Weyl tensor* given in a background with dimension n$>$ 2 by

$$W_{\mu\nu}{}^{\rho\sigma} = B_{\mu\nu}{}^{\rho\sigma} - \frac{4}{\mathrm{n}-2}\, g_{[\mu}{}^{[\rho} B_{\nu]}{}^{\sigma]} + \frac{2}{\text{(n-2)(n-1)}} B g_{[\mu}{}^{[\rho} g_{\nu]}{}^{\sigma]}\ , \tag{6.11}$$

and of the trace free *extrinsic conformation tensor* defined[7] for an embedding of dimension p by

$$C^{\mu}{}_{\nu}{}^{\rho} = K^{\mu}{}_{\nu}{}^{\rho} - \frac{1}{\mathrm{p}}\eta^{\mu}{}_{\nu} K^{\rho}\ , \tag{6.12}$$

the cruder version (6.8) is replaceable by the more refined formula

$$\Omega_{\mu\nu}{}^{\rho}{}_{\sigma} = 2C_{[\mu}{}^{\tau\rho} C_{\nu]\tau\sigma} + \eta^{\alpha}{}_{\mu}\eta^{\beta}{}_{\nu} W_{\alpha\beta}{}^{\kappa}{}_{\lambda} \gamma^{\rho}{}_{\kappa} \gamma^{\lambda}{}_{\sigma}\ . \tag{6.13}$$

for the (itself trace free and conformally invariant) outer curvature tensor.

REFERENCES

[1] J. Stachel, *Phys. Rev.* **D21**, 2171 (1980); **D12**, 2182 (1980).

[2] W. Kopczynski, *Phys. Rev.* **D36**, 3582 (1987); **D36**, 3589 (1987).

[3] D. H. Hartley, R.W. Tucker, in *Geometry of Low Dimensional Manifolds, 1* , L.M.S. Lecture Note Series **150**, ed. S. Donaldson, C. Thomas (Cambridge U.P., 1990).

[4] B. Carter, in *Formation and Evolution of Cosmic Strings,* ed. G. Gibbons, S. Hawking, T. Vachaspati, pp143-178 (Cambridge U.P., 1990).

[5] B. Boisseau, C. Barrabes, *Physics Letters* **B279**, 259 (1992).

[6] B. Boisseau, P.S. Letelier, *Phys. Rev.* **D46**, 1721 (1992).

[7] B. Carter, *J. Geom. Phys.* **8**, 53 (1992).

[8] B. Carter, *J. Class. Quantum Grav.* **9**, 19 (1992).

[9] J. Garriga, A. Vilenkin, *Phys. Rev.* **D47**, 3265 (1993).

[10] J. Guven, *Phys. Rev.* **D48**, 4464 (1993).

[11] B. Carter, *Phys. Rev.* **D48**, 4835 (1993).

[12] J. Guven, *Phys. Rev.* **D48**, 5563 (1993).

[13] A.L. Larsen, V. Frolov, *Nucl. Phys.* **B414**, 129 (1994).

[14] R. Capovilla, J. Guven, *Phys. Rev.* **D51**, 6736 (1995).

[15] R. Capovilla, J. Guven, *Phys. Rev.* **D52**, 1072 (1995).

[16] R.A. Battye, B. Carter, *Phys. Lett.* **B357**, 29 (1995).

[17] J.A. Schouten, *Ricci Calculus* (Springer, Heidelberg, 1954).

[18] S. Kobayashi, K. Nomizu *Foundations of Differential Geometry* (Intersciences, New York, 1969).

[19] B.Y. Chen, *Geometry of Submanifolds* (Dekker, New York, 1973).

[20] M. Spivak, *Differential Geometry* (Publish or Perish, Berkeley, 1979).

[21] Y. Choquet-Bruhat, C. DeWitt-Morette *Analysis, Manifolds, Physics II* (North Holland, Amsterdam, 1989).

[22] L.P. Eisenhart, *Riemannian Geometry* (Princeton U.P., 1926, reprinted 1960).

DÉPARTEMENT D'ASTROPHYSIQUE RELATIVISTE ET DE COSMOLOGIE, CNRS UPR-176, OBSERVATOIRE DE PARIS-MEUDON, 92195 MEUDON, FRANCE.

Contemporary Mathematics
Volume **203**, 1997

Spaces Admitting a Foliation by Isotropic Hypersurfaces

T.PAPAKOSTAS

ABSTRACT. All spaces solutions of Einstein's field equations admitting a foliation by isotropic (null) hypersurfaces (here after referred under the acronymus "ishyps") are determined in a geometric way and the general metric for these spaces is given.

1. Introduction

The study of submanifolds with a degenerate matric,immersed in a pseudo-Reimannian (Lorentzian) manifold has been discarded in general in spite the great importance of isotropic submanifolds in the context of the General Theory of Relativity.

In fact,gravitational waves in General Relativity can be assimilated to "ishyps" and it's important to find out the spaces, solutions of Einstein's field equations admitting a foliation by such hypersurfaces. Then it will be possible to construct models for the propagation of gravitational waves.

In previous publications [1] [2] the conditions which permit to construct the bundle of adapted frames over an ishyps, have been found, giving raise to the so called radiative structures or G-structures.

The purpose of this paper is to present the general metric of the spaces of General Relativity,admitting a foliation by isotropic hypersurfaces, to determine the Petrov type of the corresponding Weyl tensor and the properties of the null vector generating the ishyps,we do not impose any particular energy-momentum tensor.

1991 *Mathematics Subject Classification.* Primary 53C22 ; Secondary 57R30..

2. Bundle of adapted frames over an ishyps

Let $(V_{3,1}$ g) be a Lorentzian space time, the metric g being of signature -2. Let b denotes the degenetate metric induced by g, on the isotropic hypersurface S and [n] the line field, generating the kernel of b i.e.

$$[n] = \{kn, k \in R^{\cdot}/b(n) = 0\}.$$

One can suppose that n is the restriction to S of an isotropic (null) vector field in $V_{3,1}$ also denoted by n ,so g(n,n) = 0. At each point x of S, the cone of the isotropic directions of x , has a first order contact with S, but the tangent scape T_xS, does not contain any time-like vector, the future light cone at x being entirely on one side of S and the past light cone at x entirely on the other side.

We shall work in the moving frames of the complex vectorial formalism [3] [4] [5]. A covariant null tetrad of one forms h^a (a = 1,2,3,4) is defined, in which the metric can be written under the form:

$$ds^2 = g = h^1 \otimes h^2 + h^2 \otimes h^1 - h^3 \otimes h^4 - h^4 \otimes h^3 \tag{1}$$

In local coordinates system (X^i), we have (i=1,2,3,4)

$$h^1 = n_i dx^i \quad , \quad h^2 = l_i dx^i \quad , \quad h^3 = -\bar{m}_i dx^i \quad , \quad h^4 = -m_i dx^i \tag{2}$$

Where l,n are real and null and $\bar{m}$,m complex conjugate and null.

In the space of complex self dmal 2-forms, with basis:

$$Z^1 = h^1 \wedge h^2 \ , \ Z^2 = h^1 \wedge h^2 - h^3 \wedge h^4 \ , \ Z^3 = h^4 \wedge h^2 \quad a, b, c = 1, 2, 3 \tag{3}$$

the complex connection 1-forms $s^a{}_b$ are defined by:

$$dZ^a = s^a{}_b \wedge Z^b \tag{4}$$

related to the vectorial connection 1-forms s_a by:

$$s_a = (1/8) e_{abc} c^{cd} s^b{}_d \tag{5}$$

e_{abc} is the 3-permutation symbol and c^{cd} the metric components in the space of the complex self dual 2-forms.

The tetrad components s_α defined by $s_\alpha = s_{\alpha a} h^a$ are 12 complex valued functions, which are none others than the Newman-Penrose (N-P) spin coefficient.

$$C_{\alpha a} = \begin{pmatrix} j & t & s & q \\ e & c & b & a \\ p & m & l & k \end{pmatrix} \tag{6}$$

The complex curvate 2-forms $S^a{}_b$ are defined by:

$$ds^a{}_b - s^a{}_c \wedge s^c{}_b = S^a{}_b \tag{7}$$

and the vectorial curvature 2-forms by:

$$S_a = (1/8)e_{abc}c^{cd}S^b{}_d \tag{8}$$

On expanding S_a in basis $[Z^a, \bar{Z}^a]$ one obtains

$$S_a = (C_{ab} - 1/6Rc_{ab})F^b + E_{a\bar{b}}\bar{Z}^b \tag{9}$$

where the quantities C_{ab} and $E_{a\bar{b}}$ are related to the NP curvature components X_a and V_{ab} as follows:

$$C_{ab} = \begin{pmatrix} X_0 & X_1 & X_2 \\ X_1 & X_2 & X_3 \\ X_2 & X_3 & X_4 \end{pmatrix}, \quad V_{ab} = \begin{pmatrix} V_{00} & V_{01} & V_{02} \\ V_{10} & V_{11} & V_{12} \\ V_{21} & V_{20} & V_{22} \end{pmatrix} \tag{10}$$

C_{ab} and $E_{a\bar{b}}$ are the irreducible components of the Riemann Tensor.

If we assume that the curves tangent to n are given to us "a priori" by the geometry, the "natural" choice of the tetrad (1) is restricted. The local frame is determined in the subgroup G_n of the Lorentz group:

$$\begin{aligned} \tilde{h}^1 &= \exp^a h^1 \\ \tilde{h}^2 &= \exp^{-a}(q\bar{q}h^1 + h^2 + qh^3 + \bar{q}h^4) \\ \tilde{h}^3 &= \exp^{ib}(h^3 + \bar{q}h^1) \\ \tilde{h}^4 &= \exp^{-ib}(h^4 + qh^1) \end{aligned} \tag{11}$$

q=c+id ,a,b,c,d are functions of the coordinates.

If we identify n the null vector field which generates the ishyps with the null vector n of the tetrad (1), then the invariance of this isotropic direction by the action of the structural group G acting on the bundle of the pseudo-orthonormal frames, which is the group $O(3,1)$ permit , to write G as the semidirect product: $(R \otimes O(2) \times R^2)$, the consequences of this decomposition in the level of the connection are summarized by the following proposition [1]:

PROPOSITION 1. : *Any moving frame adapted to an ishyps possesses the property:*

$$p = m = l = k = 0 \tag{12}$$

The annulation of these four spin coefficients will permits us to characterize the null vector n and to find the metric tensor satisfying (12).

3. General metric of spaces admitting a foliation by ishyps

Relations (12), substituted in the Cartan's structure equation and the Frobenius theorem imply the existance of a local coordinate system (u,v,x,y) such that:

$$\begin{aligned} h^1 &= Rdu \\ h^2 &= Mdu + Ndv + Kdx + Qdy \\ h^3 &= (Edu + Fdx + Gdy) + i(Hdu + Pdx + Ldy) \\ h^4 &= (Edu + Fdx + Gdy) - i(Hdu + Pdx + Ldy) \end{aligned} \tag{13}$$

where E,F,G,H,K,L,M,N,O,P,Q,R are real valued functions of the coordinates.

We can use the transformations (11) to get another null frame:

$$\begin{aligned}\tilde{h}^1 &= du \\ \tilde{h}^2 &= \tilde{M}du + \tilde{N}dv \\ \tilde{h}^3 &= (\tilde{E}du + \tilde{F}dx + \tilde{G}dy) + i(\tilde{H}du + \tilde{P}dx + \tilde{L}dy) \\ \tilde{h}^4 &= (\tilde{E}d\acute{u} + \tilde{F}dx + \tilde{G}dy) + i(\tilde{H}du + \tilde{P}dx + \tilde{L}dy)\end{aligned} \tag{14}$$

The transformations which makes $\tilde{h}^1$ a closed form is the transformation permitting to choose a of (11) in such way that the equation:

$$n^j \nabla_j n_i = pn_i \tag{15}$$

which is valid because of (12) and states that the null vector n is geodesic, becomes:

$$n^j \nabla_j n_i = 0 \tag{16}$$

If we drop , the metric corresponding to the choice (14) can be written:

$$\begin{aligned}ds^2 = g = 2du(Mdu + Ndv) - 2(Edu + Fdx + G + dy)^2 - \\ -2(Hdu + Pdx + Ldy)^2\end{aligned} \tag{17}$$

In this tetrad the NP spin coefficients satisfy the conditions:

$$p = m = l = k = 0 \tag{18}$$

$$\bar{a} + b = c + \bar{c} = q - \bar{q} = 0 \tag{19}$$

Substitution of (18),(19) in the NP field equations gives us some conditions on the curvature components X_A and V_{ab}:

$$X_3 = X_4 = X_2 - \bar{X}_2 = 0 \tag{20}$$

$$V_{02} = V_{20} = V_{12} = V_{21} = V_{22} = 0 \tag{21}$$

But the relations (18),(19) permit us to determine the functional dependance of some of the functions in (17).Calculating the spin coefficients of metric (17) and implying conditions (18),(19) we get the following differential equations:

$$PG_v + GP_v - LF_v = FL_v \tag{22}$$

$$PG_v - GP_v - LF_v = -FL_v \tag{23}$$

$$FG_v + FP_v + GF_v = -PL_v \tag{24}$$

Where

$$G = \frac{\partial G}{\partial v}\ ,\ P_v = \frac{\partial P}{\partial v}\ \ etc$$

This system is satisfied if

$$G_v = -\frac{L}{G}L_v\ ,\ P_v = \frac{F}{G}L_v\ ,\ F_v = -\frac{P}{G}L_v \tag{25}$$

Then expanding the terms of the metric (17) we get:

$$ds^2 = g = 2\{du(Mdu + Ndv) - [(E^2 + H^2)du^2 + 2(EF + HP)dudx+$$

$$+2(EG + HL)dudy + (F^2 + P^2)dx^2 + 2(FG + PL)dxdy + (G^2 + L^2)dy^2]\} \quad (26)$$

We can prove by straight forward calculation,using relations (24), that the metric components:

$$g_{xx} = F^2 + P^2 \ , \ g_{xy} = FG + PL \ , \ g_{yy} = G^2 + L^2 \quad (27)$$

are independant of the coordinate u, that is:

$$\frac{\partial g_{xx}}{\partial v} = \frac{\partial g_{xy}}{\partial v} = \frac{\partial g_{yy}}{\partial v} = 0. \quad (28)$$

Then we can rewrite the metric (26) in a different null frame, this transformations leaving (18)-(21) invariant:

$$ds^2 = 2du(Mdu + Ndv + Pdx + Qdy) - 2(Fdx + Gdy)^2 - 2(Edx + Hdy)^2 \quad (29)$$

where

$$F = F(u, x, y) \ , \ G = G(u, x, y) \ , \ E = E(u, x, y) \ , \ H = H(u, x, y).$$

Then the consequences of (18)-(21),the equation (16) and the equation giving the decomposition of a geodesic null vector field [7], can be summarized, in the following theorem:

THEOREM 1. : *Spaces admitting a foliation by isotropic hypersurfaces are algebraically special in the Petrov classification. The null vector generating the ishyps is geodesic, shear free and non diverging and the general metric is given by (29).*

This line element is more convenient for calculations and it will be used in the future for the construction of models for gravitational radiation. This metric belongs to Kundt's class of solutions of Einstein's equations [7]. The metric of the ishyps is deduced by (29) putting u=constant and it is degenerate.

Two subcases arise from this general metric [6]:

The first corresponds to the additional assumption that n is also covariantly constant,this is the G_1–connection case and the corresponding spaces are well studied [$P = Q = G = E = 0$, $F = H$ in (29)] [7]

The second is obtained imposing the condition that the ambiant space-time connection induces a unique connection over the ishyps. This unique connection has been called "generalized Robinson–Bertotti spaces" because the corresponding metric, solution of Einstein–Maxwell with non singular electomagnetic field is the "Robinson–Bertotti universe". [$M = P = Q = G = E = 0$, $F = H$ in (29)]. In this case the metric is the product of two 2-dimensional spaces of constant curvature and hence admits a six parameter group of motions [8].

4. Concluding Remarks

We have obtained the general metric of spaces of General Relativity admitting a foliation by isotropic hypersurfaces, which includes two very interesting subcases already presented [6]. It is worth noting that the case of G_0 - connection has been used for the quantization of the gravitational field [9].

Two projects seem to be promising:

The first is to contrust a model describing the propagation of gravitional radiation, far from the source.

The second is to use the foliation by isotropic hypersurfaces to construct a Hamiltonian formalism for General Relativity,á la Ashtekar, using the coordinate u as time coordinate, this will not be the first attempt (see for example [10]) but in the others tentatives there are no clear arguments about the metric supposed to describe the foliation of space by isotropic hypersurfaces.

References

[1]. G. Burdet, M. Perrin, *Les ondes gravitationnelles*, Math. and Fund. Applications **50** (1992).

[2]. G. Burdet, M. Perrin, *Gravitational Waves Without Gravitons*, Lett. Math. Phys. **25** (1992), 39-45.

[3]. M. Cahen, R. Debever, L. Defrise, *A complex vectorial formalism in general relativity*, Jour. Math. Mechanics **16** (1967), 761-785.

[4]. R. Debever, *Le rayonnement gravitationnel, le tenseur de Riemann en relativité générale*, Can. Phys. **118** (1964), 303-349.

[5]. R. Debever, R. G. McLenaghan, *Orthogonal transitivity, invertibility and null geodesic separability in type D electrovac solutions of Einstein's field equations with cosmological constant*, J. Math. Phys. **22** (1981), 1711-1726.

[6]. G. Burdet, T. Papacostas, M. Perrin, *Lorentzian manifolds admitting isotropic hypersurfaces solutions of Einstein's field equations*, Int. J. of Mod. Physics **D 3** (1994), 163-166.

[7]. D. Kramer, H. Stephani,M. Maccalum,E. Hertl, Exact Solutions of Einstein's Flield Equations, VEB, 1980.

[8]. G. Burdet, T. Papacostas, M. Perrin, *Sapce-time metric inducing the radiation connection*, Gen. Rel. Grav. **26** (1994), 225-232.

[9]. G. Burdet, M. Perrin, *Speculation sur la cosmologie quantique primordiale*, Lett in Math Physics **30** (1994), 317-325.

[10].J. N. Goldberg,D. C. Robinson,C. Soteriou, *Null Hypersurfaces and New Variables*, Class. Quant. Grav. **9** (1992), 1309-1328.

Technical University of Crete,Department of General studies, Polytechnioupolis,73100 CHANIA ,GREECE

E-mail address: taxiar@orion.physics.uch.gr

Contemporary Mathematics
Volume **203**, 1997

An alternative to Inflation

R. TRIAY

ABSTRACT. While Inflation is understood as a necessary ingredient of the standard world model, the short extragalactic distance scale suggested by recent measurements of the Hubble constant forces us to modify drastically such a viewpoint. The present speculation shows that a spatially closed geometry is a possible issue which reconciles the estimate age of the universe with realistic estimates of the density parameter. It turns out that the Horizon and the Flatness problems can be overcome within such a geometry.

1. Introduction

From Clifford's *Lectures and Essays* one can read "The subject of science is the human universe; that is to say, everything that is, or has been, or may be related to man. Let us then, taking several topics in succession, endeavor to make out in what cases thought about them is scientific, and in what cases not". One can easily agrees that Nature is understood by modeling virtual experiences (which are supposed to be as close as possible to reality), the number and the accuracy of predictions ensure the credibility of the related model, and Mathematics help us for having a deeper understanding when mathematical structures (e.g., symmetries) are present in the model. In the past, Cosmology was a typical example where the object of the investigation is unique, and such a weakness may explain why sensible ideas went forward warily among numerous speculations. Nevertheless, such a step provided us with the standard picture of the universe. Today, this science is getting more and more interests because of the increasing quality of observational devices. The richness of problems motivates fruitful collaborations between scientific communities, from pure observation to mathematical physics, although the difference of languages is a source of possible misunderstandings (or religious beliefs). With this in mind, I speculate on a quite controversial improvement of the standard picture which was given in the 80's by Inflation[**28, 1**]. Such an scenario was motivated by quantum field theory, in support of the SU(5) approach of grand unification theory (GUT), which sought to unify the strong and the electroweak interactions. While this bet turned out to be unsuccessful[**6**] (e.g., the proton mean life-time $> 1.6\,10^{16}$ Gyr[**46**] exceeding the age of the universe, when it was supposed to decay, see e.g. [**40**]), Inflation is still understood as a necessary

1991 *Mathematics Subject Classification.* Primary 83F05; Secondary 85A40 .

ingredient of the standard picture. The main reasons are the interpretation of both the isotropy of the cosmological background radiation (CBR), and the smallness of the spatial curvature. However, this scenario provides us with a flat geometry (the Einstein de-Sitter model) which can be questioned by recent observations. On the other hand, it turns out that the above features can be understood within a spatially closed universe[**71, 72**], as provided by the Standard world model with a non vanishing cosmological constant Λ. Actually, such a world model was already proposed according to neo-classical tests applied on quasars[**21**], and on brightest cluster galaxies[**8, 9**], it shows the possibility to restore a likely Baryon symmetric cosmology[**22, 70**] with regard to observations[**15**]. It is interesting to emphasize the fact that the universe has a globally zero electric charge (within an accuracy of 10^{-40} otherwise the electrostatic repulsion makes the gravitational attraction imperceptible) is a fundamental observation which can be easily understood within a baryon symmetric cosmologies[**64**]. The aim of the present paper is to give enlightening on such an alternative.

Section 2 introduces the standard world model, and explains why the isotropy and the homogeneity are still reliable working hypotheses for the standard picture, when present observations shows large scale structures in the distribution of galaxies. Section 3 is a polemic which describes how Inflation came on the scene. Section 4 introduces the Alternative.

2. The standard world model

The standard picture of the universe accounts for an homogeneous and isotropical distribution of dust as source of gravity, it is given by Friedmann's model[**24**]. Because the solution of Einstein's equation is a spacetime metric of Robertson-Walker (RW) type[**56, 74**], it is also known as FRW-model. Although such a model provides us with a quite limited understanding of cosmic structures, it turns out to be a sensible basis to Modern Cosmology, which begun with Lemaitre[**41**] interpretation of the Hubble law[**36**] as a general expansion of the universe. The extrapolation down to earlier epochs justified the Gamov's hot big-bang world model[**25**] and the treatment of the *Origin problem* [**51**]. Such investigations improved the standard picture by several steps. The former describes a radiation dominated universe in thermal equilibrium, and provides us with the interpretation of the cosmological background radiation (CBR) and the primordial nucleosynthesis[**5**]. The later are speculations, with no direct links with observations, which shows that an *initial singularity* must have occurred in the past[**33**]. Nowadays, contrary to working hypotheses on the homogeneity and the isotropy (of the space) which were settled at hand, the observations show that the galaxies are not as regularly distributed as that. Indeed, the presence of large scale structures[**14**], cosmic velocities fields[**68**], gravitational lenses or arcs[**60**], angular scale CBR temperature fluctuations[**58**], ..., force us to ask whether such an approach still provides us with a likely description of the universe. However, by using a geometrical interpretation of the CBR isotropy together with the recession of galaxies, it has been proved[**63**] that the space-time V_4 has a RW-metric $g_{\mu\nu}$, which rehabilitate the Standard picture defined above[1]. The key-point of this proof is that the CBR shows a blackbody spectrum. Being characteristic of a thermodynamics equilibrium, it can be defined

[1]Observations suggest that the largest scale up to which inhomogeneities appear is smaller than 100 Mpc.

by a Planckian temperature $\vec{\theta}_{\rm Pl}$ (it is a future-oriented 4-vector of length $1/{\rm k}T$, where k is the Boltzmann constant and T is the temperature). The construction of this vectors field assumes that any observer, which is located on a test particle of the cosmic fluid, is able to find a direction where to move for observing the CBR as an isotropical blackbody radiation[2] (according to Cosmological Principle). Thermodynamics tells us that $\vec{\theta}_{\rm Pl}$ is a conformal Killing vector for the space-time metric (i.e., the Lie-derivative satisfies $\mathcal{L}_{\vec{\theta}_{\rm Pl}} g_{\mu\nu} \propto g_{\mu\nu}$), and thus V_4 reads as the product of an homogeneous 3-dimension manifold V_3, where the galaxies have fixed positions, (the *comoving space*) and the cosmic time t (or the temperature T).

The RW-metric can be written as follows

$$\mathrm{d}s^2 = a^2(t)\left(\mathrm{d}\tilde{t}^2 - \mathrm{d}\sigma^2\right), \tag{1}$$

where $a(t) \propto 1/{\rm k}T$ is the (dimensionless) *expansion parameter*, $\mathrm{d}\tilde{t} = \mathrm{d}t/a(t)$ is the conformal time element and $\mathrm{d}\sigma^2$ is the metric element of V_3, one can choose $a_\circ = 1$ today ($t = t_\circ$), the past corresponds to $a < 1$ and the future to $a > 1$, since the universe is expanding. Since Friedmann's model do not account for epochs prior to decoupling, when the electrons combined with ions to form neutral H-atom at $t = t_{\rm dec} \approx 10^{14}$sec, the standard picture is given by Gamov's model, which accounts for an homogeneous and isotropical distribution of relativistic particles in thermal equilibrium. Such a picture can be improved if the stress-energy tensor $T_{\mu\nu}$ takes into accounts both CBR and dust like matter (although not interacting), and if the cosmological constant is not supposed a priori to be zero, such a solution is called Friedmann-Lemaitre-Gamov (FLG) model. At earlier epochs this model coincides to Gamov model, and the extrapolation of $\vec{\theta}_{\rm Pl}$ down to $t = t_{\rm Pl}$ could be used for describing an ideal gas, but however with some necessary caution.

The expansion law in term of a is given by the variation of the Hubble constant

$$H(t) = \frac{\dot{a}}{a} = H_\circ\sqrt{\lambda_\circ - \frac{k_\circ}{a^2} + \frac{\Omega_\circ}{a^3} + \frac{\alpha_\circ}{a^4}}, \tag{2}$$

where $H_\circ$, resp. the coefficients $\lambda_\circ$, $k_\circ$, $\Omega_\circ$ and $\alpha_\circ$, denote the present values of the Hubble constant, resp. the cosmological parameters, which specifies the candidate solution within the FLG model. The cosmological parameters are dimensionless values defined as follows:

- $\lambda = \frac{1}{3}\Lambda H^{-2}$ is the reduced cosmological constant;
- $k = KH^{-2}$ is the curvature parameter, where K is the curvature scalar of the comoving space V_3, its sign (invariant) allows us to choose between a Riemannian ($K > 0$), Euclidian ($K = 0$) or Lobatchevski ($K < 0$) geometry for V_3;
- $\Omega = \rho/\rho_{\rm c}$ is the density parameter, where ρ is the specific density of massive particles, and $\rho_{\rm c} = \frac{3}{8}\pi^{-1}{\rm G}^{-1}H^2$ is the critical energy density, G is the Newton's constant;
- $\alpha = \frac{8}{45}\pi^3{\rm G}({\rm k}T)^4\hbar^{-3}H^{-2}$ is the radiation parameter, which accounts for the CBR photon, where $\hbar$ is the Planck constant.

These parameters verify the normalization condition

$$1 = \lambda - k + \Omega + \alpha, \tag{3}$$

[2]In practice, the peculiar velocity of the Galaxy is obtained by removing the dipole anisotropy due to Doppler shift.

which can be interpreted as a scale free formulation of eq. (2).

The chronology $t \mapsto a$ is obtained by integrating the differential equation in term of t given in eq. (2), which provides us with the relation $a \mapsto t$. It is specified by measuring the values of cosmological parameters from observations, either by direct estimations or by fitting simulations, which represents several experience in essence. The behaviors of cosmological parameters with time (or equivalently with a) are easily derived from little algebra. As usual, one can say that the dynamics of the expansion is dominated either by the vacuum, the curvature, the matter or the radiation, provided the related cosmological parameter shows the larger value.

2.1. Observational constraints. The quest of cosmological parameters was the aim of observational cosmology during decades. The present state of the art provides us with the following results :

- Estimations of the cosmological constant give $\Lambda < 2\,10^{-55}\,\mathrm{cm}^{-2}$ from dynamics in clusters[**48**], and $-2\,10^{-56}\,\mathrm{cm}^{-2} \leq \Lambda < 4\,10^{-56}\,\mathrm{cm}^{-2}$ from the minimum age of the universe and the existence of high redshift objects[**10**], which reads $-1.75h^{-2} < \lambda_\circ \leq 1.14h^{-2}$, where $h = H_\circ/(100\,\mathrm{km\,sec}^{-1}/\mathrm{Mpc})$;
- A consensus tells us that in the neo-classical tests (based on the luminosity of sources, their size ...), evolutionary effects of sources masks the geometry. Hence, there is not yet a common agreement on an direct estimate of the curvature parameter k, even on its sign.
- Current estimates from dynamics of dark halo of galaxies up to scales $10h^{-1}$ Mpc give[**50**] a density parameter $\Omega_\circ \leq 0.3$.
- Finally, the only well known parameter is the present CBR temperature[**75**] $T_\circ = 2.73 \pm 0.03°\mathrm{K}$, which provides us with a radiation parameter $\alpha_\circ \approx 5\,10^{-5}h^{-2}$.

2.2. About the cosmological constant. The cosmological constant Λ has long been wrongly interpreted (almost unanimously) as a free parameter of Einstein's equations[3]. According to *Principles of General Relativity*[**20, 62**], Einstein's equation tells us how the spacetime adapts its geometry to physical constraints. The former is described by a tensor $S_{\mu\nu} = A\,g_{\mu\nu} + B\left(R_{\mu\nu} - \frac{1}{2}R\,g_{\mu\nu}\right) + \dots$, which is defined by means of invariant geometrical quantities, in such a way that the higher the order of the expansion the smaller the scale of the description, and the later are described by a (vanishing riemannian divergence) stress-energy tensor $T_{\mu\nu}$ according to usual principles in mechanics. The terms A and B are coupling constants which identify to fundamental constants $\Lambda = A/B$ and $\mathrm{G} = (8\pi B)^{-1}$, which have to be measured by observations. Since a positive Λ competes against Newton gravity, quantum fields theory suggests that $\Lambda g_{\mu\nu}/8\pi\mathrm{G}$ accounts for the quantum vacuum energy, when it must be noted that such a quantity lies solely in the tensor $T_{\mu\nu}$ and intervenes at Planck scale. Additional arguments[**65**] shows that a quantum status for Λ looks more like a decoy than derived from geometrical principles in quantum mechanics.

2.3. Boundary conditions. The present values of cosmological parameters can be classified according to behaviors of the world model, regarded as a dynamical system. Fig. 1 shows the related $\lambda_\circ$-$\Omega_\circ$ domain of acceptance, cosmologies

[3] It has been introduced solely for obtaining a static universe, when the Hubble expansion could be predicted.

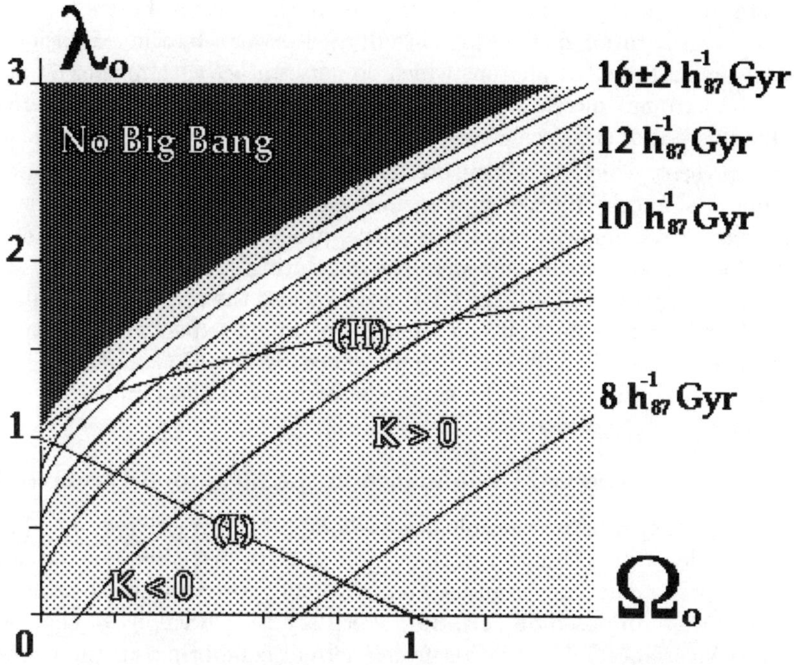

FIGURE 1. *World models in agreement with observational constraints.* Theoretical and observational constraints provide $0 < \Omega_\circ < 0.3$ and $0 \leq \lambda_\circ \leq 1.425$. Zero curvature world models are situated on the straight line (I), isocurvature lines are parallels to this line, with a positive curvature above. The white area provides us with an Age of the universe of $t_\circ = 16 \pm 2\,\mathrm{Gyr}$, curves corresponding to an Age of $t_\circ = 14, 10, 8\,\mathrm{Gyr}$ are also given ($h_{87} = H_\circ/87\mathrm{km/s}Mpc^{-1}$. universes with no Big-Bang are located in the dark area. The world models about the curve (II) show a Last Scattering Surface within a unique causal cell, which provides us with an issue to the Horizon problem.

without *big-bang* (see black region) or with a *big-crunch* (the related domain is located within the width of $\Omega_\circ$-axis) are excluded. The initial values of cosmological parameters define constraints that must be satisfied by the candidate world model at $t < t_{\mathrm{dec}}$, where the reasons why the universe is homogeneous and isotropic could be found. It has been argued[**28**] "... the standard model requires the assumption of initial conditions (taken at Planck time $t_{\mathrm{Pl}} = 5.38 \times 10^{-44}\,\mathrm{s}$) which are very implausible for two reason : • Causally disconnected regions are assumed to be nearly identical; in particular they are simultaneously at the same temperature (the *Horizon problem*). • For a fixed initial temperature, the initial value of the Hubble constant must be fine tuned to extraordinary accuracy to produce a universe which is as flat as the one we observe (the *Flatness problem*).". However, one might ask whether sensible investigations of these problems requires one to be aware of the degree of accuracy of the FLG description of the universe before $t_{\rm dec}$. Indeed, nowaday the gravitational structures are diluted enough by the expansion[4] so that the large amount of CBR photons which do not interact with matter let us know about the background metric. But the observed CBR temperature fluctuations $\delta T/T$ (due to seeds of large scale structures) tell us that irregularities of the cosmic fluids were present, which do not ensures the existence of a Planckian temperature $\vec{\theta}_{\rm Pl}$ at primordial epochs. Indeed, in addition of non trivial physics (baryons genesis, ...), one has to emphasize that dissipative processes[5] are not taken into account by the above model, when they were present during standard physics[6]. The witnesses of such a physics are both the CBR isotropy and the presence of small angular scale CBR temperature fluctuations. Hence, one might expect that the true chronology of the expansion is different of that defined by the FLG model, despite the effect can be weak (because of the smallness of ratio of the number of baryons to the number of photons $\approx 10^{-10}$).

3. The GUT's influence in the 80's upon Cosmology

Baryon symmetric cosmologies has long been attractive[**4, 47**], but there is no clear evidence of antimatter in the visible universe[**67**]. In order to understand why matter predominates over antimatter in the universe, three basic principles of baryogenesis were proposed[**59, 19, 39**], which are : – : breaking of charge and charge-parity invariance; – : deviation from thermal equilibrium in the primeval plasma; – : baryonic charge non conservation. Such a scenario was likely within 80's GUT which provided us with two main predictions : • the production of monopoles[**18**] when the grand unified symmetry breaks (at a temperature $\sim 10^{15}$ GeV) in the early universe; • the proton decay. In addition of the fact that no monopole was observed, the first feature was a disaster because the expected density of monopoles is high enough to make the universe collapsing [**76, 54**]. Hence, the end justifying the means, a possible solution for preserving GUT from death was to introduce an Inflationary epoch[**31**] at date 10^{-35}s for expanding the universe enough (during $\sim 10^{-32}$s) to dilute the density of monopoles. As resulting from a (first order) phase transition due to a symmetry breaking, the Higgs field was at the origin of the inflationary expansion. Subsequent investigations pointed out the advantages of such a mechanism. The first one is that it could increase the causal scale of the universe[**38**], which overcomes the Horizon problem. The second one is to decrease the spatial curvature down to zero[**28**], which provides us with a solution to the Flatness problem. However, the model was discarded because the inflationary expansion could not ended without producing an inhomogeneous final state[**32**]. Such a failure showed that homogeneity is not a priori ensured as easily as that, as required by Friedmann's model (the *Homogeneity problem*). Hence, the Higgs potential well was chosen to overcome this problem, which provided us with the *New Inflation*[**42, 3**]. Although producing a scale-free Harison-Zel'dovich spectrum for the density fluctuations, it turned out that the final result was still suffering of similar defects[**34, 29, 66**]. Hence, a possible solution was to not base Inflation on the Coleman-Weinberg potential, but on supersymmetric theories[**7**]. Then an

[4]The gravitational lensing effects do not dominate in the sky, no apparent shear is present, ...

[5]Let us remind that such effects are responsible of the trend toward equilibrium.

[6]For example, at the nucleosynthesis epoch $t_{\rm ns} \approx 1$sec, the density ρ of matter is typical of the mean solar density.

additional scalar field (*Inflaton*) was added to GUT with the sole purpose of performing a huge expansion[**61, 52, 30**] and preserving low enough fluctuations[**2**]. Recently, according to the scenario of *Chaotic Inflation*[**43, 44**], Inflation is interpreted as a natural consequence of chaotic initial conditions in the early universe. Although totally disconnected, analogous ideas can be found in anterior literatures about exponential expansion due to vacuum energy[**26, 27**]. It is a matter of fact that with Inflation in mind and a quantum status for the cosmological constant, the constraint $\Lambda = 0$ forces a part of the community to interpret the *Dark-Matter problem* in a different way than observations suggest[7], i.e., to believe on $\Omega_\circ = 1$. In such a case, the candidate value for the Hubble constant is $H_\circ \approx 50 \mathrm{km\,s^{-1}Mpc^{-1}}$, according to the age of globular clusters $\approx 15\mathrm{Gyr}(= \frac{2}{3}h^{-1}9.78\mathrm{Gyr})$.

4. The Alternative

Nowadays, there are considerable grounds for doubting the assertion that we live in a critical density universe[**49, 55, 12**]. The short extragalactic distance scale obtained from recent measurements[**37, 53, 23**] of the Hubble constant $H_\circ = 87 \pm 7$ $\mathrm{km\,s^{-1}Mpc^{-1}}$ suggests that a spatially closed geometry is the unique issue[**65**] which reconciles an age $t_\circ = 16.5 \pm 2\mathrm{Gyr}$ with realistic estimates of the density parameter $\Omega_\circ \approx 0.2$, see Fig. 1. Since this issue does not agree with Inflation, it is sensible to ask whether such a scenario is the unique solution to the Horizon and the Flatness problems. Although the following speculations assume implicitly reasonable values ranges for the cosmological parameters, see Sec. 2, let us mention however that they are not conventional.

4.1. The Flatness problem. Contrary to the popular belief which says that "the universe was flat when it was born", it must be noted that the scalar curvature $K = k_\circ H_\circ^2 a^{-2}$ becomes huge by approaching the singularity (which shows that one has little constraint on the primordial geometry). The Flatness problem is the puzzle of why the curvature parameter k is not orders of magnitude different from zero today (which can be interpreted as a *dynamical flatness*), although most of usual speculations account for primordial epochs. Namely, unless k is a very tiny number at the Planck time then the universe will recollapse or go into a curvature-dominated expansion (before it reaches the present age. Although, we easily understand that a *weak* anthropic principle forces us to consider a limited domain of initial conditions, such a problem is not so clear cut as that. Indeed, it must be noted that the property of having a primordial behavior close to Einstein–de Sitter model is necessarily ensured by the FLG model (regardless the present values of the curvature and the density parameters $k_\circ$ and $\Omega_\circ$), since (little algebra shows that) one has $k \sim k_\circ \alpha_\circ^{-1} a^2$ when $a \approx 0$. Moreover, by noting that $\Omega_\circ$ has no effect on the dynamics at primordial epochs, as well as $\alpha_\circ$ has no effect at present, one understands that any speculation which uses the Dark Matter problem as a premise is spurious. With these details in mind, we may however investigate the dynamical flatness at a primordial epoch, and to the question on whether $k_\circ$ could inherited of a peculiar small value (which could suggest that it is actually zero), we test "the nearness of k to zero at the origin of time" within a probabilistic viewpoint. The only parameter which does not depend on the choice of the date

[7]The Dark-Matter Problem is that the dynamical mass of extragalactic sources exceeds by more than one magnitude their luminous mass. The former is derived from the rotation curves of spiral galaxies[**57**] and from the velocity dispersion of galaxies in rich clusters.

and which accounts for the behavior of k near the singularity, is the slope of the curve $a^2 \mapsto k$, or equivalently the angle β (i.e., $\tan\beta = k_\circ \alpha_\circ^{-1}$). It turns out that the present estimates of the cosmological parameters provide us with a value in the range $-89.99° < \beta < 89.99°$. Hence, because of the present accuracy on the cosmological parameters, one understands that no reasonable probability distribution tells us that such an angle coincides with zero. Moreover, it is clear that realistic evidences that k is fine tuned with zero at a primordial epoch requires $|k_\circ| \ll \alpha_\circ$, which is not yet ensured by present observations, see Sect. 2.1.

In the way around, the flatness has been suggested by speculations on the present expansion dynamics, which point out anthropic problems called the *Dicke Coincidences*[**16, 17**] . A speculation which avoids such problems (if any) suggests that the present cosmological parameters have to be close to their final values (which are derived with $a \to +\infty$). Such a requirement provides us with two possible issues (depending on whether the cosmological constant is zero) :

- if $\Lambda \neq 0$ then one has $\lambda_\circ \approx 1$ and $k_\circ \approx \Omega_\circ$, which gives a spatially close universe with a vacuum-dominated expansion.
- if $\Lambda = 0$ then one has the choice between an empty ($\Omega_\circ = 0$) or a flat ($k_\circ = 0$) universe. It is clear that the later is more likely because of the presence of matter in the universe, which is the solution proposed by Inflation.

4.2. The Horizon problem. The events related to the emission of CBR photons when they decoupled at $t_{\rm dec}$ from the primeval plasma, define a sphere centered on our position on V_3, called *Last Scattering Surface*. Let this sphere be denoted by $\mathcal{S}_{\rm LSS}$, its comoving radius $\tau_{\rm rec}$ is calculated by using eq. (2), with $a_{\rm rec} = (1 + z_{\rm rec})^{-1}$, where $z_{\rm rec}$ is the redshift of CBR photons. Calculations[**50**] show that its present value depends weakly on cosmological parameters, it varies from $z_{\rm rec} = 1360$ for a baryon density $\Omega_B h^2 = 0.013$ up to $z_{\rm rec} = 1520$ for $\Omega_B h^2 = 1$. The location on V_3 of all event which is causally connected to a particular emission event lies within a sphere $\mathcal{S}_{\rm dec}$. Such a sphere is called *particle horizon* of the above event, its conformal radius is given by $\tilde{t}_{\rm dec} = \tilde{t}_\circ - \tau_{\rm rec}$, where $\tilde{t}_\circ$ is the conformal age of the universe. On one hand, a drawing on the space – conformal time diagram shows that the Horizon problem appears at conformal time $\tilde{t} = 2\tilde{t}_{\rm dec}$ (since CBR photons observed toward opposite directions in the sky come from regions which were not causally connected), and the calculation shows that $\tilde{t}_\circ > 2\tilde{t}_{\rm dec}$. On the other hand, it is clear that if the RW-metric is ensured from the origin of time then the universality of physics does not required causal link between CBR emission events. Actually, the problem is to understand how the cosmic fluid became as homogeneous and isotropical as observed by COBE, and one uses a the minimum number of 2D causal cells which is required for a regular covering of $\mathcal{S}_{\rm LSS}$, by assuming the FLG world model as geometrical support. Hence, because the size of cells depends on cosmological parameters, one understands that the smaller this number the more likely these values (since the larger this number the more efficient the smoothing process). It turns out that the issue depends on the curvature parameter :

- if $k_\circ \leq 0$ then this number is so large ($\geq 10^4$) that it may be difficult to propose a realistic smoothing process. Hence, within the Inflation scenario, the spheres $\mathcal{S}_{\rm dec}$ were enlarged enough during the inflationary epoch to contain $\mathcal{S}_{\rm LSS}$ at decoupling.

- if $k_\circ > 0$ then the manifold V_3 identifies to the hypersphere S_3, and the center of $\mathcal{S}_{\rm LSS}$ is located at our antipode. Hence, the Horizon problem can be canceled by choosing values of cosmological parameters such that the sphere $\mathcal{S}_{\rm LSS}$ becomes smaller than $\mathcal{S}_{\rm dec}$. The line (II) in Fig. 1 defines the world models for which $\mathcal{S}_{\rm LSS}$ is confined within a single point. Although the number of causal regions within $\mathcal{S}_{\rm LSS}$ may be still large, they have time to synchronize before the CBR photons are observed.

It is also interesting to investigate the perennially of these solutions, although observations do not indicate that the CBR isotropy is permanent. In the case $k_\circ \leq 0$, the calculation shows that if $\Lambda = 0$ then $\mathcal{S}_{\rm LSS}$ increases with time. Thus, if the amount of expansion factor due to Inflation is finite then both the Homogeneity and the Horizon problems will emerge in future, which shows that they are inherent to $k_\circ \leq 0$ geometries. Such a feature indicates that if the space is infinite then it must exist before the matter for ensuring the homogeneity (since V_3 contains an infinite number of spheres $\mathcal{S}_{\rm dec}$), and the later has to be created all over the space at the same date. On the other hand, if $k_\circ > 0$ then the universe has been generated from a *single initial event* (e.g., before $t_{\rm Pl}$), which makes more easier of rendering the cosmic fluid with appropriated properties. In this case, the condition $\Lambda > 0$ is required for ensuring an eternal expansion, which makes the conformal age of the universe bounded $t < \tilde{t}_\infty$. Hence, the Horizon problem can be canceled in eternity for particular values of cosmological parameters. Therefore, a copernican principle applied to such a problem should force us to prefer a positive curvature universe, which has a finite mass equal to $M = \frac{3}{4}\pi\Omega_\circ k_\circ^{-3/2}(\mathrm{G}H_\circ)^{-1}$.

5. Conclusion

The present speculation shows that the Standard world model still provides us with a likely geometrical frame, when galaxies redshifts surveys shows large scale structures and cosmic velocities fields. The Horizon and the Flatness problems are investigated within such a framework, and it turns out that a spatially closed geometry is not excluded. On the other hand, a polemic on the way how Inflation came on the scene, in addition of the short extragalactic distance scale suggested by recent measurements of the Hubble constant forces us to question such a solution.

ACKNOWLEDGMENTS. I express my gratitude to the organizers of the Conference, in particular Hanna Nencka and Philippe Combe. Thanks to Monty and Roy Chisholm for having provided me with materials on W.K. Clifford.

References

[1] L.F. Abbott, S.-Y. Pi, *Inflationary Cosmology* (World Scientific, 1986).

[2] Adams, Freese, A.H. Guth, *Constraints on the scalar-field potential in inflationary models*, Phys. Rev. D **43**, 965 (1991)

[3] A. Albrecht, P.J. Steinardt, *Cosmology for Grand Unified Theories with Radiatively Induced Symetry Breacking*, Phys. Rev. Lett. **48**, 1220 (1982)

[4] H. Alfven, *Antimatter and the Development of the Metagalaxy*, Rev. Mod. Phys. **37**, 652 (1965).

[5] R. Alpher, H. Bethe and G. Gamov, *The Origin of Chemical Elements*, Phys. Rev. **73**, 803 (1948)

[6] U. Amaldi, W. De Boer, H. Furstenau, *Comparison of grand unified theories with electroweak and strong coupling constants measured at LEP*, Phys. Lett. B **260**, 447 (1991).

[7] J.M. Bardeen, P.J. Steinardt, M.S. Turner, *Spontaneous Creation of Almost Scale-Free Density Perturbations in an Inflationary UNiverse*, Phys. Rev. D **28**, 679 (1983)

[8] G. Bigot, H.H. Fliche, R. Triay, *Determination of q_o by a technique of null-correlation*, Astron. & Astrophys. **206**,1 (1988).

[9] G. Bigot, R. Triay, *A null correlation technique to determine q_o*, in *The quest for the fundamental constants in Cosmology*, 9th Moriond Astrophysics meeting, eds J. Audouze and J. Trân Thanh Vân, Ed. Frontières (1989).

[10] S.M. Carroll, W.H. Press, E.L. Turner, *The cosmological constant* , Annu. Rev. Astron. Astrophys. **30**, 499 (1992).

[11] H.T. Cho, R. Kantowski, *Measure on a subspace of FRW solutions and "the flatness problem" of standard cosmology*, Phys. Rev. D **50**, 6144 (1994)

[12] P. Coles, G. Ellis, *The case for an open Universe*, Nature **370** (1994) 609.

[13] D.H. Coule, *Canonical measure and the Flatness of a FRW universe*, Class. Quantum Grav **12**, 455 (1995).

[14] V. de Lapparent, M.J. Geller, J.P. Huchra *A Slice of the Universe*, Astrophys. J. Lett. **302**, L1 (1986).

[15] F.X. Desert, E. Schatzman, *Physical constraints on the bipartition model of the Universe*, Astron. & Astrophys. **158**, 135 (1986).

[16] R.H. Dicke, *Gravitation and the Universe*, (The Jayne lectures for 1969, American Philosophical Society, Philadelphia, 1970), p. 62. (1970)

[17] R.H. Dicke, P.J.E. Peebles, in *General Relativity: An Einstein Centenary Survey*, ed. S.W. Hawking & W. Israel (Cambridge University Press, London, 1979) (1979).

[18] P. Dirac, ., Proc. Roy. Soc. London A **133**, 60 (1931)

[19] A.D. Dolgov, *Antimatter in the Universe*, in Proc. of *Antihydrogen Workshop*, ed. J. Eades, (J.C. Baltzer Ed., Munich, 1993), p. 97 (1992)

[20] A. Einstein, ., Ann. der Phys. **49**, 769 (1916).

[21] H.H. Fliche, J.M. Souriau, *Quasars and Cosmology*, Astron. & Astrophys. **78**, 87 (1979).

[22] H.H. Fliche, J.M. Souriau, R. Triay, *A Possible Large Scale Structure of the Universe*, Astron. & Astrophys. **108**, 256 (1982).

[23] W.L. Freedman, B.F. Madore, J.R. Mould, R. Hill, L. Ferrares, R.C. Kennicutt Jr, A. Saha, P.B. Stetson, J.A. Grahan, H. Ford, J.G. Hoessel, J. Huchra, S.M. Hughes and G.D. Illingworth, *Distance to the Virgo cluster galaxy M100 from the Hubble Space Telescope observations of Cepheids*, Nature **371**, 757 (1994).

[24] A. Friedmann, ., Z. Phys. **10**, 377 (1922).

[25] G. Gamov, *Expanding Universe and the Origin of Elements*, Phys. Rev. **70**, 572 (1946)

[26] E.B. Gliner, ., Sov. Phys. JETP **22**, 378 (1966)

[27] E.B. Gliner, I.G. Dymnikova, ., Sov. Phys. Astr. Lett. **1**, 93 (1975)

[28] A.H. Guth, *Inflationary Universe : A Possible solution to the Horizon adn Flatness Problems* , Phys. Rev. D **23**, 347 (1981).

[29] A.H. Guth, S-Y. Pi, *Fluctuatios in the New Inflationary Universe* , Phys. Rev. Lett. **49**, 1110 (1982)

[30] A.H. Guth, P.J. Steinhardt, *The Inflationary Universe*, Scientific American, May, 90 (1984)

[31] A.H. Guth, S-H.H. Tye, *Phase Transitions and Magnetic Monopole Production in the Very Early Universe*, Phys. Rev. Lett. **44**, 631 (1980)

[32] A.H. Guth, E.J. Weinberg, *Cosmological consequences of a first-order phase transition in the SU(5) grand unified model*, Phys. Rev. D **23**, 876 (1981)

[33] S.W. Hawking, R. Penrose, *The singularities of gravitational collapse and cosmology*, Proc. Roy. Soc. London A **314**, 529 (1970)

[34] S.W. Hawking, ., Phys. Lett. B **115**, 295 (1982).

[35] Yue Hu, M.S. Turner, E.J. Weinberg, *Dynamical solutions to the horizon and flatness problems*, Phys. Rev. D **49**, 3830 (1994)

[36] E.P. Hubble, . , Proc. Nat. Acad. Sci. **15**, 168 (1927).

[37] G.H. Jacoby, D. Branch, R. Ciardullo, R.L. Davies, W.E. Harris, M.J. Pierce, C.J. Pritchet,J.L. Tonry, D.L. Welch, , Pub. Astron. Soc. Pac. **104**, 599 (1992).

[38] D. Kazanas, *Dynamics of the Universe and Spontaneous Symmetry Breacking*, Astrophys. J. Lett. **241**, L59 (1980).

[39] E.W. Kolb, M.S Turner, *Grand Unified Theories and the Origin of the Baryon Asymmetry*, Ann. Rev. Nucl. Part. Sci. **33**, 645 (1983).

[40] P. Langacker, *Grand Unified theories and Proton Decay*, Phys. Rep. **72** (4), 187 (1970).

[41] G. Lemaitre, ., Ann. Soc. Sci. Bruxelles **A47**, 49 (1927).

[42] A.D. Linde, *A New Inflationary Universe Scenario : A Possible Solution of the Horizon, Flatness, Homogeneity, Isotropy and Primordial Monopole Problems*, Phys. Lett. B **108**, 389 (1982).

[43] A.D. Linde, *Chaotic Inflation*, Phys. Lett. B **129**, 177 (1983)

[44] A.D. Linde, *Lectures on Inflationary Cosmology* pprint hep-th/9410082 (based on lectures and talks given in 1994).

[45] A.G. Mirzabekian, G.A. Vilkovisky, *The model-independent approach to quantum gravity: implications for asymptotic flatness*, Phys. Lett. B **317**, 517 (1993).

[46] L. Montanet, et al. *Review of Particles Properties*, Phys. Rev. D **50,2**, 81 (1994).

[47] R. Omnes, ., Phys. Rep. **3**, 1 (1970).

[48] J.V. Peach, ., Astrophys. J. **159**, 753 (1970).

[49] P.J.E. Peebles, *Tests of Cosmological Models Cosntrained by Inflation* , Astrophys. J. **284**, 439 (1984).

[50] P.J.E. Peebles, *Principles of Physical Cosmology.* (Princeton Univ. Press, 1993).

[51] R. Penrose, *Gravitational collapse and space-time singularities*, Phys. Rev. Lett. **14**, 57 (1965)

[52] S-Y. Pi, *Inflation without Tears : A Realistic Cosmological Model*, Phys. Rev. Lett. **52**, 1725 (1984)

[53] M.J. Pierce, D.L. Welch, R.D. McClure, S. van den bergh, R. Racine, P. B. Stetson, *The Hubble constant and Virgo cluster distance from observations of Cepheid variables*, Nature **371**, 385 (1994).

[54] J.P. Preskill, *Cosmological Production of Superheavy Magnetic Monopoles*, Phys. Rev. Lett. **43**, 1365 (1979)

[55] M.J. Rees, *Is the Universe Flat ?*, J. Astrophys. Astr. **5**, 331 (1984).

[56] H.P. Robertson, ., Astrophys. J. **82**, 284 (1935).

[57] V.C Rubin, W. K. Ford, Jr., N. Thonnard, *Rotational Properties of 23 Sb galaxies*, Astrophys. J. **261**, 439 (1982).

[58] R.K. Sachs, A.M. Wolfe, *Perturbations of a Cosmological Model and Angular Variations of the Microwave Background*, Astrophys. J. Lett. **147**, 73 (1967).

[59] A.D. Sakharov, , Piz'ma. Zh. Eksp. Teor. Fiz. **5**, 32 (1967).

[60] P. Schneider, *Cosmological Applications of Gravitational Lensing*, in Proc. Laredo Advanced Summer School (1995), astro-ph/0512047 at Sissa

[61] Q. Shafi, A.Vilenkin, *Inflation with SU(5)*, Phys. Rev. Lett. **52**, 691 (1984)

[62] J.M. Souriau, *Géométrie et relativité.* (Hermann ed., 1964).

[63] J.M. Souriau, *Mécanique Statistique, Groupe de Lie et Cosmologie*, in *Géométrie symplectique et physique mathématique.* Coll. Int. CNRS **237**, 59 (1974).

[64] J.M. Souriau, *Un modèle d'Univers confronté aux observations*, *Dynamics and Processes* Springer Lect. Notes in Math. **1031**, 114 (1983).

[65] J.M. Souriau, R. Triay, *The Age Problem in positive curvature world models.* pprint CPT-94/P3137

[66] A.A. Starobinsky, *Dynamics of phase transition in the New Inflationary Universe scenario and Generation of Perturbations*, Phys. Lett. B **117**, 175 (1982).

[67] S.M. Steigman, *Observational tests of Antimatter Cosmologies*, Annu. Rev. Astron. Astrophys. **14**, 339 (1976).

[68] M.A. Strauss, J.A. Willick, *The Density and Peculiar Velocities Fields of Nearby Galaxies*, Phys. Rep. **261**(5&6), 271 (1994).

[69] F.R. Tangherlini, ., Nuovo Cimento **108**, 1253 (1993).

[70] R. Triay, *Possible Large Scale Structures in the Universe*, in *Large Scale Structures and Peculiar Motions in the Universe* (1989). ed. D.W. Latham and L.N. da Costa, ASP conference series.

[71] R. Triay, *Is the Flatness Problem real ?*, pprint CPT-95/P3226

[72] R. Triay, *Geometry of the Horizon Problem*, pprint CPT-95/P3227

[73] D.A. Van den Bergh, *The formation and the Evolution of Star Clusters* ed. K. Janes, PASP Conf. Ser. (Astr. Soc. Pacif., San Francisco) **13**, 183 (1991)

[74] A.G. Walker, ., Proc. London Math. Soc. **42**, 90 (1936)

[75] D. Wilkinson, *Physical Cosmology.* – Proceedings 25th anniversary of the Cosmic Background Radiation Discovery, ed. A. Blanchard, L. Celnikier, M. Lachièze-Rey and J. Trân Thanh Vân, (Ed. Frontières,1991) "Rencontres de Blois 1990", p. 97

[76] Ya.B. Zel'dovich, M.Yu. Kholpov, *On the concentration of relic magnetic monopoles in the Universe*, Phys. Lett. B **79**, 239 (1978).

Université de Provence & Centre de Physique Théorique–CNRS, Luminy Case 907, F-13288 Marseille, France

E-mail address: triay@cpt.univ-mrs.fr

Part VI

SYMPLECTIC GEOMETRY AND SELF-SIMILAR STRUCTURES

Contemporary Mathematics
Volume **203**, 1997

A class of homogeneous symplectic manifolds

P. BIELIAVSKY, M. CAHEN, S. GUTT

ABSTRACT. Symmetric symplectic manifolds are homogeneous manifolds characterized by purely symplectic properties. They admit a unique preferred symplectic connection and are affine symmetric spaces relative to this preferred connection. We prove that such a manifold M, when simply connected, is the product of elementary pieces which are uniquely determined by M. We determine all compact ones and we describe the coadjoint orbits of semi simple groups which are covered by symmetric symplectic spaces. We end up by giving a list of those spaces which are not Kähler.

Introduction

Symmetric spaces have been extensively studied in the riemannian [**1**] or the pseudo riemannian [**2**] context. An interesting challenge is to study them in a purely symplectic context. We thus define a symmetric symplectic manifold as a symplectic manifold, whose symplectic structure is invariant by a set of symmetries. The axioms on symmetries are adapted from a beautiful book by O. Loos [**3**]. It turns out that a symplectic symmetric manifold M admits a unique symplectic connection ∇, such that (M, ∇) is an affine symmetric space. It is then easy to reduce the study of these spaces to a purely algebraic problem.

Simply connected pseudo riemannian symmetric spaces can be decomposed into a product of weakly irreducible simply connected symmetric spaces [**4**]; two such decompositions are conjugate by an isometry [**2**]. The notion of weak irreducibility is thus the good notion to define elementary pieces in the pseudo riemannian setting. No such idea works in the symplectic framework and our definition of the elementary pieces may seem a little crude; nevertheless it gives a decomposition theorem and two decompositions in elementary pieces are conjugate by an automorphism of the symmetric space. Furthermore, in the pseudo hermitian case, our definition gives the same result as the usual definition.

The paper is divided in 3 sections. In the first we give the definition of symmetric symplectic manifolds, we construct the preferred symplectic connection and we show how to reduce the study of these manifolds to algebra. The second section, which is the most important one, contains the definition of indecomposability and the theorem which asserts that any simply connected symmetric symplectic

1991 *Mathematics Subject Classification.* Primary 53C15; Secondary 53A10,53C35 .

manifold is the product of indecomposable ones and that two such decompositions are conjugate by an automorphism. The third section recalls some of our previous results [**5**] on coadjoint orbits of semi simple groups and adds a list of the symplectic symmetric spaces with simple transvection group which are not Kähler. This answers a question of Alekseevsky.

1. Symmetric symplectic manifolds: definitions and elementary properties

DEFINITION 1.1. A **symmetric symplectic manifold** (denoted hereafter S. S. M.) is a triple (M, ω, s) where M is a smooth connected manifold, where ω is a smooth symplectic form on M and where s is a smooth map $M \times M \to M$: $(x, y) \to s(x, y) \underset{\text{not}}{=} s_x(y)$ such that:

(i) For all x in M, s_x is an involutive symplectic diffeomorphism of (M, ω), called the symmetry at x,
(ii) For all x in M, x is an isolated fixed point of s_x,
(iii) For all x, y in M one has $s_x s_y s_x = s_{s_x(y)}$.

DEFINITION 1.2. The symmetric symplectic manifolds (M, ω, s) and (M', ω', s') are **isomorphic** if there exists a symplectic diffeomorphism $\varphi : (M, \omega) \to (M', \omega')$ such that for all x in M, $\varphi \circ s_x = s'_{\varphi(x)} \circ \varphi$. We shall denote $\mathrm{Aut}(M, \omega, s)$ the **automorphism group** of (M, ω, s).

DEFINITION 1.3. An affine connection ∇ on (M, ω, s) is said to be an **admissible connection** if:

(i) $\nabla \omega = 0$.
(ii) For all x in M, s_x is an affine transformation of (M, ∇).

THEOREM 1.4. *A S. S. M. (M, ω, s) has a unique admissible connection ∇. Furthermore:*

(i) *If X, Y, Z are smooth vector fields on M one has, for any x in M:*

$$\omega_x(\nabla_X Y, Z) = \frac{1}{2} X_x \omega(Y + s_{x\star} Y, Z)$$

(ii) *The admissible connection ∇ has no torsion and is thus a symplectic connection.*
(iii) *The automorphism group of (M, ω, s) is the intersection of the affine group of (M, ∇), Aff(M, ∇), and the symplectic diffeomorphism group, Symp (M, ω).*

PROOF. One first observes that $(s_{x\star x})^2 = \mathrm{id}_{|M_x}$ and $(s_{x\star})$ has thus eigenvalues ± 1. Choose an auxiliary riemannian metric g, for which s_x is an isometry; if $s_{x\star x}$ admits the eigenvalue $+1$, there exists a geodesic γ of (M, g) passing through x, which is pointwise stabilized by s_x. This contradicts the fact that x is an isolated fixed point of s_x. As $s_{x\star x}$ is an orthogonal transformation and admits -1 as the only eigenvalue, $s_{x\star x} = -\mathrm{id}_{|M_x}$.
Assume there exists an admissible connection ∇; then:

$$\begin{aligned} \omega_x(\nabla_x Y, Z) &= X_x \omega(Y, Z) - \omega_x(Y, \nabla_x s_{x\star} Z) \\ &= X_x(\omega(Y, Z) - \omega(Y, s_{x\star} Z)) - \omega_x(\nabla_x Y, Z) \end{aligned}$$

Hence:

$$\begin{aligned}\omega_x(\nabla_x Y, Z) &= \frac{1}{2}X_x(\omega(Y,Z) - \omega(Y, s_{x\star}Z))\\ &= \frac{1}{2}X_x(\omega(Y,Z) - s_x^*(\omega(s_{x\star}Y, Z)))\\ &= \frac{1}{2}X_x(\omega(Y,Z) + \omega(s_{x\star}Y, Z)).\end{aligned}$$

This proves unicity and (i). One checks readily that formula (i) defines an affine connection, and that relative to this connection ω is parallel. Furthermore if $\varphi \in \mathrm{Aut}(M,\omega,s)$:

$$\omega(\varphi_\star Y + s_{x\star}\varphi_\star Y, Z) = \varphi^{-1\star}(\omega(Y + s_{\varphi^{-1}(x)_\star}Y, \varphi_\star^{-1}Z))$$

and thus:

$$\begin{aligned}\omega_x(\nabla_{\varphi_\star X}\varphi_\star Y, Z) &= \frac{1}{2}(\varphi_\star X)_x\varphi^{-1\star}(\omega(Y + s_{\varphi^{-1}(x)_\star}Y, \varphi_\star^{-1}Z))\\ &= \frac{1}{2}X_{\varphi^{-1}(x)}\omega(Y + s_{\varphi^{-1}(x)\star}Y, \varphi_\star^{-1}Z)\\ &= \omega_x(\varphi_\star\nabla_X Y, Z).\end{aligned}$$

This shows that φ is an affine transformation; in particular all symmetries are affine transformations, which proves that ∇ is an admissible connection.
If T denotes the torsion tensor of ∇:

$$\begin{aligned}s_{x\star}T_x(Y,Z) = -T_x(Y,Z) &= s_{x\star}(\nabla_Y Z - \nabla_Z Y - [Y,Z])_x\\ &= T_x(s_{x\star}Y, s_{x\star}Z) = T_x(Y,Z) = 0.\end{aligned}$$

We have already proven that

$$\mathrm{Aut}(M,\omega,s) \subset \mathrm{Aff}(M,\nabla) \cap \mathrm{Symp}(M,\omega).$$

For the reverse inclusion, notice that if ψ is both affine and symplectic, $\psi \circ s_x \circ \psi^{-1}$ is an affine transformation stabilizing $\psi(x)$, whose differential at $\psi(x)$ is $-\mathrm{id}_{|M_{\psi(x)}}$. As M is connected, $\psi s_x \psi^{-1} = s_{\psi(x)}$ and ψ is an automorphism of (M,ω,s).
From now on we shall always endow the S. S. M., (M,ω,s) with the admissible connection ∇, which is canonical. □

PROPOSITION 1.5. The S. S. M.,(M,ω,s) is an affine symmetric space.

PROOF. The symmetry s_x, being affine, stabilizes the geodesics of (M,∇) passing through x. As $s_{x\star|M_x} = -\mathrm{id}_{|M_x}$, there exists a neighborhood of x where s_x coincides with the geodesic symmetry at x. Hence the conclusion. □

We can now use the well known theory of affine symmetric spaces; we recall hereunder standard definitions and state as facts classical propositions whose proofs may be found in [**1, 6**].

DEFINITION 1.6. The **transvection group** $G(M)$ of (M,ω,s) is the group generated by products of an even number of symmetries. Clearly $G(M)$ is a normal subgroup of $\mathrm{Aut}(M,\omega,s)$.

From now on we shall choose a base point o in M. The involutive automorphism $\tilde{\sigma}$ of $\mathrm{Aut}(M,\omega,s)$ defined by $\tilde{\sigma}(g) = s_0 g s_0$ $(g \in \mathrm{Aut}(M,\omega,s))$ is simply called the involutive automorphism of $\mathrm{Aut}(M,\omega,s)$. **Facts**:

(i) The automorphism group $\mathrm{Aut}(M,\omega,s)$ is a Lie transformation group of M; it acts transitively on M. The transvection group $G(M)$ is the smallest subgroup of $\mathrm{Aut}(M,\omega,s)$ stable by $\tilde{\sigma}$ and acting transitively on M; it is a connected Lie subgroup of $\mathrm{Aut}(M,\omega,s)$.

(ii) Let K denote the stabilizer of o in $G(M)$ and let $G^{\tilde{\sigma}}$ (resp $G_o^{\tilde{\sigma}}$) denote the group of fixed points of $\tilde{\sigma}$ in $G(M)$ (resp its connected component). Then $G^{\tilde{\sigma}} \supseteq K \supseteq G_o^{\tilde{\sigma}}$. The representation of K on $M_o : k \to k_{\star o}$ is faithful.

(iii) We shall identify $M = G/K$ and denote by π the canonical projection $G \to M : g \to g \cdot o \cdot$. The holonomy group of (M,∇) at o is isomorphic to K, and we can view $G \xrightarrow{\pi} M$ as the holonomy bundle.

(iv) If $\mathcal{G}$ (resp $\mathcal{K}$) is the Lie algebra of $G(M)$ (resp K) and if $\sigma = \tilde{\sigma}_{\star e}$ (e = neutral element of G), $\mathcal{K}$ is the subalgebra of $\mathcal{G}$ of fixed points of σ. We shall denote by $\mathcal{P}$ the subspace of $\mathcal{G} = \{x \in \mathcal{G} \mid \sigma(x) = -x\}$. Then $\mathcal{G} = \mathcal{K} \oplus \mathcal{P}$ and $\pi_{\star e|\mathcal{P}} : \mathcal{P} \to M_o$ is a linear isomorphism, which allows us to identify M_o with $\mathcal{P}$. Furthermore $[\mathcal{P},\mathcal{P}] = \mathcal{K}$.

(v) The curvature tensor of (M,∇) is parallel ($\nabla R = 0$). If X, Y, Z belong to $\mathcal{P}$, one has - with the identification of $\mathcal{P}$ with M_o - $R_o(X,Y)Z = -[[X,Y],Z]$. ($[\ ,\ ]$ denoting the bracket in $\mathcal{G}$).

DEFINITION 1.7. A **symmetric symplectic triple** (S. S. T.) is a triple $t = (\mathcal{G},\sigma,\Omega)$ where

(i) $\mathcal{G}$ is a finite dimensional real Lie algebra

(ii) σ is an involutive automorphism of $\mathcal{G}$; if we write $\mathcal{G} = \mathcal{K} \oplus \mathcal{P}$ with $\sigma = \mathrm{id}_{|\mathcal{K}} \oplus -\mathrm{id}_{|\mathcal{P}}$, then

$$[\mathcal{P},\mathcal{P}] = \mathcal{K}$$

and the action of $\mathcal{K}$ on $\mathcal{P}$ is faithful.

(iii) Ω is a $\mathbb{R}$- valued Chevalley 2 - cocycle of $\mathcal{G}$ with respect to the trivial representation of $\mathcal{G}$ on $\mathbb{R}$. Moreover for any $k \in \mathcal{K}$ and any $X \in \mathcal{G}$, $\Omega(k,X) = 0$ and $\Omega\,|_{\mathcal{P}\times\mathcal{P}}$ has maximal rank (ie: $(\mathcal{P},\Omega\,|_{\mathcal{P}\times\mathcal{P}})$ is a symplectic vector space).

DEFINITION 1.8. The symmetric symplectic triples $(\mathcal{G},\sigma,\Omega)$, $(\mathcal{G}',\sigma',\Omega')$ are **isomorphic** if there exists a Lie algebra isomorphism $\varphi : \mathcal{G} \to \mathcal{G}'$ such that $\varphi \circ \sigma = \sigma' \circ \varphi$ and $\varphi^\star\Omega' = \Omega$.

The following basic result is standard:

THEOREM 1.9. *There is a bijective correspondence between the isomorphism classes of simply connected S. S. M.,(M,ω,s) and the isomorphism classes of S. S. T.,$(\mathcal{G},\sigma,\Omega)$. The algebra $\mathcal{G}$ is the Lie algebra of the transvection group $G(M)$; σ is the differential of the involutive isomorphism $\tilde{\sigma}$ restricted to $G(M)$; Ω is obtained from $\Omega\,|_{\mathcal{P}\times\mathcal{P}}$ by extending by o in $\mathcal{K}$.*

As a symmetric symplectic manifold (M,ω,s) is a symplectic homogeneous space of the group $G(M)$, it seems natural, when possible, to relate (M,ω,s) to a coadjoint orbit of $G(M)$ in $\mathcal{G}^\star$. Recall first the two following definitions:

DEFINITION 1.10. **[7]** Let G be a Lie group of transformations of the symplectic manifold (M,ω) preserving the symplectic form ω. The **action** is said to be

hamiltonian if for any X in $\mathcal{G}$ (= Lie algebra of $\mathcal{G}$) the fundamental vector field $X^\star$ on M, associated to X by the action, ($X_x^\star = \frac{d}{dt} exp - tX.x \mid_{t=0}$; $x \in M$) is hamiltonian; i. e. there exists a smooth function f_X on M such that $i(X^\star)\omega = df_X$. This is always the case if M is simply connected.

DEFINITION 1.11. [**7**][**10**] The symplectic action of G on (M, ω) is said to be **strongly hamiltonian** if

(i) it is hamiltonian
(ii) the map $\mathcal{G} \to \mathcal{C}^\infty(M) : X \to f_X$ is a homomorphism of Lie algebras. The Lie algebra structure on $\mathcal{C}^\infty(M)$ is given by the Poisson bracket; if f, g belong to $\mathcal{C}^\infty(M)$, $\{f, g\} = -\omega(X_f, X_g)$ and $i(X_f)\omega = df$.

The map $X \to f_X$ may always be assumed to be linear; but it is not always a homomorphism.

PROPOSITION 1.12. Let $(\mathcal{G}, \sigma, \Omega)$ be a S. S. T. and let (M, ω, s) be an associated simply connected symmetric symplectic manifold; the action of the transvection group $G(M)$ on M is strongly hamiltonian if and only if Ω is a coboundary ($\Omega = \partial\xi$). In this situation (M, ω, s) is a $G(M)$equivariant, symplectic covering of θ_ξ (= coadjoint orbit of ξ in $\mathcal{G}^\star$). Denote G_ξ the stabilizer of ξ in $G(M)$ and by N the group of ineffectiveness (i.e $N = \{g \in G(M) \mid g.\eta = \eta \forall \eta \in \theta_\xi\}$) Then the orbit θ_ξ is a symmetric symplectic manifold if and only if G_ξ/N is contained in the group of fixed points of $\tilde{\sigma}$ in $G(M)/N$, where $\tilde{\sigma}$ is the involutive automorphism of $G(M)/N$ induced by σ.

PROOF. Assume the action of $G(M)$ on M is strongly hamiltonian. Then by the theorem of Kostant (M, ω) is a $G(M)$ equivariant, symplectic covering of a certain coadjoint orbit θ of $G(M)$ in $\mathcal{G}^\star$. The form of the symplectic form in θ shows that Ω must be a coboundary.
Conversely assume that $\Omega = \partial\xi$ and let $L : M = G/K \to \mathcal{G}^\star : x = gK \to g \cdot \xi$; we have denoted by G the connected, simply connected group with algebra $\mathcal{G}$ and by K the (necessarily connected)[**8**] group of fixed points of $\tilde{\sigma}$ (= the involutive automorphism of G induced by σ).
As $\xi \circ \mathrm{ad}\, k = 0, k \in \mathcal{K}$ (= Lie algebra of K), ξ is invariant by the coadjoint action of K; as $\xi \circ \mathrm{ad}\, p = 0$ (for a certain $p \in \mathcal{P}$) implies $p = 0$ one sees that the Lie algebra $\mathcal{G}_\xi$ of the stabilizer G_ξ of ξ in G is equal to $\mathcal{K}$; hence the connected component of G_ξ is K. One checks that L is the moment map for the action of G on M and that this action is strongly hamiltonian.
As ξ vanishes on $[\mathcal{K}, \mathcal{G}]$ one may assume, without restriction, that ξ vanishes on $\mathcal{P}$; hence $\sigma^\star\xi \underset{\text{def}}{=} \xi \circ \sigma = \xi$. This observation implies that G_ξ is $\tilde{\sigma}$ stable. To define possible symmetries on the orbit θ_ξ one computes $L(s_x y)$ $(x, y \in M)$. Denote by $\pi : G \to M$ the canonical projection and let $x = \pi(g_1)$, $y = \pi(g_2)$;

$$
\begin{aligned}
& < L(s_x y), X > = < L(\pi(g_1\tilde{\sigma}(g_1^{-1}g_2)), X > \\
= \quad & < g_1\tilde{\sigma}(g_1^{-1}g_2)\xi, X > = < \xi, \mathrm{Ad}\,(g_2^{-1}g_1)\mathrm{Ad}\,\tilde{\sigma}(g_1^{-1})\sigma(X) >
\end{aligned}
$$

Hence the "symmetry" at the point $L(x)$, $u_{L(x)}$ should read:

$$u_{L(x)}L(y) = \sigma^\star\tilde{\sigma}(g_1)g_1^{-1}L(y).$$

It has all the properties of symmetries; but to be well defined it is necessary (and sufficient) that for any a in G_ξ, any g_1, g_2 in G:

$$\sigma^\star\tilde{\sigma}(g_1 a)a^{-1}g_1^{-1}g_2\xi = \sigma^\star\tilde{\sigma}(g_1)g_1^{-1}g_2\xi$$

which means that $\tilde{\sigma}(a)a^{-1}$ fixes all points of the orbit for any $a \in G_\xi$, i. e. $\tilde{\sigma}(a)a^{-1} \in N$. As N is a normal subgroup of G, stable by $\tilde{\sigma}$, we have θ_ξ diffeomorphic to $G/N/G_\xi/N$ and $\tilde{\sigma}$ induces an involutive automorphism $\hat{\sigma}$ of G/N. The condition hence reads that $\hat{\sigma}$ fixes all points of G_ξ/N. □

2. Block decomposition of simply connected symmetric symplectic manifolds

The decomposition of simply connected symmetric symplectic manifolds in "elementary" factors presented here, may appear at first sight very different from the well known one for pseudo hermitian symmetric spaces. It nevertheless agrees with it in this particular context. We shall first describe the decomposition in "elementary factors" of symmetric symplectic triples; the corresponding decomposition of the manifold is an easy consequence.

DEFINITION 2.1. A S. S. T. $(\mathcal{G}, \sigma, \Omega)$ is said to be **flat** if $K = 0$. In this case $[\mathcal{P}, =PP] = 3D0=20$ and thus $\mathcal{G}$ is an abelian algebra.

DEFINITION 2.2. Let $(\mathcal{G}_i, \sigma_i, \Omega_i)$ $(i = 1, 2)$ be S. S. T.'s. Then the triple $(\mathcal{G}_1 \oplus \mathcal{G}_2, \sigma_1 \oplus \sigma_2, \Omega_1 \oplus \Omega_2)$ is a S. S. T. called the **direct sum** of the S. S. T.'s $(\mathcal{G}_i, \sigma_i, \Omega_i)$ $(i = 1, 2)$. A S. S. T. $(\mathcal{G}, \sigma, \Omega)$ is said to be **decomposable** if there exist S. S. T.'s $(\mathcal{G}_i, \sigma_i, \Omega_i)$ $(i = 1, 2)$ such that $(\mathcal{G}, \sigma, \Omega)$ is isomorphic to $(\mathcal{G}_1 \oplus \mathcal{G}_2, \sigma_1 \oplus \sigma_2, \Omega_1 \oplus \Omega_2)$. A S. S. T.. $(\mathcal{G}, \sigma, \Omega)$ is said to be **indecomposable** if it is not decomposable.

THEOREM 2.3. *Let $\mathcal{T} = (\mathcal{G}, \sigma, \Omega)$ be a S. S. T.. and let $\mathcal{T}_\alpha$ (resp. $\mathcal{T}'_\beta$) $(0 \le \alpha \le p; 0 \le \beta \le q) = (\mathcal{G}_\alpha, \sigma_\alpha, \Omega_\alpha)$ (resp. $(\mathcal{G}'_\beta, \sigma'_\beta, \Omega'_\beta)$ be a S. S. T. such that :*

(i) $(\mathcal{G}, \sigma, \Omega) = \bigoplus_{\alpha=0}^{p}(\mathcal{G}_\alpha, \sigma_\alpha, \Omega_\alpha) = \bigoplus_{\beta=0}^{q}(\mathcal{G}'_\beta, \sigma'_\beta, \Omega'_\beta)$

(ii) $(\mathcal{G}_0, \sigma_0, \Omega_0)$ *(resp. $(\mathcal{G}'_0, \sigma'_0, \Omega'_0)$) is flat*

(iii) *for all $\alpha \ge 1$ (resp. $\beta \ge 1$) $\mathcal{T}_\alpha$ (resp. $\mathcal{T}'_\beta$) is indecomposable and non flat*

Then $p = q$ and there exists a permutation $\eta \in S_p$ and an automorphism φ of $\mathcal{T}$ such that

(a) $\varphi\mathcal{T}_0 = \mathcal{T}'_0$

(b) $\varphi\mathcal{T}_\alpha = \mathcal{T}'_{\eta(\alpha)}$

The proof will be divided in a sequence of easy lemmas. Clearly $\mathcal{G}_0$ (resp. $\mathcal{G}'_0) = \mathcal{P}_0$ (resp. $\mathcal{P}'_0$) is contained in the center of $\mathcal{G}$; the intersection of $\overline{\mathcal{G}}_0 \underset{\text{not}}{=} \oplus_{\alpha\ge1}\mathcal{G}_\alpha$ (resp. $\overline{\mathcal{G}}'_0 = \oplus_{\beta\ge1}\mathcal{G}'_\beta$) with the center of $\mathcal{G}$ is completely isotropic. Indeed if u, v are elements of $\overline{\mathcal{G}}_0 \cap Z(\mathcal{G})$ such that $\Omega(u, v) = 1$, one may assume that they both belong to a certain $\mathcal{G}_\alpha$ and then

$$\mathcal{G}_\alpha = (\mathbb{R}u \oplus \mathbb{R}v) \oplus (\mathbb{R}u + \mathbb{R}v)^\perp \cap \mathcal{P}_\alpha \oplus \left[(\mathbb{R}u \oplus \mathbb{R}v)^\perp \cap \mathcal{P}_\alpha, (\mathbb{R}u + \mathbb{R}v)^\perp \cap \mathcal{P}_\alpha\right]$$

and $\mathcal{G}_\alpha$ would be either flat or decomposable. Hence $\overline{\mathcal{G}}_0 \cap Z(\mathcal{G}) = \overline{\mathcal{G}}'_0 \cap Z(\mathcal{G}) =$ radical of $Z(\mathcal{G})$.

LEMMA 2.4. *Let $\mathcal{G} = \mathcal{G}_0 \oplus \overline{\mathcal{G}}_0 = \mathcal{G}'_0 \oplus \overline{\mathcal{G}}'_0$. Then there exists an automorphism φ of $\mathcal{T} = (\mathcal{G}, \sigma, \Omega)$ such that $\varphi(\mathcal{G}'_0) = \mathcal{G}_0$ and $\varphi(\overline{\mathcal{G}}'_0) = \overline{\mathcal{G}}_0$.*

PROOF. Let π (resp. $\overline{\pi}$) be the projection of $\mathcal{G}$ onto $\mathcal{G}_0$ (resp. $\overline{\mathcal{G}}_0$). The map $\pi_{|\mathcal{P}'_0} : \mathcal{P}'_0 \to \mathcal{P}_0$ is injective hence a linear isomorphism. If not, there exists $0 \neq x \in \mathcal{P}'_0 \cap \overline{\mathcal{P}}$; choose $y \in \mathcal{P}'_0$ such that $\Omega(x, y) = 1$; but $\Omega(x, y) = \Omega(x, \overline{\pi}(y)) = 0$ since x and $\overline{\pi}(y) \in Z(\mathcal{G}) \cap \overline{\mathcal{P}}_0$.
The space $\overline{\pi}(\mathcal{P}'_0)$ is central, totally isotropic and contained in $\overline{\mathcal{P}}'_0$. Observe that :

$$\begin{aligned} \overline{\pi}(\mathcal{P}'_0)^{\perp} \cap \overline{\mathcal{P}}_0 &= (\overline{\pi}(\mathcal{P}'_0) + \mathcal{P}_0)^{\perp} = (\mathcal{P}_0 + \mathcal{P}'_0)^{\perp} = \overline{\mathcal{P}}_0 \cap \overline{\mathcal{P}}'_0 \\ &= \overline{\pi}(\mathcal{P}'_0)^{\perp} \cap \overline{\mathcal{P}}'_0. \end{aligned}$$

Choose a totally isotropic subspace $\overline{V} \subset \overline{\mathcal{P}}'_0$, in duality with $\overline{\pi}(\mathcal{P}'_0)$ and such that :

$$\overline{\mathcal{P}}'_0 = \overline{\mathcal{P}}'_0 \cap \overline{\mathcal{P}}_0 \oplus \overline{V}$$

Observe that $\overline{\pi}_{|\overline{V}} : \overline{V} \to V (\subset \overline{\mathcal{P}}_0)$ defines a subspace V in duality with $\overline{\pi}(\mathcal{P}'_0)$. Indeed

$$\Omega(\overline{\pi}(\overline{v}), \overline{\pi}(p'_o)) = \Omega(\overline{v}, \overline{\pi}(p'_0))$$

Hence we may write :

$$\overline{\mathcal{P}}_0 = \overline{\mathcal{P}}'_0 \cap \overline{\mathcal{P}} \oplus V$$

Let $l : V \to \overline{\pi}(\mathcal{P}'_0) \oplus V$ be a linear injection such that $l(V)$ is a totally isotropic subspace in duality with $\overline{\pi}(\mathcal{P}'_0)$. Observe finally that $l \circ \overline{\pi}(\overline{v}) - \overline{\pi}(\overline{v})$ is an element contained in the radical of $Z(\mathcal{G})$. Then the map $\varphi : \mathcal{G} \to \mathcal{G}$ defined by

$$\varphi_{|\mathcal{P}'_0} = \pi_{|\mathcal{P}'_0} \qquad \varphi_{|\overline{\mathcal{P}}_0 \cap \overline{\mathcal{P}}'_0} = \mathrm{id}_{|\overline{\mathcal{P}}_0 \cap \overline{\mathcal{P}}'_0}$$
$$\varphi_{|\overline{V}} = l \circ \overline{\pi}_{|\overline{V}}$$

is the desired automorphism. □

Lemma 2.4 shows that one may assume in theorem 2.3 that the center $Z(\mathcal{G})$ is totally isotropic subspace of $\mathcal{P}$. The following notations are convenient for the rest of the proof

$$\begin{aligned} \overline{\mathcal{T}}_\alpha &= \bigoplus_{\beta \neq \alpha} \mathcal{T}_\beta = (\overline{\mathcal{G}}_\alpha, \overline{\sigma}_\alpha, \overline{\Omega}_\alpha) = \overline{\mathcal{K}}_\alpha \oplus \overline{\mathcal{P}}_\alpha \\ \pi_a &: \mathcal{G} \to \mathcal{G}_\alpha \ ; \ \overline{\pi}_a : \mathcal{G} \to \overline{\mathcal{G}}_\alpha. \end{aligned}$$

where π_α (resp. $\overline{\pi}_\alpha$) is the canonical projection parallely to $\overline{\mathcal{G}}_\alpha$ (resp. $\mathcal{G}_\alpha$). So are the parallel primed notations.

LEMMA 2.5. *For all $\alpha \geq 1$,* **there exists $\beta \geq 1$ such that** $\mathcal{P}_\alpha \cap \mathcal{P}'_\beta \neq 0$.

PROOF. One simply observes that $0 \neq [\mathcal{K}_\alpha, \mathcal{P}] = \bigoplus_\beta [\mathcal{K}_\alpha, \mathcal{P}'_\beta]$. □

We choose a couple α, β such that $U = \mathcal{P}_\alpha \cap \mathcal{P}'_\beta \neq 0$ and let $W = \pi_\alpha(\mathcal{P}'_\beta)$.

LEMMA 2.6. *If W is a symplectic subspace of $\mathcal{P}_\alpha$, then $W = \mathcal{P}_\alpha$ and furthermore $\mathcal{P}_\alpha \cap \overline{\mathcal{P}}'_\beta = 0$, $\pi_\alpha(\overline{\mathcal{P}}'_\beta) \subset Z(\mathcal{G}_\alpha)$; also β is the unique index such that $\mathcal{P}_\alpha \cap \mathcal{P}'_\beta \neq 0$ and $\dim \mathcal{P}'_\beta \geq \dim \mathcal{P}_\alpha$.*

PROOF. Observe that $\Omega(W^\perp\cap\mathcal{P}_\alpha, \pi_\alpha(P'_\beta)\oplus\overline{\pi}_\alpha(\mathcal{P}'_\beta)) = 0$ implies that $\Omega(W^\perp\cap\mathcal{P}_\alpha, \mathcal{P}'_\beta) = 0$, i. e. $W^\perp\cap\mathcal{P}_\alpha \subset \mathcal{P}'^{\perp}_\beta$; hence

$$[W, W^\perp\cap\mathcal{P}_\alpha] \subset \pi_\alpha[\mathcal{P}'_\beta, W^\perp\cap\mathcal{P}_\alpha] = 0.$$

As $\mathcal{P}_\alpha = W\oplus^\perp W^\perp\cap\mathcal{P}_\alpha$, the indecomposability of $\mathcal{T}_\alpha$ and the fact that $0\neq U$ imply that $W^\perp\cap\mathcal{P}_\alpha = 0$ or $W = \mathcal{P}_\alpha$. We then remark that

$$\Omega(\mathcal{P}_\alpha, \mathcal{P}_\alpha\cap\overline{\mathcal{P}}'_\beta) = \Omega(\pi_\alpha(\mathcal{P}'_\beta), \mathcal{P}_\alpha\cap\overline{\mathcal{P}}'_\beta) = \Omega(\mathcal{P}'_\beta, \mathcal{P}_\alpha\cap\overline{\mathcal{P}}'_\beta) = 0$$

Hence $\mathcal{P}_\alpha\cap\overline{\mathcal{P}}'_\beta = 0$. The facts that π_α is a homomorphism and $W = \mathcal{P}_\alpha$ show that $\pi_\alpha(\overline{\mathcal{P}}'_\beta) \subset Z(\mathcal{G}_\alpha)$. □

LEMMA 2.7. *If for all pairs γ,δ such that $\mathcal{P}_\gamma\cap\mathcal{P}'_\delta\neq 0$, $\pi_\gamma(\mathcal{P}'_\delta)$ (resp. $\pi'_\delta(\mathcal{P}_\gamma)$) is a symplectic subspace of $\mathcal{P}_\gamma$ (resp. $\mathcal{P}'_\beta$), then the theorem is true.*

PROOF. With these assumptions, the lemma above implies $\dim\mathcal{P}_\alpha = \dim\mathcal{P}'_\beta$. One observes that $\overline{\pi}_\alpha(\mathcal{P}'_\beta)$ is central; indeed for any $p'_\beta\in\mathcal{P}'_\beta$ $[\overline{\pi}_\alpha(p'_\beta), \overline{\mathcal{P}}_\alpha] = [p'_\beta, \overline{\mathcal{P}}_\alpha] \subset \mathcal{K}'_\beta\cap\overline{\mathcal{K}}_\alpha$. Furthermore if $k\in[\mathcal{K}'_\beta\cap\overline{\mathcal{K}}_\alpha]$, $[k,p]\neq 0$ only if $p\in\mathcal{P}'_\beta\cap\overline{\mathcal{P}}_\alpha$; but $\mathcal{P}_\alpha\cap\overline{\mathcal{P}}'_\beta = 0$ implies $\overline{\mathcal{P}}_\alpha+\mathcal{P}'_\beta = \mathcal{P}$ and by dimension argument the sum is direct.

The map $\varphi:\mathcal{G}\to\mathcal{G}$ defined by $\varphi_{|\mathcal{P}'_\beta} = \pi_{\alpha|\mathcal{P}'_\beta}$, $\varphi_{|\overline{\mathcal{P}}'_\beta} = \overline{\pi}_{\alpha|\overline{\mathcal{P}}'_\beta}$, $\varphi_{|\mathcal{K}} = \mathrm{id}_{|\mathcal{K}}$ is, in view of the above, a linear isomorphism. It is clearly a Lie algebra isomorphism and $\varphi_{|\mathcal{P}}$ is symplectic as the center is totally isotropic. □

To conclude the proof of the theorem it is enough to prove that the assumptions of the above lemma are always satisfied.

This will be done by induction on the dimension of $W\cap W^\perp$, using an analogue of theorem 2.3 for symmetric pairs. We give the relevant definitions and state the theorem 2.9 whose proof will be given after the conclusion of the proof of theorem 2.3.

DEFINITION 2.8. A **symmetric couple** briefly a S.C., $(\mathcal{G},\sigma)$, is a couple where $\mathcal{G}$ is a finite dimensional real Lie algebra, σ is an involutive automorphism of $\mathcal{G}$. If $\mathcal{G} = \mathcal{K}\oplus\mathcal{P}$ is the corresponding decomposition of $\mathcal{G}$, one assumes that $[\mathcal{P},\mathcal{P}] = \mathcal{K}$ and that the representation of $\mathcal{K}$ on $\mathcal{P}$ is faithful. A S.C. $(\mathcal{G},\sigma)$ is said to be **flat** if $\mathcal{K} = 0$.
A symplectic symmetric triple $(\mathcal{G},\sigma,\Omega)$ has an underlying symmetric couple $(\mathcal{G},\sigma)$. The notions of direct sum of S.C., of indecomposability and decomposability are defined as for S. S. T..

THEOREM 2.9. *Let $\mathcal{C}$ be a symmetric couple and let*

$$\mathcal{C} = \bigoplus_{\alpha=0}^{p}\mathcal{C}_\alpha = \bigoplus_{\beta=0}^{q}\mathcal{C}'_\beta$$

two decompositions such that :

(i) *$\mathcal{C}_0$ and $\mathcal{C}'_0$ are flat*
(ii) *$\forall\alpha,\beta\geq 1$ $\mathcal{C}_\alpha$ and $\mathcal{C}'_\beta$ are indecomposable and non flat.*

Then $p = q$ and there exists an element η of the symmetric group S_p and an automorphism φ such that $\varphi(\mathcal{C}_0) = \mathcal{C}'_0$ and $\varphi(\mathcal{C}_\alpha) = \mathcal{C}'_{\eta(\alpha)}$.

PROOF. **of theorem 2.3** Choose as before α, β such that $0 \neq U = \mathcal{P}_\alpha \cap \mathcal{P}'_\beta$ and assume that $0 \neq r \in X = \text{rad}\,(\pi_\alpha(\mathcal{P}'_\beta))$ $(= \{y \in \pi_\alpha(\mathcal{P}'_\beta) |$ such that $\Omega(y,z) = 0$, for all $z \in \pi_\alpha(\mathcal{P}'_\beta)\})$. Apply theorem 2.9 to $\overline{\mathcal{P}}'_\beta$ and write $\underline{\overline{\mathcal{C}}'_\beta} = \bigoplus_{l=0}^{\overline{q}\beta} \overline{\mathcal{C}}'_{\beta,l}$, where $\underline{\overline{\mathcal{C}}'_\beta}$ means the symmetric couple underlying $\overline{\mathcal{T}}'_\beta$. There exists $p \in \overline{\mathcal{P}}'_{\beta,l}$ $(l \geq 1)$ such that $\Omega(r,p) = 1$; indeed r can not be orthogonal to $\overline{\mathcal{P}}'_\beta$, because it would imply that Ω is degenerate on $\mathcal{P}$; hence there exists $p \in \overline{\mathcal{P}}'_\beta$ such that $\Omega(r,p) = 1$. This element may not belong to $\overline{\mathcal{C}}'_{\beta,0}$ which is central as this would contradict the isotropy of the center of $\mathcal{G}$; indeed, we shall see below that r is also central.
Having chosen r and p we observe the following useful facts

(i) $\overline{\pi}_\alpha(p) \neq 0$, indeed if it were 0, $p \in \mathcal{P}_\alpha \cap \overline{\mathcal{P}}'_\beta$ and thus

$$\Omega(r,p) = \Omega(\pi_a(p'_\beta), p) = \Omega(p'_\beta, p) = 0;$$

a contradiction.

(ii) $\pi_\alpha(p) \neq 0$; indeed if not, $\Omega(r,p) = \Omega(r, \pi_\alpha(p)) = 0$; a contradiction.

(iii) $r \in \overline{\mathcal{P}}'_\beta$ and is central; indeed $\Omega(r, \pi_\alpha(p'_\beta)) = 0$, $\forall p'_\beta \in \mathcal{P}'_\beta$ implies $r \in \overline{\mathcal{P}}'_\beta$. This in turn implies $[r, p'_\beta] = 0$, $\forall p'_\beta \in \mathcal{P}'_\beta$. Now write $[r, \overline{p}'_\beta] = [\pi_\alpha(p'_\beta), \overline{p}'_\beta] = [\pi_\alpha(p'_\beta), \pi_\alpha(\overline{p}'_\beta)] = 0$ and r is central. In particular, the center being totally isotropic this shows that $\pi_\alpha(p)$ is not central.

(iv) $\overline{\pi}_\alpha(p)$ is central. To prove this point we decompose the indecomposable symmetric symplectic triples into indecomposable symmetric couples; i.e we write :

$$\mathcal{G} = \mathcal{G}_0 \oplus \bigoplus_{\alpha=1}^{p} \mathcal{G}_\alpha = \mathcal{G}'_0 \oplus \bigoplus_{\beta=1}^{q} \mathcal{G}'_\beta$$

$$\mathcal{C}_\alpha = \bigoplus_{a=0}^{q_\alpha} \mathcal{C}_{\alpha,a} \quad \mathcal{C}'_\beta = \bigoplus_{b=0}^{q'_s} \mathcal{C}'_{\beta,b} \quad \overline{\mathcal{C}}'_\beta = \bigoplus_{c=0}^{\overline{q}'_\beta} \overline{\mathcal{C}}'_{\beta,c}$$

By choice $p \in \overline{\mathcal{P}}'_{\beta,c}$. Hence by theorem 2.9 there exists a unique pair (μ, a) such that $\overline{\mathcal{P}}'_{\beta,c} \cap \mathcal{P}_{\mu,a} \neq 0$. Write :

$$p = \pi_{\mu,a}(p) + \overline{\pi}_{\mu,a}(p)$$

The proof of theorem 2.9 shows that $\overline{\pi}_{\mu,a}(p)$ is central. If μ would be different from α, $\pi_\alpha(p) = \pi_\alpha \overline{\pi}_{\mu,a}(p)$ would be central; a contradiction. Hence $\mu = \alpha$ and $\overline{\pi}_\alpha(p) = \overline{\pi}_\alpha \overline{\pi_{\alpha,a}}(p)$ is central.

We now use r and p to construct subspaces $\overline{\mathcal{D}}'_\beta$ (resp. $\mathcal{D}'_\beta$) of $\overline{\mathcal{P}}'_\beta$ (resp. $\mathcal{P}'_\beta$). Define $\mathcal{D}'_\beta = \{x \in \mathcal{P}'_\beta | \Omega(\pi_\alpha(p), x) = 0\}$ and choose y in $\mathcal{P}'_\beta$ such that $\pi_\alpha(y) = r$. Then :

$$\mathcal{P}'_\beta = \mathbb{R}y \oplus \mathcal{D}'_\beta$$

The 2-dimensional subspace of $\pi_\alpha(\overline{\mathcal{P}}'_\beta)$ spanned by r and $\pi_\alpha(p)$ has an orthogonal supplement Y in $\pi_\alpha(\overline{\mathcal{P}}'_\beta)$. Define $\overline{\mathcal{D}}'_\beta$ by :

$$\begin{aligned}\overline{\mathcal{P}}'_\beta &= \mathbb{R}r \oplus \mathbb{R}p \oplus \overline{\mathcal{D}}'_\beta \\ \pi_\alpha(\overline{\mathcal{D}}'_p) &= Y\end{aligned}$$

Use these decompositions of $\mathcal{P}'_\beta$ and $\overline{\mathcal{P}}'_\beta$ to construct an automorphism of the symmetric couple $(\mathcal{G},\sigma)$. Define $A : \mathcal{G} \to \mathcal{G}$ by :

$$\begin{aligned}A_{|\mathcal{K}\oplus\mathcal{D}'_\beta\oplus\mathbb{R}r\oplus\overline{\mathcal{D}}'_\beta} &= \mathrm{id}_{|\mathcal{K}\oplus\mathcal{D}'_\beta\oplus\mathbb{R}r\oplus\overline{\mathcal{D}}'_\beta} \\ Ap &= \pi_\alpha(p) \\ Ay &= \overline{\pi}_\alpha(y)\end{aligned}$$

We first prove that A is a linear isomorphism. Observe that $A_{|\mathcal{P}'_\beta}$ is injective; indeed if it were not there would exist $d'_\beta \in \mathcal{D}'_\beta$, $a \in \mathbb{R}$ such that $d'_\beta + a\overline{\pi}_\alpha(y) = 0$; this would imply $\overline{\pi}_\alpha(y) \in \mathcal{P}'_\beta$ and as $y \in \mathcal{P}'_\beta$, $\pi_\alpha(y) = r$ would belong to $\mathcal{P}'_\beta$; a contradiction. We next observe that $A_{|\overline{\mathcal{P}}'_\beta}$ is injective; indeed if it were not there would exist $\overline{d}'_\beta \in \overline{\mathcal{D}}'_\beta$, $a, b \in \mathbb{R}$ $(b \neq 0)$ such that $\overline{d}'_\beta + ar + b\pi_\alpha(p) = 0$; this would imply $\pi_\alpha(p) \in \overline{\mathcal{P}}'_\beta$

$$\Omega(r,p) = \Omega(r,\pi_\alpha(p)) = \Omega(\pi_\alpha(p'_\beta),\pi_\alpha(p)) = \Omega(p'_\beta,\pi_\alpha(p)) = 0$$

if $r = \pi_\alpha(p'_\beta)$; hence a contradiction.

We then show (i) that $A(\mathcal{P}'_\beta)$ is a symplectic subspace of $\mathcal{P}$, (ii) that $A_{|\mathcal{P}'_\beta} : \mathcal{P}'_\beta \to A(\mathcal{P}'_\beta)$ is a symplectic map and (iii) that $A(\overline{\mathcal{P}}'_\beta) \subset A(\mathcal{P}'_\beta)^\perp$. Hence by dimensionality we have $A(\overline{\mathcal{P}}'_\beta) = A(\mathcal{P}'_\beta)^\perp$ and A is a linear isomorphism.

For (i, ii) we have :

$$\begin{aligned}&\Omega(d'_{1\beta} + a_1\overline{\pi}_\alpha(y), d'_{2\beta} + a_2\overline{\pi}_\alpha(y)) \\ =\ &\Omega(d'_{1\beta} + a_1 y - a_1 r, d'_{2\beta} + a_2 y - a_2 r) \\ =\ &\Omega(d'_{1\beta} + a_1 y, d'_{2,\beta} + a_2 y)\end{aligned}$$

For (iii) we have similarly :

$$\begin{aligned}&\Omega(d'_\beta + a\overline{\pi}_\alpha(y)), (\overline{d}'_\beta + br) + c\pi_\alpha(p)) \\ =\ &\Omega(a\overline{\pi}_\alpha(y), (\overline{d}'_\beta + br)) \quad \text{(using the definition of } \mathcal{D}'_\beta) \\ =\ &\Omega(a\overline{\pi}_\alpha(y), \overline{d}'_\beta) \\ =\ &-a\Omega(r, \overline{d}'_\beta) = 0 \quad \text{(using the definitions of } \overline{\mathcal{D}}'_\beta)\end{aligned}$$

Using the facts that r and $\overline{\pi}_\alpha(p)$ are central one checks that, for any $p_1, p_2 \in \mathcal{P}$,

$$[Ap_1, Ap_2] = A[p_1 p_2]$$

To conclude that A is a Lie algebra isomorphism we observe

(i) that

$$[\mathcal{K}, \mathcal{P}'_\beta] \subset \mathcal{P}_\alpha \cap \mathcal{P}'_\beta + \overline{\mathcal{P}}_\alpha \cap \mathcal{P}'_\beta$$

(ii) that

$$[\mathcal{K}, \overline{\mathcal{P}}'_\beta] \subset \mathcal{P}_\alpha \cap \overline{\mathcal{P}}'_\beta + \overline{\mathcal{P}}_\alpha \cap \overline{\mathcal{P}}'_\beta$$

(iii) and that $\mathcal{P}_\alpha \cap \mathcal{P}'_\beta \subset \mathcal{D}'_\beta$, $\overline{\mathcal{P}}_\alpha \cap \mathcal{P}'_\beta \subset \mathcal{D}'_\beta$, $\mathcal{P}_\alpha \cap \overline{\mathcal{P}}'_\beta \subset \mathbb{R}r \oplus \overline{\mathcal{D}}'_\beta$, $\overline{\mathcal{P}}_\alpha \cap \overline{\mathcal{P}}'_\beta \subset \overline{\mathcal{D}}'_\beta$.

Hence that $A_{|[\mathcal{K},\mathcal{P}]} = \mathrm{id}_{|[\mathcal{K},\mathcal{P}]}$.
We now use A to construct a new decomposition of the given symmetric symplectic triple. Define :

$$\begin{aligned}\tilde{\mathcal{G}}' &= \mathcal{K}'_\beta \oplus A\mathcal{P}'_\beta \\ \overline{\tilde{\mathcal{G}}}' &= \overline{\mathcal{K}}'_\beta \oplus A\overline{\mathcal{P}}'_\beta\end{aligned}$$

and observe that the triple $(\tilde{\mathcal{G}}', \sigma_{|\tilde{\mathcal{G}}'}, \Omega_{|\tilde{\mathcal{G}}'})$ is indecomposable non flat and isomorphic to the triple $(\mathcal{G}'_\beta, \sigma_{|\mathcal{G}'_\beta}, \Omega_{|\mathcal{G}'_\beta})$. For this triple $(\tilde{\mathcal{G}}', \sigma_{|\tilde{\mathcal{G}}'}, \Omega_{|\tilde{\mathcal{G}}'})$ we have the following properties

(i) $\tilde{\mathcal{P}} \cap \mathcal{P}_\alpha \supset \mathcal{P}'_\beta \cap \mathcal{P}_\alpha = U$
(ii) $\pi_\alpha(\tilde{\mathcal{P}}') \subset \pi_\alpha(\mathcal{P}'_\beta)$
(iii) $\dim \mathrm{rad}\pi_\alpha(\mathcal{P}') < \dim \mathrm{rad}\pi_\alpha(\mathcal{P}'_\beta)$.

Indeed $\tilde{\mathcal{P}}' = A\mathcal{P}'_\beta \supset U$; which implies (i). For the second statement

$$\pi_\alpha(A\mathcal{P}'_\beta) = \pi_\alpha(\mathcal{D}'_\beta + \mathbb{R}\overline{\pi}_\alpha(y)) = \pi_\alpha(\mathcal{D}'_\beta) \subset \pi_\alpha(\mathcal{P}'_\beta).$$

For the last (and crucial property), we have $\mathrm{rad}\pi_\alpha(A\mathcal{P}'_\beta) = \mathrm{rad}\pi_\alpha(\mathcal{D}'_\beta)$ and $\mathrm{rad}(\pi_\alpha\mathcal{P}'_\beta) \supset \mathrm{rad}\pi_\alpha(\mathcal{D}'_\beta) \oplus \mathbb{R}r$.

Iterating this construction we build a triple isomorphic to $(\mathcal{G}'_\beta, \sigma_\beta, \Omega_{|\mathcal{G}'_\beta})$ of the form $(\mathcal{K}'_\beta \oplus \tilde{\mathcal{P}}^{(k)}, \sigma, \Omega)$ such that

$$\begin{aligned}&\tilde{\mathcal{P}}^{(k)} \cap \mathcal{P}_\alpha \supset U \\ &\dim \mathrm{rad}\pi_\alpha(\tilde{\mathcal{P}}^{(k)}) = 0 \\ &\pi_\alpha(\tilde{\mathcal{P}}^k) \subset \pi_\alpha(\mathcal{P}'_\beta)\end{aligned}$$

But then lemma 2.6 implies $\pi_\alpha(\tilde{\mathcal{P}}^k) = \mathcal{P}_\alpha$; hence $\pi_\alpha(\mathcal{P}'_\beta) = \mathcal{P}_\alpha$ and $\mathrm{rad}\pi_\alpha(\mathcal{P}'_\beta) = 0$ which is a contradiction and ends the proof of theorem 2.3. □

PROOF. **of theorem 2.9** The notations are essentially the same as for theorem 2.3.
One first observes that $\forall \alpha \geq 1$, there exists $\beta \geq 1$ such that $\mathcal{P}_\alpha \cap \mathcal{P}'_\beta \neq 0$. Indeed $[\mathcal{K}_\alpha, \mathcal{P}] \subset \bigoplus_{\beta \geq 1}[K_\alpha, \mathcal{P}'_\beta] \subset \bigoplus_{\beta \geq 1} \mathcal{P}_\alpha \cap \mathcal{P}'_\beta$; choose such a couple (α, β) and let $U = \mathcal{P}_\alpha \cap \mathcal{P}'_\beta$.
One must have $\pi_\alpha(\mathcal{P}'_\beta) = \mathcal{P}_\alpha$ and $\mathcal{P}_\alpha \cap \overline{\mathcal{P}}'_\beta = 0$; indeed there exists $C_1 \subset \pi_\alpha(\mathcal{P}'_\beta)$, $D_1 \subset \pi_\alpha(\overline{\mathcal{P}}'_\beta)$ two subspaces such that $U \subset C_1$, $\mathcal{P}_\alpha \cap \overline{\mathcal{P}}'_\beta \subset D_1$, and $\mathcal{P}_\alpha = C_1 \oplus D_1$. These subspaces are stable by $\mathcal{K}$, and commute as π_α is a homomorphism. The indecomposability of $\mathcal{C}_\alpha$ and $U \neq 0$, implies $D_1 = 0$ and $C_1 = \mathcal{P}_\alpha$. We then conclude that $\dim \mathcal{P}'_\beta \geq \dim \mathcal{P}_\alpha$ and that there is only one $\beta \geq 1$ such that $\mathcal{P}_\alpha \cap \mathcal{P}'_\beta \neq 0$; by symmetry this says that $\dim \mathcal{P}_\alpha = \dim \mathcal{P}'_s$ and that $p = q$. The correspondance between α and β given by $\mathcal{P}_\alpha \cap \mathcal{P}'_\beta \neq 0$ defines an element η of the symmetric group S_p $(\beta = \eta(\alpha))$. This also implies $\dim \mathcal{P}_0 = \dim \mathcal{P}'_0$. Remark finally that $\overline{\pi}_\alpha(\mathcal{P}'_\beta)$ is central; indeed if $p'_\beta \in \mathcal{P}'_\beta$,

$$[\overline{\pi}_\alpha(p'_\beta), \overline{\mathcal{P}}_\alpha] = [p'_\beta, \overline{\mathcal{P}}_\alpha] \subset \mathcal{K}'_\beta \cap \overline{\mathcal{K}}_\alpha$$

But if $k \in \mathcal{K}'_\beta \cap \overline{\mathcal{K}}_\alpha$, $[k, \mathcal{P}] \subset \mathcal{P}'_\beta \cap \overline{\mathcal{P}}_\alpha = 0$; hence $\mathcal{K}'_\beta \cap \overline{\mathcal{K}}_\alpha = 0$; i.e. $[\overline{\pi}_\alpha(p'_\beta), \mathcal{P}] = 0$. Define then $\varphi : \mathcal{G} \to \mathcal{G}$ by :

$$\varphi = \psi_{|\mathcal{P}'_o} \oplus \bigoplus_{\alpha=1} \pi_{\alpha|\mathcal{P}'_{\eta(\alpha)}} \oplus \mathrm{id}_{|\mathcal{K}}$$

where ψ is a linear isomorphism $\mathcal{P}'_0 \to \mathcal{P}_0$; one checks that φ is the automorphism of $\mathcal{C}$ one was looking for. □

Theorems 2.3 and 2.9 present two decompositions of symmetric symplectic triples : the one given by theorem 2.3 is in indecomposable symmetric symplectic triples; the second given by theorem 2.9 is in indecomposable symmetric couples. In this second decomposition one forgets the role of the 2-cocycle Ω. The following proposition describes a situation when the two decompositions coincide.

PROPOSITION 2.10. Let $\mathcal{T} = (\mathcal{G}, \sigma, \Omega)$ be a symmetric symplectic triple. Assume Ω is a coboundary. Then

(i) The Lie algebra $\mathcal{G}$ has no center
(ii) The decompositions given by theorems 2.3 and 2.9 coincide
(iii) The decomposition given by theorem 2.3 is unique up to a permutation of the factors.

PROOF. Let α be an element of $\mathcal{G}^*$ such that $\Omega = \delta\alpha$; as $\Omega_{|\mathcal{P}\times\mathcal{P}}$ is non degenerate, $\mathcal{G}$ has no center. Now let V, W be subspaces of $\mathcal{P}$ such that $\mathcal{P} = V \oplus W$ and $[V, W] = 0$. Then V and W are symplectic, orthogonal and $\mathcal{K}$ stable. Indeed $\Omega(v, w) = -\alpha([v, w]) = 0$; so one simply has to check $\mathcal{K}$ stability. Now

$$\Omega(w, [k, v]) = 0 \quad \forall v \in V,\ w \in W,\ k \in \mathcal{K}.$$

Indeed if $k = [w_1, w_2]$ one applies Jacobi identity and if $k = [v_1, v_2]$ one applies $\mathcal{K}$ invariance of Ω and Jacobi identity. Point (iii) comes from the proof of theorem 2.9 which shows that $\overline{\pi}_\alpha(\mathcal{P}'_\beta) = 0$; hence $\mathcal{P}'_\beta = \mathcal{P}_\alpha$. □

The result analoguous to theorem 2.3 for symmetric symplectic manifolds is now easy to formulate.

DEFINITION 2.11. Let (M_i, ω_i, s_i) $(i = 1, 2)$ be a symmetric symplectic manifold. The triple $(M_1 \times M_2, s_1 \times s_2, \omega_1 \otimes \omega_2)$ is called the **product** of the given symmetric symplectic manifolds. A S.S.M is said to be **decomposable** if it is the product of two non trivial S.S.M's; it is called **indecomposable** if it is not decomposable.

The following theorem is now an easy consequence of theorem 2.3.

THEOREM 2.12. *Let (M, ω, s) be a simply connected symmetric symplectic manifold. Then there exist simply connected symmetric symplectic manifolds (M_i, ω_i, s_i) $(0 \le i \le r)$ such that (M_0, ω_0, s_0) is a certain $\mathbb{R}^{2n}$ with its standard symplectic form and its standard symmetries, (M_i, ω_i, s_i) $(1 \le i \le r)$ is indecomposable and $(M, \omega, s) = \prod_{i=0}^{r}(M_i, \omega_i, s_i)$. Furthermore two such decompositions are conjugate by an automorphism of (M, ω, s).*

3. Holonomy, complete reducibility

Let (M, ω, s) be a symmetric symplectic manifold; let ∇ be its canonical connection and let $G(M)$ be its transvection group. Choose a base point o in M and let K be the stabilizer of o in $G(M)$. As (M, ∇) is an affine symmetric space K is isomorphic to the linear holonomy group of (M, ∇) at o.

DEFINITION 3.1. (M, ω, s) is said to be **completely reducible** if the action of K on M_0 is completely reducible.

PROPOSITION 3.2. If (M, ω, s) is completely reducible, the group $G(M)$ is reductive. Conversely if $G(M)$ is reductive, the symmetric space (M, ω, s) is completely reducible.

PROOF. Let $(\mathcal{G}, \sigma, \Omega)$ be the S.S.T associated to (M, ω, s). If (M, ω, s) is completely reducible the action of $\mathcal{K}$ on $\mathcal{P}$ is completely reducible and conversely. One may choose a Levi decomposition of $\mathcal{G}$, stable by σ :

$$\mathcal{G} = \mathcal{S} \oplus \mathcal{R} \quad \sigma\mathcal{S} = \mathcal{S} \quad \sigma\mathcal{R} = \mathcal{R}$$

where $\mathcal{R}$ = radical of $\mathcal{G}$ and $\mathcal{S}$ is a Levi factor. We shall write $K_{\mathcal{S}} = \mathcal{K} \cap \mathcal{S}$, $K_{\mathcal{R}} = \mathcal{K} \cap \mathcal{R}$, $\mathcal{P}_{\mathcal{S}} = \mathcal{P} \cap \mathcal{S}$, $\mathcal{P}_{\mathcal{R}} = \mathcal{P} \cap \mathcal{R}$. Observe that $K_{\mathcal{R}}$ is contained in the nil radical of $\mathcal{G}$ and has thus a nilpotent action on $\mathcal{P}$.
Assuming the action of $\mathcal{K}$ on $\mathcal{P}$ to be completely reducible, one may write $\mathcal{P} = \bigoplus_{\alpha} \mathcal{P}_\alpha$ where the $\mathcal{P}_\alpha$'s are $\mathcal{K}$ irreducible invariant subspace. If $K_{\mathcal{R}} \neq 0$, let $W_\alpha = \{p \in \mathcal{P}_\alpha \mid [K_\alpha, p] = 0\}$; this subspace of $\mathcal{P}_\alpha$ is $\mathcal{K}$ stable, different from 0 and hence $W_\alpha = \mathcal{P}_\alpha$, $\forall \alpha$. Hence $K_a = 0$, a contradiction. If $K_a = 0$, $\mathcal{P}_{\mathcal{R}} = Z(\mathcal{G})$ and $\mathcal{G} = \mathcal{S} \oplus Z$. Hence $\mathcal{G}$ is reductive. The inverse is obvious using the fact that $K_{\mathcal{S}}$ is completely reducible [9]. □

DEFINITION 3.3. A symmetric space (M, ω, s) is called **semi simple** (resp. **simple**) if $G(M)$ is semi simple (resp. simple).

PROPOSITION 3.4. Let (M, ω, s) be a simply connected symmetric space which is completely reducible. The $(M, \omega, s) = (M_0, \omega_0, s_0) \times (M_1, \omega_1, s_1)$ where (M_0, ω_0, s_0) is the standard flat symplectic space and (M_1, ω_1, s_1) is a semi simple symmetric space.

The semi simple symmetric spaces have been described elsewhere [5]. We simply recall here the basic results.

PROPOSITION 3.5. Let (M, ω, s) be a simply connected semi simple symmetric symplectic space. Then $(M, \omega, s) = \prod_{i=1}^{r} (M_i, \omega_i, s_i)$ where (M_i, ω_i, s_i) is a simply connected simple symmetric symplectic space.

Let $(\mathcal{G}, \sigma, \Omega)$ be a S.S.T with $\mathcal{G}$ simple; write $\mathcal{G} = \mathcal{K} \oplus \mathcal{P}$ and denote by $\beta$the Killing form of $\mathcal{G}$. The subspaces $\mathcal{K}$ and $\mathcal{P}$ are β-orthogonal; in particular the restriction of β to $\mathcal{P} \times \mathcal{P}$ (resp. $K \times K$) is not degenerate. If X (resp. ξ) is an element of $\mathcal{G}$ (resp. $\mathcal{G}^*$) we shall denote X^b (resp. $\xi^\#$) the element of $\mathcal{G}^*$ (resp. $\mathcal{G}$) defined by $X^b = \beta(X, \cdot)$ (resp. $\beta(\xi^\#, \cdot) = \xi$).

PROPOSITION 3.6. Let $(\mathcal{G}, \sigma, \Omega)$ be a symmetric symplectic triple with $\mathcal{G}$ simple. Then

(i) There exists $0 \neq u \in Z(K)$ such that $\Omega = \partial u^b$, where ∂ is the Chevalley coboundary operator. Conversely if $0 \neq u \in Z(K)$ $\partial u^b_{|\mathcal{P}\times\mathcal{P}}$ is a symplectic 2 form.

(ii) if $\mathcal{G}$ admits an almost complex structure $\mathcal{J}$, commuting with the adjoint action of $\mathcal{G}$, $\mathcal{J}$ commutes with σ.

If $\mathcal{G}$ is a real simple Lie algebra; its complexified algebra $\mathcal{G}^C$ is either simple or the direct sum of two isomorphic simple complex ideals; in the first case $\mathcal{G}$ is said to be absolutely simple; in the second case $\mathcal{G}$ may be considered as a simple complex algebra, viewed as a real algebra. If $(\mathcal{G},\sigma,\Omega)$ is a simple S.T, and $\mathcal{G}$ is absolutely simple, $(\mathcal{G}^C,\sigma^C,\Omega^C)$ is a simple complex symmetric symplectic triple. If on the other hand $\mathcal{G}$ is a complex algebra, viewed as a real Lie algebra, $(\mathcal{G},\sigma,\Omega)$ may be viewed as a simple complex symmetric triples. The simple complex symmetric triples are described by

THEOREM 3.7. *Let $(\mathcal{G},\sigma,\Omega)$ be a simple complex S.T and let u be the element of the center of $\mathcal{K}$ such that $\Omega = \partial u^b$. Then u is a semi simple element. Its centralizer in $\mathcal{G}$ coincides with $\mathcal{K}$; hence $\mathcal{G}$ and $\mathcal{K}$ have the same rank. Furthermore the dimension of the center of $\mathcal{K}$ is one. Finally it is isomorphic to a simple complex triple $(\mathcal{G},\sigma_1,\Omega_1)$ constructed as follows. Let $\mathcal{H}$ be a Cartan subalgebra of $\mathcal{G}$; let ϕ be the system of roots of $\mathcal{G}$ relative to $\mathcal{H}$; let Δ be a basis of ϕ and let ϕ^+ be the corresponding system of positive roots. Assume α is an element of Δ which appears in the maximal root μ of ϕ with coefficient 1. Denote by ξ the element of $\mathcal{H}_{\mathbb{R}}$ such that $\alpha(\xi) = 1$ and if $\alpha' \in \Delta$, $\alpha' \neq \alpha$, $\alpha'(\xi) = 0$. Then $\sigma_1 = \exp \pi i \; ad\, \xi$ and there exists a non zero complex number a such that $u_1 = a\xi$ and $\Omega_1 = \partial u_1^b$.*

DEFINITION 3.8. A S.S.T $(\mathcal{G},\sigma,\Omega)$ is said to be pseudo **Kähler** if there exists an almost complex structure $\mathcal{J}$ on $\mathcal{P}$ commuting with the adjoint action of $\mathcal{K}$ and leaving $\Omega_{|\mathcal{P}\times\mathcal{P}}$ stable.It is said to be Kähler if the symmetric bilinear form $\Omega(J,.)$ is positive definite.

PROPOSITION 3.9. Let (M,ω,s) be a compact symmetric symplectic space. Then M is completely reducible. In particular it is kähler.

PROOF. The transvection group of a compact symmetric space is compact. This implies the first statement and proposition 3.4 implies the second statement. □

With the help of proposition 3.4 it is easy to give a list of symmetric symplectic triples which are simple and do not admit kähler structure. We simply indicate the algebra $\mathcal{G}$ and the subalgebra $\mathcal{K}$.It is said Kähler if the symmetric bilinear form $\Omega(J,.)$ is positive definite

1. Classical algebras

$$\begin{array}{ll}
\mathcal{G} = sl(n,\mathbb{R}) & \mathcal{K} = sl(p,\mathbb{R}) \oplus sl(n-p,\mathbb{R}) \oplus \mathbb{R}\ 0<p<n \\
\mathcal{G} = su^*(2n) & \mathcal{K} = su^*(2p) \oplus su^*(2n-2p) \oplus \mathbb{R}\ 0<p<n \\
\mathcal{G} = su(n,n) & \mathcal{K} = sl(n,\mathbb{C}) \oplus \mathbb{R} \\
\mathcal{G} = so^*(4n) & \mathcal{K} = su^*(2n) \oplus \mathbb{R} \\
\mathcal{G} = so(p,q) & \mathcal{K} = so(p-2,q) \oplus \mathbb{R} \\
\mathcal{G} = so(n,n) & \mathcal{K} = sl(n,\mathbb{R}) \oplus \mathbb{R} \\
\mathcal{G} = sp(n,n) & \mathcal{K} = su^*(2n) \oplus \mathbb{R} \\
\mathcal{G} = sp(n,\mathbb{R}) & \mathcal{K} = sl(n,\mathbb{R}) \oplus \mathbb{R}
\end{array}$$

2. Exceptional algebras

$$\begin{array}{ll} \mathcal{G} = E_6^1 & \mathcal{K} = so(5,5) \oplus \mathbb{R} \\ \mathcal{G} = E_6^4 & \mathcal{K} = so(1,9) \oplus \mathbb{R} \\ \mathcal{G} = E_7^1 & \mathcal{K} = E_6^1 \oplus \mathbb{R} \\ \mathcal{G} = E_7^3 & \mathcal{K} = E_6^4 \oplus \mathbb{R} \end{array}$$

References

[1] E. Cartan *Sur une classe remarquable d'espaces de Riemann.* Bull. Soc. Math. France **54**, 1926, pp 214-264 and **55**, 1927, pp 114-134.
See also S. Helgason **Differential geometry, Lie groups and symmetric spaces.** Ac. Press 1978 and the references therein.
[2] M. Cahen, M. Parker *Pseudo riemannian symmetric spaces.* Mem. Ann. Math. Soc., 1980, **24, 229**, pp 1-108.
[3] O. Loos **Symmetric spaces I, II.** Benjamin 1969.
[4] H. Wu *On the de Rham decomposition theorem.* Ill. J. Math., **8**, 1964, pp 291-311.
[5] P. Bieliavsky, M. Cahen, S. Gutt *Symmetric symplectic manifolds.* in Modern group theoritical methods in Physics, Math. Studies, 18 pp 63-75
[6] S. Kobayashi, K. Nomizu **Foundations of differential geometry.** Vol II, , Interscience 1969
[7] N. Wallach, V. *Symplectic Geometry and Fourier Analysis*, Math. Sc. Press, 1977
[8] S. Koh *On affine symmetric spaces.* Trans. Am. Math. Soc. **119**, 1965, pp 291-309
[9] C. Chevalley, *Groupes algébriques*, Hermann 1968
[10] J. M. Souriau, *Structure dres systèmes dynamiques*, Dunod 1970

Département de Mathématiques. Université Libre de Bruxelles. Campus Plaine CP 218. 1050 Bruxelles. Belgium

E-mail address: `sgutt@ulb.ac.be`

Current address: Department of Mathematics. University of California at Berkeley. Berkeley, CA 94720. U.S.A.

Current address: Département de Mathématiques. Université Libre de Bruxelles.Campus Plaine CP 218. 1050 Bruxelles. Belgium

E-mail address: `sgutt@ulb.ac.be`

Current address: Département de Mathématiques. Université de Metz. Ile du Saulcy. F-57045 Metz cedex. France.

Contemporary Mathematics
Volume **203**, 1997

r^{th} Order Conditionally Convergent Series of Fractal Domains

Jenny Harrison

ABSTRACT. We study integration of a p-form ω defined over a domain written as an infinite series of p-dimensional simplicial chains $\sum_{i=1}^{\infty} S_i$. Term-by-term integration is valid as long as the series of chains converges in the λ- norm [H2] and ω is sufficiently smooth. In particular, if $\sum_{i=1}^{\infty} S_i$ converges in the λ-norm and ω is a p-form of class C^r where r $\geq \lambda - p - 1 + Lip$ if $\lambda \in \mathbb{Z}$ and r $\geq \lambda$ -p if $\lambda \in \mathbb{R}/\mathbb{Z}$, then

$$\int_{\sum_{i=1}^{\infty} S_i} \omega = \sum_{i=1}^{\infty} \int_{S_i} \omega.$$

We define the *extrinsic dimension* dim_E of a series $\sum_{i=1}^{\infty} S_i$ which measures its degree of conditional convergence. If $\lambda > dim_E(\sum_{i=1}^{\infty} S_i)$ then $\sum_{i=1}^{\infty} S_i$ is a domain of integration for forms of class C^r.

Self-similar fractals such as the Van Koch snowflake S may be written as an infinite sum of finite simplicial chains S = $\sum_{i=1}^{\infty} S_i$. Its extrinsic dimension is $\frac{ln4}{ln3}$. Thus forms of class $C^\alpha, \alpha \geq \frac{ln4}{ln3} - 1$, may be integrated directly over S and term by term integration is possible.

We show this result is sharp by defining in $\mathbb{R}^2$ a 1-form ω of class C^α such that $\omega(x, 0) = 0$ for all x; and a series $\sum_{i=1}^{\infty} S_i$ which converges to the unit interval I contained in the x-axis in the λ-norm for $\lambda > 1 + \alpha$ with $lim \sum_{i=1}^{\infty} \int_{S_i} \omega = -1$ but $\int_{\Sigma S_i} \omega = \int_I \omega = 0$.

Introduction

Mathematicians encountering a 1-form ω defined in a neighborhood of a non-smooth set such as a fractal curve γ in $\mathbb{R}^3$ may wish to integrate ω over γ. A common approach is to integrate the form over approximating smooth curves and take limits. Herein lie two basic difficulties which are often dismissed or ignored. Does the limit exist and is it unique? Other simple, related questions have been unanswered until recently. How does the integral vary when a domain is perturbed? Is there a natural way to determine how far apart are two oriented curves γ_1 and γ_2? The Hausdorff metric may come to mind but it does not take into account orientation and multiplicity and is therefore too limited. For example, if γ_1 is an oriented closed curve and γ_2 is the same as γ_1 taken twice around, the Hausdorff metric gives their distance as 0 even though these curves are not the same. Another

1991 *Mathematics Subject Classification.* Primary 28A78; Secondary 28A75,28A80 .

approach is to consider the area of a minimal spanning set of the two curves, their *homological distance.* However, if one curve has Hausdorff dimension greater than two then any set spanning γ_1 and γ_2 will have infinite area.

The λ-norms defined in [H2] lead to a new metric which can deal effectively with these questions and treat issues of existence and uniqueness of limits of integrals. Here we apply techniques from that paper to analyze infinite series of chains and other domains of integration.

The λ-norms are related to the flat and sharp norms of Whitney [W]. Eells has defined a related family of norms [E]. The low differentiability version of this theory is similar to that developed in [H-N].

The λ-norm is defined here for domains in Euclidean space $\mathbb{R}^n$ of arbitrary codimension but can also be defined on smooth Riemannian n-manifolds using the inner product structure.

1. Deficiencies of the Riemann and Lebesgue integrals

It is well understood that the Riemann integral has a number of deficiencies. A sequence f_n of Riemann integrable functions may converge pointwise to f over a fixed interval I, but we can conclude nothing about the relationship between $\int_I f$ and $lim \int_I f_n$. The quantities may not exist and if they do exist they may not be equal. Of special historical interest has been the question as to whether $\int_I \sum_{n=1}^{\infty} g_n = \sum_{n=1}^{\infty} \int_I g_n$.

These questions were resolved by the introduction of Lebesgue's L^1 norm. As long as $f_n \longrightarrow f$ in the L^1 norm their integrals converge. Furthermore, the Banach space obtained upon completion of the simple functions contains functions not Riemann integrable. Elements of this Banach space are called Lebesgue integrable functions.

The following examples show that dual problems arise if one considers domains $S_n \to S$ in the sense of homology in $\mathbb{R}^n$ and a fixed form ω defined in a neighborhood of the domains in $\mathbb{R}^n$.

Examples

1. Let $\theta : \mathbb{R} \to \mathbb{R}$ be a smooth, positive function so that $\theta(t) = n$ for $\sqrt{n} \leq$ t $< \sqrt{(n+1)}, n \geq 0, n \in \mathbb{Z}$. Define ω to be the form $\omega(x,y) = \theta(y)dx$. Define $f_n : I \to \mathbb{R}$ by $f_n(x) = \sqrt{n}$ for $x \leq 1/n$ and $f_n(x) = 0$ for $x > 1/n$. Let S_n be the graph of f_n together with the vertical line $x = 1/n, 0 \leq y \leq \sqrt{n}$, oriented from left to right. Then $S_n \to I$ in the sense of homology since the subgraph of S_n has area $1/\sqrt{n} \to 0$. But $\int_{S_n} \omega = 1$ for all n, whereas $\int_I \omega = 0$.
2. Construct a similar sequence of curves, but this time, let $\theta(t) = n^2$ for $\sqrt{n} \leq$ t $< \sqrt{(n+1)}$. Then $\lim_{n \to \infty} \int_{S_n} \omega$ does not exist and $\int_I \omega = 0$

3. Let S $=\lim S_n$ denote the van Koch snowflake where the S_n are smooth curves approximating S in the sense of homology. The curve S is not a domain of integration in Lebesgue theory since it is nonrectifiable. Yet, $\lim \int_{S_n} \omega$ exists for every $\omega \in C^1$. (See below.)

As in Lebesgue theory these questions of convergence may also be clarified by a norm, but this time the norm is defined for domains of integration rather than for the integrands. As long as the domains converge in this norm, their integrals will converge.

The λ-norm for $\lambda \geq$ p is first defined on simplicial p-chains which are equivalence classes of formal sums of simplexes with real coefficients. (Roughly, two formal sums of simplexes are equivalent if and only if they have a common simplicial subdivision. See [W] or [H2] for more details.)

The elements of the Banach spaces $\mathcal{A}^{p,\lambda}$ obtained upon completing the simplicial p-chains with the λ-norm are called λ-*domains of integration* because one may extend the integral to them in a natural way: Suppose A$\in \mathcal{A}^{p,\lambda}$. A p-form ω may be integrated over A if $\omega \in C^r$ where

$$* \qquad r = r(\lambda, p) = \left\{ \begin{array}{cc} \lambda - p - 1 + Lip & \text{if } \lambda \in \mathbb{Z} \\ \lambda - p & \text{if } \lambda \in \mathbb{R}/\mathbb{Z} \end{array} \right\}$$

Let $|\omega|_r$ denote the C^r norm on forms.

DEFINITION 1. For $A \in \mathcal{A}^{p,\lambda}$ and $\omega \in C^r$ define $\int_A \omega = lim \int_{A_n} \omega$ where A_n are simplicial p-chains converging to A in the λ-norm.

The Continuity Theorem in section 3 implies that this integral is well defined.

2. Norms on domains

Analytic definition of the natural l-norm. For A a polyhedral p-chain and $\lambda > p$, define

$$|A|_\lambda = \sup \left\{ \int_A \omega : \omega \text{ a differential } p\text{-form}, |\omega|_r \leq 1 \right\}.$$

Geometric definition of the natural l-norm. The analytic definition is simple to state but difficult to use with examples. We next give an equivalent geometric definition.

The geometric λ-norm is first defined assuming that $\lambda \in \mathbb{Z}$. The norm is defined initially relative to a δ-*system of coordinates.* That is the given δ-system can be obtained from an orthogonal system by a linear change of coordinates with norm $\leq$ $1+\delta$. These constraints can be relaxed. In [H2] it is shown that the λ-norm changes to an equivalent norm under a C^r change of coordinates.

A linear system of coordinates leads to a system of projections $\Pi = \{\pi_1, \pi_2, ..., \pi_N\}$ onto the N coordinate p-planes where N $= \binom{n}{p}$. If the system of coordinates is a δ-system of coordinates then the system of projections is called a δ-*system of projections.*

Components of chains If a simplicial p-chain A $= \Sigma a_i\sigma_i$ is a sum of p-simplexes parallel to the coordinate p-planes then A is called a "simple" p-chain. The ith component A^i of a simple p-chain A is merely the chain consisting of those simplexes of A parallel to the ith coordinate p-plane. In general, one may write A as a limit of simple chains A_k in the sense of homology. The ith component A^i of A is defined as the limit of the ith components of the A_k. If A^i is a component of A relative to a δ-coordinate system then A^i is called a *δ-component* of A.

Denote $|A|^p =$ M(A) the *mass* or p-dimensional Lebesgue measure of A.

DEFINITION 2. Fix $\lambda \geq p+1$. The λ-norm $|\ \ |^\lambda$ is the largest semi-norm in the space of semi-norms for simplicial p-chains satisfying:

- $|\sigma|' \leq M(\sigma)$ for all p-simplexes σ and
- $|(\partial C)^i|' \leq |C|^\lambda$ for all simplicial (p+1)-chains C and δ-components of ∂C.

This definition can be expanded to obtain an alternate definition, useful for calculating the λ-norm in some examples. (See [H4]) We use projections and recursion to find $|A|^\lambda$ for $\lambda \geq p+1$.

If π is a projection of $\mathbb{R}^n$ with rank p, define the *π-projected mass* of A, (with multiplicity) as

$$M_\pi(A) = \inf\left\{\sum_{i=1}^k |a_i| M(\pi\sigma_i) : \sum_{i=1}^k a_i\sigma_i \text{ equivalent to } A\right\}$$

DEFINITION 3. For $\lambda \geq$ p+1 and a δ-system Π of projections define the *(λ,Π)-norm* $|A|^\lambda_\Pi$ of A recursively by

$$|A|^\lambda_\Pi = \sum_{\pi_i \in \Pi} \inf\{M_{\pi_i}(A - \partial C_i) + |C_i|^\lambda\}$$

where for each i the infimum is taken over all simplicial (p+1)-chains C_i in $\mathbb{R}^n$.

Define

$$|A|^\lambda = \inf \sum_{j=1}^k |A_j|^\lambda_{\Pi_j}$$

where the right hand side is infimized over all sums $A = \sum_{j=1}^k A_j$ and δ -systems Π_j.

It is easy to verify that $|\ \ |^\lambda$ is a semi-norm. The fact that $|A|^\lambda \neq 0$ for simplicial $A \neq 0$ requires more work and can be found in [H2]. Then $|\ \ |^\lambda$ is a norm on simplicial p-chains. It changes to an equivalent norm under a smooth change of coordinates.

For v$\in \mathbb{R}^n$ and A a simplicial p-chain in $\mathbb{R}^n$, let T_vA denote A translated by the vector v.

Examples

4. Let A $= \sigma - T_v\sigma$ where σ is a 1-simplex in the x-axis with length ϵ and v $= (0,\beta)$. Then $|A|^1 = M(A) = 2\epsilon$. From definition 3 we calculate $|A|^2$ using projections π_1 and π_2 onto the x- and y-axes, resp. Let $C_1 = \sigma\times$ v. Then $M_{\pi_1}(A - \partial C_1) = 0$ and $|C_1|^2 = M(C_1) = \varepsilon\beta$. Let $C_2 = 0$. Then $M_{\pi_2}(A - \partial C_2) = 0$. Thus $|A|^2 \leq \varepsilon\beta$.

   We can also use definition 2: $|A|^2 = |(\partial C_1)^1|^2 \leq |C_1|^2 = M(C_1) = \varepsilon\beta$.
5. Let w $= (\gamma,0)$ and consider B $= \sigma - T_w\sigma$. Again $|B|^1 = M(B) = 2\epsilon$. But $|B|^2 \leq \frac{2}{\delta}\varepsilon\gamma$. (In figure 1, sin $\theta < 1 + \delta$. One may apply the triangle inequality and obtain $|\sigma - T_w\sigma|^2 \leq \sum|\sigma_i - \sigma_{i+1}|^2$ and then estimate each term $|\sigma_i - \sigma_{i+1}|^2$ as in the preceding example.)

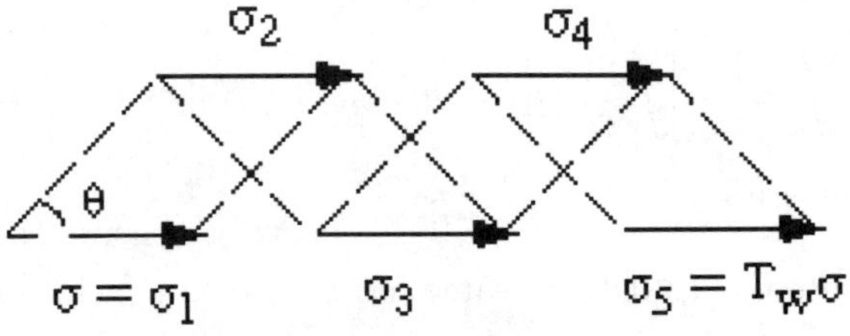

FIGURE 1. Estimating $|\sigma - T_w\sigma|^2$

6. Let S be the circle of radius ε in the xy-plane, centered at the origin and $v = (0, 0, \beta)$. Let $A = S - T_vS$. Then $|A|^1 = M(A) = 4\Pi\varepsilon$. We estimate $|A|^2$. Let D be the disk with boundary S. Using $C_1 = C_2 = D - T_vD, C_3 = 0$ we see that $|A|^2 \leq 2|C_1|^2 \leq 4|D|^2 = 4\Pi\varepsilon^2$.

   Finally we estimate $|A|^3$. As in the preceding estimate we have $|A|^3 \leq 2|C_1|^3$. So we are reduced to estimating $|C_1|^3$. Use Definition 2 and let V be the solid $D \times v$, oriented so that $\partial V = D - T_vD$. Then $|C_1|^3 = |(\partial V)^i|^3 \leq |V|^3 = M(V)$. Therefore $|A|^3 \leq 2M(V) = 2\Pi\varepsilon^2\beta$.

3. Integration on fractals

In the examples of the previous section, we see that the λ-norm may decrease as λ increases. This is quite general; it follows from the definitions that

Proposition *If A is a simplicial p-chain then* $|A|^\kappa \leq |A|^\lambda$, $p \leq \lambda \leq \kappa$.

Thus the Banach spaces $\mathcal{A}^{p,\lambda}$ form nested Banach spaces of λ-natural p-chains (or λ-chains) with $\mathcal{A}^{p,\lambda}$ a linear subspace of $\mathcal{A}^{p,\kappa}$. Define

$$\mathcal{A}^p = \bigcup_\lambda \mathcal{A}^{p,\lambda}.$$

We will call elements of $\mathcal{A}^p$ *p-dimensional domains of integration.* By the following continuity theorem the integral can be uniquely defined on all p-dimensional domains of integration for integrands of p-forms $\omega \in C^r$ (see * for the definition of r)

Continuity theorem *Let $\lambda \geq p$. If A is a simplicial p-chain and ω is a p-form of class C^r then*

$$\int_A \omega \leq |\omega|_r |A|^\lambda.$$

The proof to this uses the mapping $\Psi(\omega)\bullet A = \int_A \omega$ for ω a p-form of class C^r and A a simplicial p-chain. It is shown in [H2] that Ψ determines a 1-1 correspondence between equivalence classes of differential forms $\omega \in C^r$ and λ-cochains X and there exists a constant $k > 0$ such that $k|\omega|_r \leq |\Psi(\omega)|^\lambda \leq |\omega|_r$. Thus $|\Psi(\omega)\bullet A| \leq |\Psi(\omega)|^\lambda |A|^\lambda \leq |\omega|_r |A|^\lambda$.

7. **Example** Let S be a circle of radius ε in $\mathbb{R}^3$ with z-coordinate 0 and v = (0,0,β). Let $S' = T_v S$. Then for $\omega \in C^{1+Lip}$

$$\left|\int_S \omega - \int_{S'} \omega\right| = \left|\int_{S-S'} \omega\right| \leq |\omega|_{1+Lip}|S - S'|^3 \leq |\omega|_{1+Lip} 2\Pi\varepsilon^2\beta.$$

(See example 6.)

If we use v = (β,0,0) the same estimate with a factor of $2/\delta$ is obtained. The *boundary operator* ∂ on simplicial (p+1)-chains may be extended to all of $\mathcal{A}^{p+1,\lambda}$. This follows easily since for any p-chain A, $\sum_{i=1}^N A^i = A$ in the (p+1)-norm. Then $|\partial C|^\lambda \leq \sum_{i=1}^N |(\partial C)^i|^\lambda \leq N|C|^\lambda$. We obtain $\partial : \mathcal{A}^{p+1,\lambda} \longrightarrow \mathcal{A}^{p,\lambda}, \lambda \geq p+1$.

Generalization of Stokes' theorem *If ω is a (p-1)-form of class C^{r+1} and $A \in \mathcal{A}^{p,\lambda}$ then*

$$\int_A d\omega = \int_{\partial A} \omega.$$

We remark that in the case ω is a 0 form or function, Stokes' theorem is called the Fundamental Theorem of Calculus: If C is a smooth curve in space with endpoints p and q and f is smooth then f(q) - f(p) = $\int_C grad f \bullet dx$. If X is a smooth vector field and S a smooth surface Stokes' theorem takes the form of Green's theorem or Stokes' theorem (proper): $\int_{\partial S} X \bullet dx = \int_S (curl X)\bullet n d\sigma$. If D is a smooth volume then it becomes the Divergence Theorem or Gauss' Theorem: $\int_{\partial D} X \bullet n = \int_D (div X) dV$.

In all cases, we may replace smooth domains with λ-domains of integration if the integrand ω is of class C^{r+1}.

4. Extrinsic dimension

We next show how to define the λ-norm for $\lambda \geq p, \lambda \in \mathbb{R}$. We can then define $\mathcal{A}^{p,\lambda}$ for $\lambda \in \mathbb{R}$ and treat its elements as domains of integration as in definition 1. The above theorems for $\lambda \in \mathbb{Z}$ are all valid for $\lambda \in \mathbb{R}$, including the Continuity Theorem and the generalized Stokes' theorem.

DEFINITION 4. **(p+α)-mass** *Let Π be a δ-coordinate system and $\pi_i \in \Pi$. Let C be a simplicial (p+1)-chain and $0 < \alpha \leq 1$. We consider all expressions of C as* *an infinite sum* $\sum_{i=1}^{\infty} a_i(\sigma_i \times v_i)$ *where* σ_i *is a p-chain* $\pi_i \sigma_i$ *is parallel to* σ_i *and* v_i *is a vector with* $\pi_i(v_i) = 0$. *Define*

$$M_{\pi_i}^{n+\alpha}(C) = inf \sum_{i=1}^{\infty} |a_i| M(\sigma_i) ||v_i||^{\alpha}$$

DEFINITION 5. For $0 < \alpha \leq 1$ define

$$|A|_{\Pi}^{n+\alpha} = \sum_{\pi_i \in \Pi} inf\{M_{\pi_i}(A - \partial C_i) + M_{\pi_i}^{n+\alpha}(C_i)\}$$

where each term of the sum is infimized over all (p+1)-chains C_i.
Use recursion to define $|A|^{\lambda}$ for all real $\lambda \geq p$.

As before we deduce from the definitions monotonicity of the λ-norm as a function of $\lambda \in \mathbb{R}$. This leads to a new definition of dimension for fractal sets which is closely related to integration theory:

Extrinsic dimension and index Fix A$\in \mathcal{A}^p$, a p-dimensional domain of integration, and define $\mathcal{D}_A : [p, \infty) \longrightarrow \mathbb{R} \cup \infty$ by $\mathcal{D}_A(\lambda) = |A|^{\lambda}$. Then $\mathcal{D}_A$ is monotone decreasing. We define the *extrinsic dimension* by

$$dim_E(A) = \inf\{\lambda : \mathcal{D}_A(\lambda) < \infty\}.$$

The *index* of A is defined as $\inf\{\lambda : \mathcal{D}_A(\lambda') = \mathcal{D}_A(\lambda) \text{ for all } \lambda' \geq \lambda\}$.

Natural questions arise about these quantities and the function $\mathcal{D}_A$: If A is a simplicial chain, is the index of A finite? How is $\mathcal{D}_A(dim_E(A))$ related to Hausdorff measure if A is a self-similar set?

Conjecture *If A is self-similar then* $dim_E(A) = dim_H(A)$, *the Hausdorff dimension of A.*

The extrinsic dimension is analogous in some ways to Hausdorff dimension but they not identical. An important difference is that $dim_H(A)$ is bounded by the topological dimension n of the ambient space containing A, whereas $dim_E(A)$ may greatly exceed n.

Examples

8. Define B as in example 5, and let v = (0,β) and $C = B - T_v B$. This is a quadropole and its 2-norm is bounded by $(2/\delta)^2 \varepsilon\beta\gamma$. (See [H2] Lemma 1.14 for details.) If the quadropole is translated by a vector of length δ, the three norm of the difference is bounded by $(2/\delta)^3 \varepsilon\beta\gamma\delta$. This process can be repeated indefinitely to obtain examples in the plane of simplicial 1-chains with arbitrarily large index and extrinsic dimension 1. Infinite sums of these can be used to construct examples of one dimensional λ-chains in $\mathbb{R}^2$ with arbitrarily large extrinsic dimension. (See [H3].)
9. The Sierpinski gasket G is usually described as the infinite intersection of nested families of 2-dimensional equilateral triangles. (See Figure 2.)

   It is also described as the limit set of the three affine transformations mapping the largest triangle to each of the largest shaded triangles. From these definitions, it is not clear how best to approximate G. Is it best approximated by the sum of 2-dimensional triangles D_n or their boundaries G_n? Does G have a natural orientation and multiplicity?

FIGURE 2. The Sierpinski gasket

We use the λ-norm to find a more precise description of G as an element of $\mathcal{A}^{1,\lambda}$ where $\lambda \geq \ln 3/\ln 2 = 1+\alpha$.

The λ-norm allows us to describe G as a limit of simplicial 1-chains G_n where G_n is the sum of the positively oriented boundaries of the 3^n black triangles D_n of side length $1/2^n$. If $\beta = \alpha$ then $|G_n|^{1+\alpha} \leq 3^n/2^{n(1+\alpha)} = 1$ and if $\beta > \alpha$ then $|G_n|^{1+\beta} \to 0$ as n $\to \infty$. Hence $|G|^{1+\beta} = 0$ and it follows that G is equivalent to the 0 chain. The Sierpinski gasket is "ghostlike" for it is only visible in one Banach space $\mathcal{A}^{p,\alpha}$.

Theorem *There exists a 1-form ω of class C^α such that $\int_G \omega = 1$ and $|\omega|_\alpha = 1$ but $\int_G \nu = 0$ for any form ν of class $C^\beta, \beta > \alpha$.*

Proof . Apply the Continuity Theorem for the second part. For the first part, apply the Hahn Banach extension theorem to find a $(1+\alpha)$-cochain X so that $X \bullet G = |G|^{1+\alpha} = 1$. But X corresponds to $\omega \in C^\alpha$ with $X \bullet G = \int_G \omega$.

In particular, the integral of any C^1 form over the Sierpinski gasket G vanishes.

5. Conditional convergence of series of simplicial chains

We say that the infinite series $\sum_{n=1}^\infty S_n$ *converges in the λ-norm* if the sequence of partial sums forms a Cauchy sequence in the λ-norm. That is $|\sum_{n=j}^k S_n|^\lambda \to 0$ as j,k $\to \infty$. If $\sum_{n=1}^\infty S_n$ converges in the λ-norm then $|\sum_{n=1}^\infty S_n|^\lambda = \lim_{k\to\infty} |\sum_{n=1}^k S_n|^\lambda$

When we calculate the mass norm $|\sum_{n=1}^\infty S_n|^1$ of a series of simplicial 1-chains, we infimize over equivalence classes of sums of masses of simplexes which is analogous to the notion of absolute convergence of a series of real numbers. Here, orientation of each simplex is ignored as we are concerned only about the mass of the simplexes involved. This infimum may not be finite (as with the snowflake seen below) in which case we could consider $|\sum_{n=1}^\infty S_n|^2$. Here, we do need to know the orientations of the 1-chains S_n because this information dictates possible partial spanning 2-chains. The orientations of the 2-chains are not used to calculate $|\sum_{n=1}^\infty S_n|^2$, but their areas are needed.

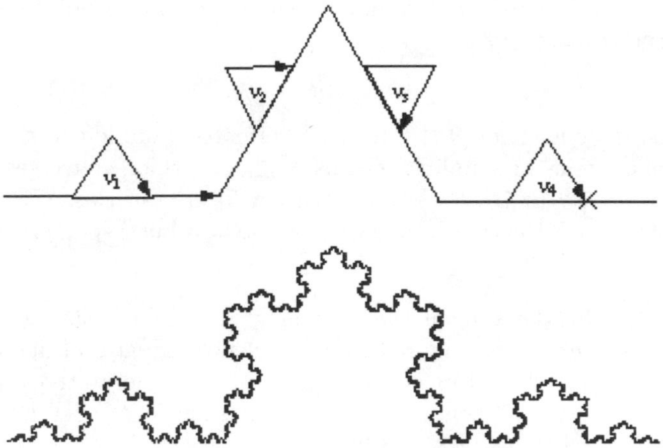

FIGURE 3. The Van Koch Snowflake

Let $\sum_{n=1}^{\infty} S_n$ be a series of simplicial p-chains. If $|\sum_{n=1}^{\infty} S_n|^p < \infty$ we say that the series $\sum_{n=1}^{\infty} S_n$ is *absolutely convergent.* If $|\sum_{n=1}^{\infty} S_n|^p = \infty$ but $|\sum_{n=1}^{\infty} S_n|^{p+1} < \infty$ the series is *first order conditionally convergent.* In general, if $|\sum_{n=1}^{\infty} S_n|^{p+r-1} = \infty$ and $|\sum_{n=1}^{\infty} S_n|^{p+r}$ then $\sum_{n=1}^{\infty} S_n$ is an rth *order conditionally convergent* series.

Theorem *If $\sum_{n=1}^{\infty} S_n$ converges in the λ-norm and $\omega \in C^r$ then*

$$\int_{\sum_{n=1}^{\infty} S_n} \omega = \sum_{n=1}^{\infty} \int_{S_n} \omega.$$

Example

10. The Van Koch Snowflake S. Consider the four affine transformations $\varphi_1, \varphi_2, \varphi_3, \varphi_4$ which have S as their limit set. In particular φ_i takes the unit equilateral triangle S_0 to the triangle V_i as depicted, $1 \leq \text{i} \leq 4$. Let S_1 be the sum of the V_i. Let S_k denote the sum of oriented images of S_0 under the composition of k of the φ_i. S_k will consist of 4^k oriented triangles of side length $1/3^k$. Let $T_n = \sum_{k=1}^{n} S_k$.

    Claim: $\{T_n\}$ is a Cauchy sequence in the λ-norm for all $\lambda > ln4/ln3$.

    Proof of claim: If $\lambda < 2$, this follows easily from the definition of the (λ-1)-mass. Consider the weighted areas of the 2-dimensional triangles $\sum_k A_k$ where $\partial A_k = S_k$. Their total weighted area is $\sum_k 4^k/3^{k\lambda} < \infty$. For $\lambda \geq 2$, apply monotonicity.

This construction offers more precision than the common draw-and-erase method and more direct geometrical interpretation than the definition as the limit set of affine transformations. Here we obtain natural orientation and multiplicity on the limit set, inherited from the original triangle S_0, and also natural approximators T_n for use in integration.

The next example shows that term by term integration is not always possible with insufficiently smooth forms.

Example

11. Let I be the one-chain [0,1] $\times\{0\}$ in $\mathbb{R}$, oriented from left to right. We find an infinite series of simplicial chains $\sum_{k=1}^{\infty} A_k$ which converges to I in the λ-norm for $\lambda > ln4/ln3$. We construct a form ω of class C^{α}, where $\alpha =$ ln4/ln3 - 1 such that $\omega|_I = 0$ and thus $\int_I \omega = 0$ but $\sum_{k=1}^{\infty} \int_{A_k} \omega = -1$.

Let S_n be the sum of the 4^n triangles of side length 3^{-n} as in the previous example. (See figure 3) Let Q_n denote squares of diameter $3^{-n}/2$ with centers at the centers of the S_n and sides parallel to the x- and y-axes. Assume these are all contained in the unit square B_1. Let B_n denote the rectangle $\{(x,y) : 0 \leq x \leq 1, 3^{-n} \leq y \leq 3^{-(n+1)}\}$. Let $L_n : \mathbb{R}^2 \longrightarrow \mathbb{R}^2$ be the affine transformation with $L_n(B_1) = B_n$. Let C_n denote the simplicial 1-chain $3^n L_n(Q_n)$. Note that the squares of Q_n appear in C_n as rectangles with length 3^{-n}, height 3^{-2n} and weight 3^n.

Step 1. Note that $|C_n|^{\lambda} = 4^n 3^{-n\lambda} \rightarrow 0$ if $\lambda > ln4/ln3$. Thus $I - C_n \rightarrow I$ in the λ-norm.

Step 2. Define $A_1 = I - C_1$ and $A_n = C_{n-1} - C_n$ for $n \geq 2$. Then $\sum_{n=1}^{\infty} A_n \rightarrow I$ in the λ-norm.

Step 3. Define a 1-form $\omega_n = f_n dy$ on each C_n as the dual to the vector field tangent to the vertical 1-vectors v comprising C_n but with norm $3^{-n\alpha}$ so that $\int_{C_n} \omega_n = 1$. The Whitney Extension Theorem can be used to extend ω to a form defined on all of $\mathbb{R}^2$ of class C^{α}. Thus $\lim_{n\to\infty} \sum_{k=1}^{n} \int_{A_k} \omega = \lim_{n\to\infty} \left(\int_{I-C_n} \omega \right) = -1$. Since $\omega = 0$ on I then $\int_I \omega = 0$.

Conclusion The methods in this paper may be used to gain geometrical insights into currents and distributions. One may consider naturally arising smooth or simplicial approximators A_n to a potential domain A in the sense of homology and show these converge in the λ-norm. Then the question of whether $\int_{A_n} \omega \rightarrow \int_A \omega$ reduces to showing that $A_n \rightarrow A$ in the λ-norm as long as ω is sufficiently smooth.

References

E: Eells, James, *Geometric aspects of currents and distributions*, Proc. Nat. Ac. of Sce. 41 (1955), 493-496.

H1: Harrison, Jenny, *Stokes' theorem on nonsmooth chains*, Bulletin AMS Oct 1993.

H2: Harrison, Jenny, *Continuity of the integral as a function of the domain.* UC Berkeley Center for Pure and Applied Mathematics preprint, Oct. 1995.

H3: Harrison, Jenny, *Dimension of domains of integration*, in preparation.

H4: Harrison, Jenny, *Whitney's missing boundary operator, a comparison of the flat, sharp and natural norms*, in preparation.

H-N:1: Harrison, Jenny and Norton, Alec, *Geometric integration on fractal curves in the plane* Indiana Journal, Vol. 40 (1991), pp. 567-594.

H-N:2: Harrison, Jenny and Norton, Alec, *The Gauss-Green theorem for fractal boundaries* Duke Journal of Mathematics, Vol. 67, No. 3, (1992), pp. 575-588, .

W: Whitney, Hassler, Geometric Integration Theory. Princeton University Press, 1957.

DEPARTMENT OF MATHEMATICS, UNIVERSITY OF CALIFORNIA, BERKELEY
E-mail address: harrisson@math.berkeley.edu

Part VII

FIELD THEORY

Contemporary Mathematics
Volume **203**, 1997

Chern-Simons Vortices

C. DUVAL AND P. A. HORVÁTHY

ABSTRACT. In $(2+1)$ dimensions, the Maxwell term $-\frac{1}{4}F_{\alpha\beta}F^{\alpha\beta}$ can be replaced by the Chern-Simons three-form $(\kappa/4)\epsilon^{\alpha\beta\gamma}A_\alpha F_{\beta\gamma}$, yielding a novel type of 'electromagnetism'. This has been proposed for studying the Quantum Hall Effect as well as High-Temperature Superconductivity. The gauge field can be coupled to a scalar field either relativistically or non-relativistically. In both cases, one admits finite-energy, vortex solutions.

1. Introduction

The phenomenological description of 'ordinary' superconductivity is provided by Landau-Ginzburg theory [1]: the Cooper pairs formed by the electrons are represented by a scalar field, whose charge is twice that of the electron. The scalar fields interact through their electromagnetic fields, governed by the Maxwell equations. The theory admits static, vortex-like solutions [2].

Landau-Ginzburg theory is non-relativistic; a relativistic version is provided by the Abelian Higgs model [3], which admits again static and purely magnetic vortex-type solutions.

In the eighties, however, new phenomena were discovered, namely the Quantum Hall Effect [4] and High-Temperature Superconductivity [5]. None of these is described by Landau-Ginzburg theory. Hence the need for a new theory.

Both phenomena mentioned above are essentially planar. The mathematical room for a modification is provided by the three-form

$$\frac{\kappa}{4}\epsilon^{\alpha\beta\gamma}A_\alpha F_{\beta\gamma}, \tag{1.1}$$

introduced by Chern and Simons [6] in their study of secondary characteristic classes. This form, which only exists in $(2+1)$ dimensions, can be added to the conventional Maxwell term $-\frac{1}{4}F_{\alpha\beta}F^{\alpha\beta}$ [7]. The use of the Chern - Simons term has in fact been proposed in the above contexts [8], [9].

1991 *Mathematics Subject Classification.* 81T13, 81T8O; 35Q51, 35Q55.

Since the Chern - Simons term is of lower-order in derivatives, for large distances it dominates the Maxwell term. Thus, for describing long-range phenomena, the Maxwell term can even be dropped: the dynamics of the 'electromagnetic' field will be described by the Chern-Simons term (1.1) alone.

Below, we review various aspects of Chern-Simons gauge theory. First, we explain the construction of relativistic topological vortices. The non-relativistic limit is physically relevant owing to the intrinsically non-relativistic character of condensed matter physics. It also provides an explicitly solvable model. Finally, we consider spinorial models. Again, explicit solutions are found.

2. Relativistic Chern-Simons vortices

Let us consider a complex scalar field, ψ, coupled to an abelian gauge field, described by the vector potential [1-form], $A_\mu dx^\mu$, defined on (2+1)-dimensional Minkowski space with the metric $(g_{\mu\nu}) =$ diag $(1,-1,-1)$, the coordinates being $x^0 = t$ and $(x_i) = \vec{x}$. The action $S = \int d^3x\, \mathcal{L}$ is given by [10]

$$\mathcal{L} = \tfrac{1}{2} D_\mu\psi (D^\mu\psi)^* + \tfrac{1}{4}\kappa\, \epsilon^{\alpha\beta\gamma} A_\alpha F_{\beta\gamma} - U(\psi), \tag{2.1}$$

where $D_\mu\psi = \partial_\mu\psi - ieA_\mu\psi$ denotes the covariant derivative, the constant e is the electric charge of the field ψ, and κ is a constant. The potential $U(\psi)$ is chosen, for later convenience, to be

$$U(\psi) = \frac{\lambda}{4}|\psi|^2\big(|\psi|^2 - 1\big)^2. \tag{2.2}$$

The Euler-Lagrange equations of the action (2.1) read

$$\begin{cases} \tfrac{1}{2} D_\mu D^\mu \psi = -\dfrac{\delta U}{\delta\psi^*} \equiv -\dfrac{\lambda}{4}\big(|\psi|^2 - 1\big)\big(3|\psi|^2 - 1\big)\psi, \\[2mm] \tfrac{1}{2}\kappa\, \epsilon^{\mu\alpha\beta} F_{\alpha\beta} = e j^\mu, \end{cases} \tag{2.3}$$

where $F_{\alpha\beta} = \partial_\alpha A_\beta - \partial_\beta A_\alpha$ is the 'electromagnetic' field, and $j^\mu \equiv (\varrho, \vec{\jmath})$ is the current

$$j^\mu = \frac{1}{2i}\big[\psi^* D^\mu\psi - \psi (D^\mu\psi)^*\big]. \tag{2.4}$$

The first of the equations in (2.3) is the Non-Linear Klein-Gordon equation (NLKG) familiar from the Abelian Higgs model [3]; the second, called *Field-Curent Identity* (FCI), replaces the Maxwell equations.

It follows from the Bianchi identity, $\epsilon^{\alpha\beta\gamma}\,\partial_\alpha F_{\beta\gamma} = 0$, that the current (2.4) is conserved, $\partial_\mu j^\mu = 0$.

Due to the presence of the vector potential in the Chern-Simons term, the Lagrangian (2.1) is manifestly *not* invariant with respect to a local gauge transformation

$$\psi(x) \to e^{ie\lambda(x)}\psi(x), \qquad A_\mu \to A_\mu + \partial_\mu\lambda. \tag{2.5}$$

Using the Bianchi identity, the change can be written, however, as $\partial_\alpha\left((\kappa/4)\,\epsilon^{\alpha\beta\gamma}\,F_{\beta\gamma}\,\lambda\right)$. This is a surface term, which does not change the equations of motion. The two other terms in (2.1) are gauge-invariant.

The same conclusion can also be reached by looking at the field equations: the FCIs are invariant, since both the field strength and the current are invariant; both sides of the NLKG get in turn be multiplied by $e^{ie\lambda}$.

Finite-energy configurations

Let us consider a static field configuration (A_μ, ψ). The energy, defined as the space integral of the time-time component of the energy-momentum tensor associated with the Lagrangian (2.1), is

$$E \equiv \int d^2\vec{x}\,T^{00} = \int d^2\vec{x}\Big[\tfrac{1}{2}D_i\psi(D^i\psi)^* - \tfrac{1}{2}e^2A_0^2|\psi|^2 + \kappa A_0 B + U(\psi)\Big], \tag{2.6}$$

where $B = -F^{12}$ is the magnetic field. Note that this expression is *not* positive definite. Observe, however, that the static solutions of the equations of movement (2.3) are stationary points of the energy. Variation of (2.6) w. r. t. A_0 yields one of the equations of motion, namely

$$-e^2A_0|\psi|^2 + \kappa B = 0. \tag{2.7}$$

Eliminating A_0 from (2.6) using this equation, we obtain the *positive definite* energy functional

$$E = \int d^2\vec{r}\,\Big[\tfrac{1}{2}D_i\psi(D^i\psi)^* + \frac{\kappa^2}{2e^2}\frac{B^2}{|\psi|^2} + U(\psi)\Big]. \tag{2.8}$$

We are interested in static, finite-energy configurations. Finite energy at infinity is guaranteed by the conditions[1]

$$\begin{cases} i.) & |\psi|^2 - 1 = \mathrm{o}(1/r), \\ ii.) & B = \mathrm{o}(1/r), \qquad r \to \infty. \\ iii.) & \vec{D}\psi = \mathrm{o}(1/r). \end{cases} \tag{2.9}$$

Therefore, the $U(1)$ gauge symmetry is broken for large r. In particular, the scalar field ψ is covariantly constant, $\vec{D}\psi = 0$. This equation is solved by parallel transport,

$$\psi(\vec{x}) = \exp\left[i\int_{\vec{x}_0}^{\vec{x}} eA_i dx^i\right]\psi_0,$$

which is well-defined whenever

$$\oint eA_i dx^i = \int_{\mathbb{R}^2} d^2\vec{x}\,eB \equiv e\Phi = 2\pi n, \qquad n = 0, \pm1, \dots. \tag{2.10}$$

Thus, the magnetic flux is *quantized.*

[1] These conditions are in no way necessary; they yield the so-called *topological solitons.* Non-topological solutions are constructed in Ref. [11].

By i.),the asymptotic values of the Higgs field provide us with a mapping from the circle at infinity into the vacuum manifold, which is again a circle, $|\psi|^2 = 1$. Since the vector potential behaves asymptotically as

$$A_j \simeq -\frac{i}{e}\,\partial_j \log \psi, \tag{2.11}$$

the integer n in Eq. (2.10) is precisely the *winding number* of this mapping; n is called the *topological charge* (or *vortex number*).

Spontaneous symmetry breaking generates mass [12]. This occurs in a somewhat unusual guise, though. Expanding j^μ around the vacuum expectation value of ψ we find $j^\mu = -eA^\mu$ so that the FCI in (2.3) is approximately

$$\tfrac{1}{2}\kappa\,\epsilon^{\mu\alpha\beta}F_{\alpha\beta} \simeq -e^2 A^\mu.$$

Hence $F_{\alpha\mu} = -(e^2/\kappa)\,\epsilon_{\alpha\mu\beta}A^\beta$. Inserting here $F_{\alpha\beta}$ and deriving by ∂^α we find that the gauge field A^μ satisfies the Klein-Gordon equation

$$\Box A^\mu = -\Big(\frac{e^2}{\kappa}\Big)^2 A^\mu, \tag{2.12}$$

showing that the mass of the gauge field is

$$m_A = \frac{e^2}{\kappa}. \tag{2.13}$$

The Higgs mass is found in turn in the usual way: the expectation value of the Higgs field can be chosen as $\psi_0 = (1,0)$. Expanding ψ as $(\psi_r, \psi_\vartheta) = (1+\varphi, \theta)$ we get

$$U = \underbrace{U(1)}_{=0} \quad + \quad \underbrace{\frac{\delta U}{\delta|\psi|}\Big|_{|\psi|=1}\varphi}_{=0} \quad + \quad \underbrace{\frac{1}{2}\Big(\frac{\delta^2 U}{\delta|\psi|^2}\Big)\Big|_{|\psi|=1}}_{m_\psi^2}\varphi^2, \tag{2.14}$$

since $|\psi| = 1$ is a critical point of U. We conclude that the mass of the Higgs particle is

$$m_\psi^2 = \frac{\delta^2 U}{\delta|\psi|^2}\Big|_{|\psi|=1} = 2\lambda. \tag{2.15}$$

(This result can also be seen by considering the radial equation (2.17) below).

Radially symmetric solutions [10]

For the radially symmetric Ansatz

$$A_0 = A_0(r), \qquad A_r = 0, \qquad A_\vartheta = A(r), \qquad \psi(r) = f(r)e^{-in\vartheta}, \tag{2.16}$$

the equations of motion (2.3) read

$$
\begin{aligned}
&\frac{1}{r}\frac{dA}{dr} + \frac{e^2}{\kappa} f^2 A_0 = 0, \\
&r\frac{dA_0}{dr} + \frac{e^2}{\kappa} f^2 \left(\frac{n}{e} + A\right) = 0, \qquad (2.17)\\
&\frac{d^2 f}{dr^2} + \frac{1}{r}\frac{df}{dr} - \frac{e^2}{r^2}\left(\frac{n}{e} + A\right)^2 f + e^2 A_0^2 f = -\frac{\lambda}{4} f(1-f^2)(1-3f^2),
\end{aligned}
$$

with asymptotic conditions

$$
\begin{aligned}
\lim_{r\to\infty} A(r) &= -\frac{n}{e}, & \lim_{r\to\infty} f(r) &= 1, \\
\lim_{r\to 0} A(r) &= 0, & \lim_{r\to 0} f(r) &= 0.
\end{aligned} \qquad (2.18)
$$

This is either seen by a direct substitution into the equations, or by re-writing the energy as

$$
E = 2\pi \int_0^\infty dr \left\{ \frac{r}{2}(f')^2 + \frac{a^2}{2r} f^2 + \frac{\kappa^2}{2e^4 f^2}\frac{(a')^2}{r} + U(f) \right\}, \qquad (2.19)
$$

where $a = eA + n$. The upper equation in (2.17) is plainly the radial form of (2.7). Then variation of (2.19) with respect to a and f yields the two other equations in (2.17).

Approximate solutions can be obtained by inserting the asymptotic value, $f \simeq 1$, into the first two equations:

$$
\frac{a'}{r} + \frac{e^2}{\kappa} A_0 = 0, \qquad A_0' + \frac{e^2}{\kappa}\frac{a}{r} = 0,
$$

from which we infer that

$$
\frac{d^2 A_0}{d\rho^2} + \frac{1}{\rho}\frac{dA_0}{d\rho} - A_0 = 0, \qquad \rho \equiv (e^2/\kappa) r.
$$

This is the modified Bessel equation [Bessel equation of imaginary argument] of order zero. Hence

$$
A_0 = C K_0\left(\frac{e^2}{\kappa} r\right). \qquad (2.20)
$$

Similarly, for $a = n/e + A$ we find, putting $\alpha = a/r$,

$$
\alpha'' + \frac{\alpha'}{\rho} - \left(1 + \frac{1}{\rho^2}\right)\alpha = 0,
$$

which is Bessel's equation of order 1 with imaginary argument. Thus $\alpha = CK_1(\rho)$ so that

$$
A = -\frac{n}{e} + C\frac{e^2}{\kappa} r\, K_1\left(\frac{e^2}{\kappa} r\right). \qquad (2.21)
$$

Another way of deriving this result is to express A from the middle equation in (2.17),

$$A = -\frac{n}{e} - \frac{\kappa}{e^2} r \frac{d}{dr} A_0.$$

The consistency with (2.21) follows from the recursion relation $K_0' = -K_1$ of the Bessel functions.

An even coarser approximation is obtained by dropping the $1/r$ terms altogether. Then both equations reduce to

$$u'' = \Big(\frac{e^2}{\kappa}\Big)^2 u \qquad \Longrightarrow \qquad A_0 - a = Ce^{-m_A r}, \tag{2.22}$$

which shows that the fields approach their asymptotic values exponentially, with characteristic length determined by the gauge field mass.

The deviation of f from its asymptotic value, $\varphi = 1 - f$, is found by inserting φ into the last eqn. of (2.17); developing to first order in φ we get

$$\varphi'' + \frac{1}{r}\varphi' - 2\lambda\varphi \simeq 0 \qquad \Longrightarrow \qquad \varphi = CK_0(\sqrt{2\lambda}\, r), \tag{2.23}$$

whose asymptotic behaviour is again exponential, with characteristic length $(m_\psi)^{-1}$,

$$\varphi = Ce^{-m_\psi r}. \tag{2.24}$$

The penetration depths of the gauge and scalar fields are therefore

$$\eta = \frac{1}{m_A} = \frac{e^2}{\kappa} \qquad \text{and} \qquad \xi = \frac{1}{m_\psi} = \frac{1}{\sqrt{2\lambda}}, \tag{2.25}$$

respectively. For small r instead, inserting the developments in powers of r, we find

$$\begin{aligned} f(r) &\sim f_0 r^{|n|} + \dots, \\ A_0 &\sim \alpha_0 - \frac{e f_0^2 n}{2\kappa |n|} r^{2|n|} + \dots, \\ A &\sim -\frac{e^2 f_0^2 \alpha_0}{2\kappa(|n|+1)} r^{2|n|+2} + \dots, \end{aligned} \tag{2.26}$$

where α_0 and f_0 are constants. In summary, the fields go as

$$|\psi(r)| = f(r) \quad \propto \quad \begin{cases} r^{|n|} & r \sim 0, \\ 1 - e^{-m_\psi r} & r \to \infty, \end{cases} \tag{2.27}$$

and

$$(2.28)\qquad \begin{aligned} |E(r)| = |A_0'| \quad &\propto \begin{cases} r^{2|n|-1} & r \sim 0, \\ e^{-m_A r} & r \to \infty, \end{cases} \\ |B(r)| = \frac{|A'|}{r} \quad &\propto \begin{cases} r^{2|n|} & r \sim 0, \\ e^{-m_A r} & r \to \infty. \end{cases} \end{aligned}$$

Self-duality

Let us suppose that the fields have equal masses, $m_\psi^2 = m_A^2 \equiv m^2$ and hence equal penetration depths. Then the Bogomolny trick applies, i.e., the energy can be rewritten in the form

(2.29)

$$E = \int d^2\vec{r}\left[\tfrac{1}{2}|(D_1 \pm iD_2)\psi|^2 + \tfrac{1}{2}\left|\frac{\kappa}{e}\frac{B}{\psi} \mp \frac{e^2}{2\kappa}\psi^*(1-|\psi|^2)\right|^2\right] \mp \int d^2\vec{r}\,\frac{eB}{2}(1-|\psi|^2).$$

The last term can also be presented as

$$\mp\frac{eB}{2} \mp \tfrac{1}{2}\vec{\nabla}\times\vec{\jmath}.$$

The integrand of the B-term yields the magnetic flux; the second is transformed, by Stokes' theorem, into the circulation of the current at infinity which vanishes since all fields drop off at infinity by assumption. Its integral is therefore proportional to the magnetic flux, $\pm e\Phi/2$. Since the first integral is non-negative, we have, in conclusion,

$$(2.30)\qquad E \geq \frac{e|\Phi|}{2} = \pi|n|,$$

equality being only attained if the 'self-duality' equations

$$(2.31)\qquad D_1\psi = \mp iD_2\psi \qquad \text{and} \qquad eB = \pm\frac{m^2}{2}\,|\psi|^2(1-|\psi|^2)$$

hold. It is readily verified that the solutions of equations (2.7) and (2.31) solve automatically the non-linear Klein-Gordon equation.

For the radial Ansatz (2.16) the self-duality equations become

$$(2.32)\qquad f' = \pm\frac{a}{r}f, \qquad \frac{a'}{r} = \pm\tfrac{1}{2}m^2 f^2(f^2-1),$$

where we introduced again $a = eA + n$. Deriving the first of these equations and using the second one, we for f we get the 'Liouville - type' equation

$$(2.33)\qquad \triangle \log f = \frac{m^2}{2}\,f^2(f^2-1).$$

Another way of obtaining the first-order eqns. (2.32) is to rewrite, for

$$U(f) = (m^2/8)\,f^2\big(f^2-1\big)^2,$$

the energy as

$$\pi\int_0^\infty rdr\left\{\left[f'\mp\frac{a}{r}f\right]^2+\frac{1}{m^2f^2}\left[\frac{a'}{r}\mp\frac{m^2}{2}f^2(f^2-1)\right]^2\right\}\pm\pi(af^2)\Big|_0^\infty\mp\pi a\Big|_0^\infty.$$

The boundary conditions read

$$\begin{aligned} a(\infty) &= 0, & f(\infty) &= 1, \\ a(0) &= n, & f(0) &= 0, \end{aligned} \tag{2.34}$$

and thus $E\geq\pi|n|$ as before, with equality attained iff the equations (2.32) hold.

For $n=0$ the only solution is the vacuum,

$$f\equiv 1, \qquad A\equiv 1. \tag{2.35}$$

To see this, note that the boundary conditions at infinity are $f(\infty)=1$ and $A(\infty)=0$. Let now $f(r)$, $A(r)$ denote an arbitrary finite-energy configuration and consider

$$f_\tau(r)=f(r), \qquad A_\tau(r)=\tau A(r) \tag{2.36}$$

where $\tau>0$ is a real parameter. This provides us with a 1-parameter family configurations with finite energy

$$E_\tau=2\pi\int_0^\infty dr\left\{\frac{r}{2}\left(f'\right)^2+\tau^2\left[\frac{a^2}{2r}f^2+\frac{r}{2m^2}\left(\frac{a'}{rf}\right)^2\right]+U(f)\right\}, \tag{2.37}$$

which is a monotonic function of τ, whose minimum is at $\tau=0$ i.e. for $a\equiv 0$. Then Eq. (2.32) implies that $f'\equiv 0$ so that $f\equiv 1$ is the only possibility.

Let us assume henceforth that $n\neq 0$. No analytic solution has been found so far. To study the large-r behaviour, put $\varphi\equiv 1-f$. Inserting $f\simeq 1$, Eq. (2.32) reduces to

$$\varphi'=\mp\frac{a}{r}, \qquad \frac{a'}{r}=\mp m^2\varphi. \tag{2.38}$$

From here we get

$$\begin{aligned} \varphi''+\frac{1}{r}\varphi'-m^2\varphi=0 \quad &\Longrightarrow \quad \varphi=CK_0(mr). \\ a''-\frac{1}{r}a'-m^2a=0 \quad &\Longrightarrow \quad a=CmrK_1(mr). \end{aligned}$$

Thus, for large r,

$$f\simeq 1-CK_0(mr) \qquad \text{and} \qquad A\simeq -\frac{n}{e}+CmrK_1(mr) \tag{2.39}$$

with some constant C. For small r instead, Eq. (2.32), yields, to $\mathrm{O}(r^{5|n|+1})$,

$$\begin{aligned} f(r) &= f_0r^{|n|}-\frac{f_0^3m^2}{2(2n+2)^2}r^{3|n|+2}+\mathrm{O}(r^{5|n|+2}). \\ A &= -\frac{f_0^2m^2}{2(2|n|+2)e}r^{2|n|+2}+\frac{f_0^2m^2}{2(4|n|+2)e}r^{4n+2}+\mathrm{O}(r^{4|n|+4}). \end{aligned} \tag{2.40}$$

The result is consistent with (2.26) since the constant α_0 is now $\alpha_0 = m/2e = e/2\kappa$.

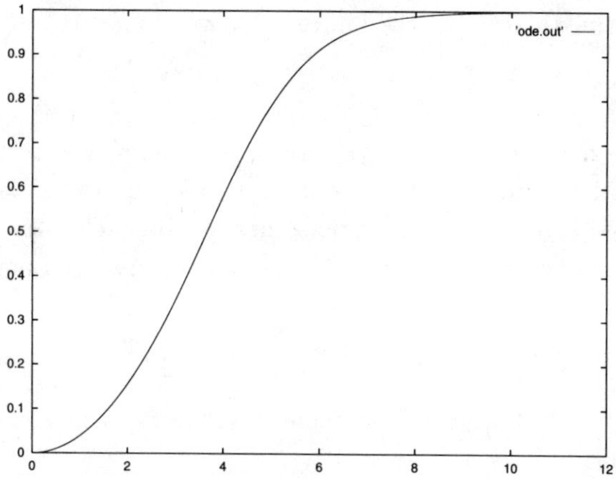

Fig. 1. *The scalar field of the radially symmetric charge-2 relativistic vortex.*

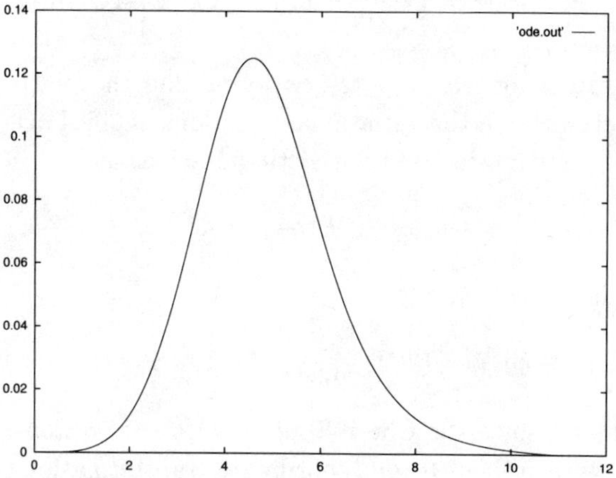

Fig. 2. *The magnetic field of the radially symmetric charge-2 relativistic vortex. Note that $B = 0$ where the scalar field vanishes: the vortex has a doughnut-like shape.*

Let us mention that Eq. (2.33) is actually valid in full generality, without the assumption of radial symmetry. Expressing in fact the vector-potential from $(D_1 \pm iD_2)\psi = 0$ as

$$e\vec{A} = \vec{\nabla}(\,\mathrm{Arg}\,\psi) \pm \vec{\nabla} \times \log|\psi| \tag{2.41}$$

and inserting into the second equation in (2.31), we get again (2.33), with $|\psi|$ replacing f. Index-theoretical calculations show that, for topological charge n,

Eqn. (2.33) admits a $2|n|$ parameter family of solutions [10].

3. Non-relativistic vortices

The non-relativistic limit of the above system is found [13,14] by setting

$$\psi = e^{-imc^2 t}\Psi + e^{+imc^2 t}\bar{\Psi},$$

where Ψ and $\bar{\Psi}$ denote the particles and antiparticles, respectively. Inserting this into the action, dropping the oscillating terms and only keeping those of order $1/c$, we see that both the particles and antiparticles are conserved. We can therefore consistently set $\bar{\Psi} = 0$. The remaining matter Lagrangian reads

$$\mathcal{L}_{matter} = i\Psi^\star D_t \Psi - \frac{|\vec{D}\Psi|^2}{2m} + \frac{\Lambda}{2}(\Psi^\star\Psi)^2, \tag{3.1}$$

where $\Lambda = e^2/mc|\kappa|$. One can show that the theory is indeed non-relativistic [15], [16]. In what follows, we put $c = 1$.

Let the constant Λ be arbitrary. Variation of $\int \mathcal{L}_{matter}$ w. r. t. $\Psi^\star$ yields the *gauged non-linear Schrödinger equation*

$$i\partial_t\Psi = \left[-\frac{\vec{D}^2}{2m} - eA_0 - \Lambda\Psi^\star\Psi\right]\Psi. \tag{3.2}$$

A fully self-consistent system is obtained by adding the matter action to the Chern-Simons action (1.1); the variational equations of this latter are clearly the FCIs in Eq. (2.3), written in non-relativistic notations as

$$B \equiv \epsilon^{ij}\partial_i A^j = -\frac{e}{\kappa}\,\varrho, \qquad E^i \equiv -\partial_i A^0 - \partial_t A^i = \frac{e}{\kappa}\epsilon^{ij}J^j, \tag{3.3}$$

where the current is now

$$j^\mu \equiv (\varrho, \vec{\jmath}) = (\Psi^\star\Psi, \frac{1}{2mi}\big[\Psi^*\vec{D}\Psi - \Psi(\vec{D}\Psi)^*\big]). \tag{3.4}$$

This is legitimate, since the Chern-Simons term is also *non-relativistic*: it is in fact invariant with respect to *any* coordinate transformation.

Self-dual vortex solutions

We would again like to find static soliton solutions. The construction of an energy-momentum tensor is now more subtle, because the theory is non-relativistic. This is nevertheless possible [15], [16]. One gets the energy functional

$$E = \int d^2x \left(\frac{|\vec{D}\Psi|^2}{2m} - \frac{\Lambda}{2}(\Psi^\star\Psi)^2\right). \tag{3.5}$$

Applying once more the Bogomolny trick, this is also written as

$$E = \int d^2x \,\frac{|(D_1 \pm iD_2)\Psi|^2}{2m} - \tfrac{1}{2}\left(\Lambda - \frac{e^2}{m|\kappa|}\right)(\Psi^\star\Psi)^2.$$

Thus, for the *specific value*

$$\Lambda = \frac{e^2}{m|\kappa|} \tag{3.6}$$

of the non-linearity, — the same as the one we obtained above by taking the non-relativistic limit of the self-dual relativistic theory — the second term vanishes. Then the energy is positive definite, whose absolute minimum — zero — is attained for "self-dual" fields i.e. for such that

$$(D_1 \pm iD_2)\Psi = 0. \tag{3.7}$$

Separating the phase as $\Psi = e^{ie\omega}\sqrt{\varrho}$, we get

$$\vec{A} = \vec{\nabla}\omega + \frac{1}{2e}\vec{\nabla} \times \log\varrho, \tag{3.8}$$

where ϱ solves the Liouville equation,

$$\triangle \log\varrho = \pm\frac{2e^2}{\kappa}\varrho, \tag{3.9}$$

cf. (2.33). Physically admissible solutions arise when the r.h.s. is negative. Hence the upper sign has to be chosen when $\kappa < 0$ and the lower sign when $\kappa > 0$. Then the general solution reads

$$\varrho = \frac{4|\kappa|}{e^2}\frac{|f'|^2}{[1+|f|^2]^2}, \tag{3.10}$$

where f is a meromorphic function. In particular, for

$$f(z) = \sum_{i=1}^{n}\frac{c_i}{z-z_i}, \tag{3.11}$$

where the c_i and the z_i are $2n$ complex numbers, we get a $4n$-parameter family of physically admissible solutions, with vortex number n.

Note that ϱ vanishes at the poles of f; these points represent the 'positions' of the vortices. The singularity in ϱ is precisely compensated by the singularity in the phase, leaving us with a regular magnetic field [15].

The number of independent parameters is in fact $4n-1$, due to a global gauge freedom; index-theory calculations [17] show that this is indeed the maximal number of independent solutions.

For $f = c/z^n$ in particular, we obtain the radially symmetric solution

$$\varrho(r) = \frac{4n^2|\kappa|}{e^2r^2}\left[\frac{|c|}{r^n} + \frac{r^n}{|c|}\right]^{-2}, \tag{3.12}$$

whose flux is *evenly* quantized,

$$\Phi = -(\,\mathrm{sg}\kappa)\frac{2h}{e} \times n. \tag{3.13}$$

A few non-relativistic vortices are shown on FIGs. 3-6 below.

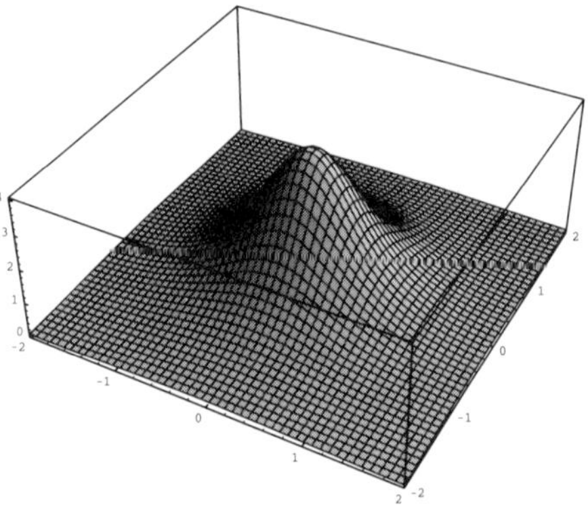

Fig. 3 *The non-relativistic radially symmetric $N = 1$ vortex has a maximum at $r = 0$.*

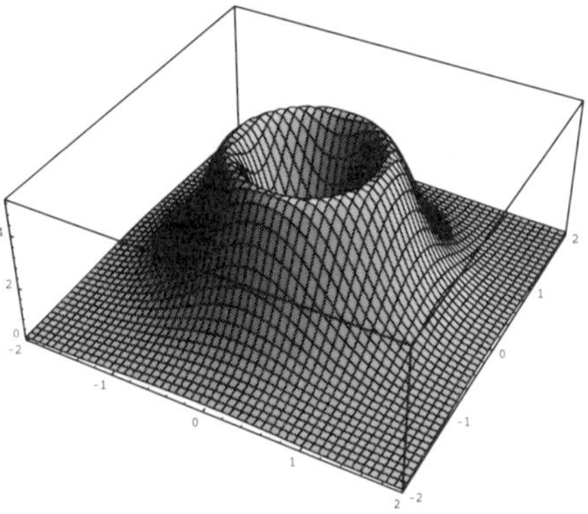

Fig. 4 *The non-relativistic radially symmetric $N = 2$ vortex has a 'doughnut-like' shape. For $N \geq 2$, the particle density vanishes at $r = 0$.*

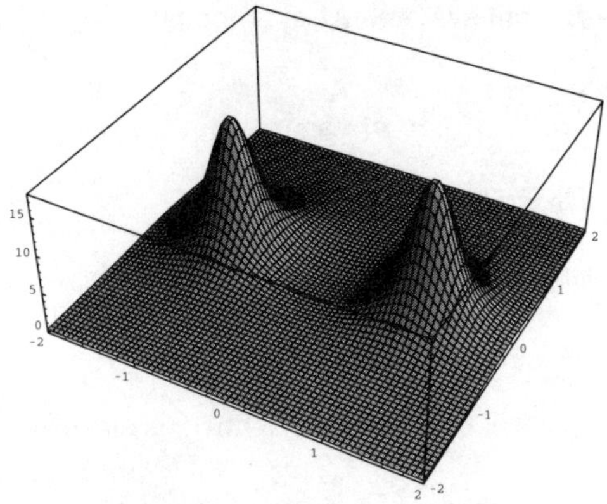

Fig. 5 *The non-relativistic* $N = 2$ *vortex representing two separated* 1*-vortices with 'positions' at* $x = \pm 1$ *and* $y = 0$.

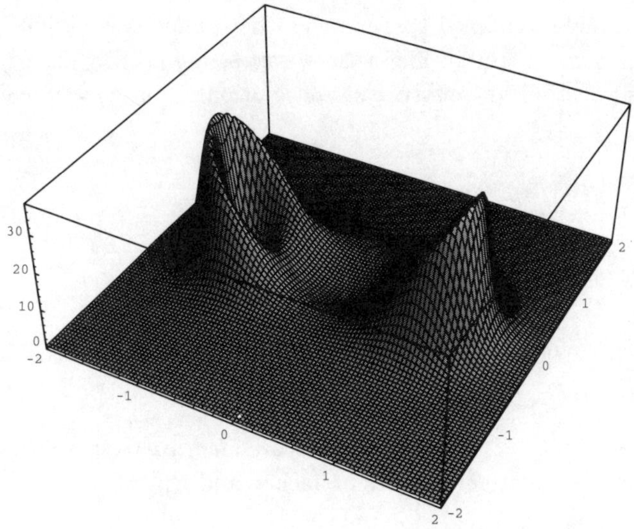

Fig. 6 *The non-relativistic* $N = 4$ *vortex representing two separated* 2*-vortices with 'positions' at* $x = \pm 1$ *and* $y = 0$.

Let us mention that the only time-dependent solutions found so far are either obtained from the static solution by the action of the Schrödinger group (the fundamental symmetry group of the theory [13, 14, 15, 16]), or are stationary [18].

4. Spinor vortices

A. Relativistic spinor vortices

In Ref. [19] Cho et al. obtain, by dimensional reduction from Minkowski space, a $(2+1)$-dimensional system. After some notational changes, their equations read

$$\tag{4.1} \begin{aligned} &\tfrac{1}{2}\kappa\epsilon^{\alpha\beta\gamma}F_{\beta\gamma} = e\big(\bar{\psi}_+\gamma_+^\alpha\psi_+ + \bar{\psi}_-\gamma_-^\alpha\psi_-\big),\\ &\big(ic\gamma_\pm^\alpha D_\alpha - m\big)\psi_\pm = 0, \end{aligned}$$

where the two sets of Dirac matrices are

$$\tag{4.2} (\gamma_\pm^\alpha) = (\pm(1/c)\sigma^3, i\sigma^2, -i\sigma^1),$$

and the $\psi_\pm$ denote the chiral components, defined as eigenvectors of the chirality operator

$$\tag{4.3} \Gamma = \begin{pmatrix} -i\sigma_3 & 0 \\ 0 & i\sigma_3 \end{pmatrix}$$

Observe that, although the Dirac equations are decoupled, the chiral components are nevertheless coupled through the Chern-Simons equation. Stationary solutions, representing purely magnetic vortices, are readily found [19]. It is particularly interesting to construct *static* solutions. For $A_0 = 0$ and $\partial_t A_i = 0$, setting

$$\tag{4.4} \psi_\pm = e^{-imt}\begin{pmatrix} F_\pm \\ G_\pm \end{pmatrix},$$

the *relativistic* system (4.1) becomes, for $c = 1$,

$$\tag{4.5} \begin{aligned} &\kappa\epsilon^{ij}\partial_i A_j = -e\big(|F_+|^2 + |G_-|^2\big),\\ &\big(D_1 + iD_2\big)F_\pm = 0, \qquad \big(D_1 - iD_2\big)G_\pm = 0. \end{aligned}$$

Now, for $F_\pm = 0$ or $G_\pm = 0$, these equations are *identical* to those which describe the *non-relativistic, self-dual* vortices of Jackiw and Pi [15] !

B. Non-relativistic spinor vortices

Non-relativistic spinor vortices can also be constructed along the same lines [20]. Following Lévy-Leblond [21], a non-relativistic spin $\frac{1}{2}$ field $\psi = \begin{pmatrix} \Phi \\ \chi \end{pmatrix}$ where Φ and χ are two-component 'Pauli' spinors, is described by the $2+1$ dimensional equations

$$\tag{4.6} \left\{ \begin{array}{cccc} (\vec{\sigma}\cdot\vec{D})\,\Phi & + & 2m\,\chi & = 0, \\ D_t\,\Phi & + & i(\vec{\sigma}\cdot\vec{D})\,\chi & = 0. \end{array} \right.$$

These spinors are coupled to the Chern-Simons gauge field through the mass (or particle) density, $\varrho = |\Phi|^2$, as well as through the spatial components of the current,

$$\vec{J} = i\big(\Phi^\dagger \vec{\sigma}\,\chi - \chi^\dagger \vec{\sigma}\,\Phi\big), \tag{4.7}$$

according to the Chern-Simons FCI (2.3). The chirality operator is still given by Eq. (4.3) and is still conserved. Observe that Φ and χ in Eq. (4.6) are *not* the chiral components of ψ; these latter are defined by $\frac{1}{2}(1 \pm i\Gamma)\psi_\pm = \pm\psi_\pm$.

It is easy to see that Eq. (4.6) splits into two uncoupled systems for ψ_+ and ψ_-. Each of the chiral components separately describe (in general different) physical phenomena in $2+1$ dimensions. For the ease of presentation, we keep, nevertheless, all four components of ψ.

Now the current can be written in the form:

$$\vec{J} = \frac{1}{2im}\Big(\Phi^\dagger \vec{D}\Phi - (\vec{D}\Phi)^\dagger\Phi\Big) + \vec{\nabla} \times \Big(\frac{1}{2m}\,\Phi^\dagger\sigma_3\Phi\Big). \tag{4.8}$$

Using $(\vec{D}\cdot\vec{\sigma})^2 = \vec{D}^2 + eB\sigma_3$, we find that the component-spinors satisfy

$$\left\{\begin{array}{lcl} iD_t\Phi &=& -\frac{1}{2m}\Big[\vec{D}^2 + eB\sigma_3\Big]\Phi, \\[2ex] iD_t\chi &=& -\frac{1}{2m}\Big[\vec{D}^2 + eB\sigma_3\Big]\chi - \frac{e}{2m}\,(\vec{\sigma}\cdot\vec{E})\,\Phi. \end{array}\right. \tag{4.9}$$

Thus, Φ solves a 'Pauli equation', while χ couples through the Darwin term, $\vec{\sigma}\cdot\vec{E}$. Expressing $\vec{E}$ and B through the FCI, (2.3) and inserting into our equations, we get finally

$$\left\{\begin{array}{ll} iD_t\Phi = & \Big[-\frac{1}{2m}\,\vec{D}^2 + \frac{e^2}{2m\kappa}\,|\Phi|^2\,\sigma_3\Big]\Phi, \\[2ex] iD_t\chi = & \Big[-\frac{1}{2m}\,\vec{D}^2 + \frac{e^2}{2m\kappa}\,|\Phi|^2\,\sigma_3\Big]\chi - \frac{e^2}{2m\kappa}\,(\vec{\sigma}\times\vec{J})\Phi. \end{array}\right. \tag{4.10}$$

If the chirality of ψ is restricted to $+1$ (or -1), this system describes non-relativistic spin $+\frac{1}{2}$ $(-\frac{1}{2})$ fields interacting with a Chern-Simons gauge field. Leaving the chirality of ψ unspecified, it describes *two* spinor fields of spin $\pm\,\frac{1}{2}$, interacting with each other and the Chern-Simons gauge field.

Since the lower component is simply $\chi = -(1/2m)(\vec{\sigma}\cdot\vec{D})\Phi$, it is enough to solve the Φ-equation. For

$$\Phi_+ = \begin{pmatrix}\Psi_+\\ 0\end{pmatrix} \qquad\text{and}\qquad \Phi_- = \begin{pmatrix}0\\ \Psi_-\end{pmatrix} \tag{4.11}$$

respectively — which amounts to working with the $\pm$ chirality components — the 'Pauli' equation in (4.10) reduces to

$$iD_t\Psi_\pm = \Big[-\frac{\vec{D}^2}{2m} \pm \lambda\,(\Psi_\pm^\dagger\Psi_\pm)\Big]\Psi_\pm, \qquad \lambda \equiv \frac{e^2}{2m\kappa}, \tag{4.12}$$

which again (3.2), but with non-linearity $\pm\lambda$, *half* of the special value Λ in (3.6) used by Jackiw and Pi. For this reason, our solutions (presented below) will be *purely magnetic*, ($A_t \equiv 0$), unlike in the case studied by Jackiw and Pi. In detail, let us consider the static system

$$\begin{cases} \Big[-\dfrac{1}{2m}(\vec{D}^2 + eB\sigma_3) - eA_t\Big]\Phi = 0, \\ \vec{J} = -\dfrac{\kappa}{e}\vec{\nabla} \times A_t, \\ \kappa B = -e\varrho, \end{cases} \tag{4.13}$$

and try the first-order Ansatz

$$\big(D_1 \pm iD_2\big)\Phi = 0. \tag{4.14}$$

Eq. (4.14) makes it possible to replace $\vec{D}^2 = D_1^2 + D_2^2$ by $\mp eB$, then the first equation in (4.13) can be written as

$$\Big[-\frac{1}{2m}eB(\mp 1 + \sigma_3) - eA_t\Big]\Phi = 0, \tag{4.15}$$

while the current is

$$\vec{J} = \frac{1}{2m}\vec{\nabla} \times \Big[\Phi^\dagger(\mp 1 + \sigma_3)\Phi\Big]. \tag{4.16}$$

Now, due to the presence of σ_3, both Eq. (4.16) and the second equation in (4.13) can be solved with a *zero* A_t and $\vec{J}$: by choosing $\Phi \equiv \Phi_+$ ($\Phi \equiv \Phi_-$) for the upper (lower) cases respectively makes $(\mp 1 + \sigma_3)\Phi$ vanish. (It is readily seen from Eq. (4.15) that any solution has a definite chirality).

The remaining task is to solve the first-order conditions

$$(D_1 + iD_2)\Psi_+ = 0, \qquad \text{or} \qquad (D_1 - iD_2)\Psi_- = 0, \tag{4.17}$$

which is done in the same way as in Section 3:

$$\vec{A} = \pm\frac{1}{2e}\vec{\nabla} \times \log\varrho, \qquad \triangle \log\varrho = \pm\frac{2e^2}{\kappa}\varrho. \tag{4.18}$$

A normalizable solution is obtained for Ψ_+ when $\kappa < 0$, and for Ψ_- when $\kappa > 0$. (These correspond precisely to having an attractive non-linearity in Eq. (4.12)). The lower components vanish in both cases, as seen from

$$\chi = -\frac{1}{2m}(\vec{\sigma} \cdot \vec{D})\Phi. \tag{4.19}$$

Both solutions only involve *one* of the $2+1$ dimensional spinor fields $\psi_\pm$, depending on the sign of κ.

The physical properties such as symmetries and conserved quantities can be studied by noting that our equations are in fact obtained by variation of the $2+1$-dimensional action given in [20], which can also be used to show that the coupled Lévy-Leblond — Chern-Simons system is, just like its scalar counterpart, Schrödinger symmetric [15].

A conserved energy-momentum tensor can be constructed and used to derive conserved quantities [20]. One finds that the 'particle number' N determines the actual values of all the conserved charges: for the radially symmetric solution, e.g., the magnetic flux, $-eN/\kappa$, and the mass, $\mathcal{M} = mN$, are the same as for the scalar soliton of [15]. The total angular momentum, however, can be shown to be $I = \mp N/2$, *half* of the corresponding value for the scalar soliton. As a consequence of self-duality, our solutions have *vanishing energy*, just like the ones of Ref. [15].

C. The non-relativistic limit

Setting

$$\psi_+ = e^{-imc^2t}\begin{pmatrix}\Psi_+\\ \widetilde{\chi}_+\end{pmatrix} \qquad \text{and} \qquad \psi_- = e^{-imc^2t}\begin{pmatrix}\widetilde{\chi}_-\\ \Psi_-\end{pmatrix}, \tag{4.20}$$

Eq. (4.1) become

$$\begin{cases} iD_t\Phi - c\vec{\sigma}\cdot\vec{D}\widetilde{\chi} = 0,\\ iD_t\widetilde{\chi} + c\vec{\sigma}\cdot\vec{D}\Phi + 2mc^2\widetilde{\chi} = 0,\end{cases} \tag{4.21}$$

where $\Phi = \begin{pmatrix}\Psi_+\\ \Psi_-\end{pmatrix}$ and $\widetilde{\chi} = \begin{pmatrix}\widetilde{\chi}_-\\ \widetilde{\chi}_+\end{pmatrix}$ In the non-relativistic limit $mc^2\widetilde{\chi} >> iD_t\widetilde{\chi}$, so that this latter can be dropped from the second equation. Redefining $\widetilde{\chi}$ as $\chi = c\widetilde{\chi}$ yields precisely our Eq. (4.6). This also explains, why one gets the same (namely the Liouville) equation both in the relativistic and the non-relativistic cases: for static and purely magnetic fields, the terms containing D_t are automatically zero.

In this paper, we only reviewed the Abelian Chern-Simons theories. Non-Abelian generalizations are studied in Refs. [14], [22] and [23].

Acknowledgement. We are indebted to our collaborators S. Nicolis (Tours) and L. Palla (Budapest), as well as to R. Jackiw (MIT) for many interesting discussions.

References

[1]]See, e.g., E. M. Lifshitz and L. PL. Pitaevski, *Statistical Physics*, Landau-Lifshitz Course of Theoretical Physics, Vol. 9. Part 2. Butterworth-Heinemann (1995).

[2] A. A. Abrikosov, *Sov. Phys. JETP* **5**, 1174 (1957).

[3] H. B. Nielsen and P. Olesen, *Nucl. Phys.* **B61**, 45 (1973).

[4] K. v. Klitzing, G. Dorda, and M. Pepper, *Phys. Rev. Lett.* **45**, 497 (1980); D. C. Tsui, H. L. Stormer, and A. C. Gossard, *Phys. Rev. Lett.* **48**, 1559 (1982).

[5] J. G. Bednorz and K. A. Müller, *Z. Phys.* **B64**, 189 (1986).

[6] S. S. Chern, *Complex Manifolds Without Potential Theory*, 2nd ed., Springer, N. Y. (1979).

[7] R. Jackiw and S. Templeton, *Phys. Rev.* **D23**, 2291 (1981); J. Schonfeld, *Nucl. Phys.* **B185**,

157 (1981); S. Deser, R. Jackiw and S. Templeton, *Phys. Rev. Lett.* **48**, 975 (1982); *Ann. Phys.* **140**, 372 (1982).

[8] S. M. Girvin, *The quantum Hall Effect*, ed. R. E. Prange and S. M. Girvin. Chap. 10. (Springer Verlag, New York (1986); S. C. Zhang, T. H. Hansson and S. Kivelson, *Phys. Rev. Lett.* **62**, 82 (1989); J. Fröhlich and U. Studer, *Rev. Mod. Phys.* **65**, 733 (1993).

[9] Y.-H. Chen, F. Wilczek, E. Witten and B. I. Halperin, *Int. Journ. Mod. Phys.* **B3** 1001 (1989).

[10] S. K. Paul and A. Khare, *Phys. Lett.* **B174**, 420 (1986); J. Hong, Y. Kim and P. Y. Pac, *Phys. Rev. Lett.* **64**, 2230 (1990); R. Jackiw and E. Weinberg, *Phys. Rev. Lett.* **64**, 2234 (1990).

[11] R. Jackiw, K. Lee and E. Weinberg, *Phys. Rev.* **D42**, 2234 (1990).

[12] S. Deser and Z. Yang, *Mod. Phys. Lett.* **A4**, 2123 (1989).

[13] R. Jackiw and S-Y. Pi, *Phys. Rev.* **D42**, 3500 (1990); *Prog. Theor. Phys. Suppl.* **107**, 1 (1992).

[14] G. Dunne, *Self-dual Chern-Simons solitons*, Springer Lecture Notes in Physics (New Series) (1995).

[15] R. Jackiw and S-Y. Pi, *Phys. Rev. Lett.* **64**, 2969 (1990).

[16] C. Duval, P. A. Horváthy and L. Palla, *Phys. Lett.* **B325**, 39 (1994).

[17] S. K. Kim, K. S. Soh, and J. H. Yee, *Phys. Rev.* **D42**, 4139 (1990).

[18] K. H. Cho, D-H. Oh, C. Rim, *Phys. Rev.* **D46**, 2709 (1992).

[19] Y. M. Cho, J. W. Kim, and D. H. Park, *Phys. Rev.* **D45**, 3802 (1992).

[20] C. Duval, P. A. Horváthy and L. Palla, *Phys. Rev.* **D52**, 4700 (1995); to be published in *Ann. Phys.* (N. Y.).

[21] J-M. Lévy-Leblond, *Comm. Math. Phys.* **6**, 286 (1967).

[22] H. J. De Vega and F. A. Schaposnik, *Phys. Rev. Lett.* **56**, 2564 (1986); B. Grossmann, *Phys. Rev. Lett.* **65**, 3230 (1990); G. V. Dunne, R. Jackiw, S-Y Pi, and C. A. Trugenberger, *Phys. Rev.* **D43**, 1332 (1991).

[23] L. Martina, O. K. Pashaev, and G. Soliani, *Phys. Rev.* **B48**, 15 787 (1993).

Contemporary Mathematics
Volume **203**, 1997

The Generalized Local Instability Criterion from the Geodesic Deviation Equation

Marek Szydłowski

ABSTRACT. The invariant criterion of local instability of geodesic motion in terms of the Riemannian (pseudoriemannian) geometry is proposed. The property of sensitive dependence on initial conditions is formulated as the local instability of nearby geodesics. In study of sectional curvature, the bivector formalism is applied.

1. Introduction

Our concentration on the study of dynamics is motivated by the following questions:

- How sensitive is the dynamics of a given system to the change of the initial conditions?
- Is a dynamical system chaotic?
- Is a dynamical system a K-system?

To answer these questions in an analytical manner a local instability criterion is proposed. In the present paper, we examine mathematical problems of this criterion related to the Riemannian and pseudoriemannian geometry.

It was Pesin [**1**] who first used the geodesic deviation equation to investigate behaviour of nearby geodesics in the ergodic theory context. In our approach, in the spirit of Pesin's theory, the local instability problem is reduced to the problem of studying the geodesic deviation equation which is written, after Arnold's concept [**2**], in the analogous form to the Newtonnian equation of motion. The considered system may be now reduced to the form of a linear harmonic oscillator with time dependent frequency, and it turns out that the negative sign of its potential energy implies its local instability.

As it is well known, the Lyapunov stability, or the boundness concept, is not invariant with respect to the coordinate transformations and time variable rescaling. It happens that the mechanical system which is stable in certain generalized coordinates is unstable in another coordinate system.

There are some reasons for studying the dynamics of systems in terms of internal Riemannian or pseudoriemannian geometry [**3**].

1991 *Mathematics Subject Classification.* 58F15; 58F17, 53C22.

This work was supported in part by the KBN research program No. 2-P-0-3D-022-10.

A Riemannian tensor or, equivalently, sectional curvature is directly derived from a metric and provides image of local behaviour of trajectories. Due to the Lobatchevsky-Hadamar theorem, from a Riemann tensor one can find global properties of systems. Local instability parameters are invariant as far as they are composed of invariant objects of a Riemannian (pseudoriemannian) geometry.

There are two natural application domains of this approach:

- Geodesic motion of a test particle in spacetimes of General Relativity.
- Geodesic motion in spaces with the Jacobi metric.

According to the Maupertuis-Jacobi principle, dynamics of systems with natural Lagrange function (mechanical systems) can be reduced to motion of geodesics on a Riemannian space with the Jacobi metric (a manifold with boundary) [**4**, **5**]. Due to the generalized Maupertuis-Jacobi principle, mechanical systems with an indefinite kinetic energy form is represented by geodesics on a pseudoriemannian space with the Jacobi metric (a manifold with a singular boundary set) [**6, 7, 8**]. In both cases the Jacobi metric is conformally related to a metric taken from the kinetic energy form. The conformal factor is a function of potential and kinetic energy of a system.

If a system (geodesic flow) is locally unstable, nearby geodesics diverge exponentially which implies a sensitive dependence on initial conditions — the key indicator of deterministic chaos. Then the long-time prediction of motion in the phase space is impossible since small initial perturbations grow arbitrarily large as the system evolves in time. In general the property of local instability seems to be necessary but not sufficient condition for the existence of deterministic chaos.

The dynamical system $(\mathcal{M}, \phi_t)$ is chaotic if it contains a chaotic invariant subset $\Lambda \subset \mathcal{M}$ such that 1) Λ is invariant under geodesic flow ϕ_t, i.e., $\phi_t(\Lambda) \subset \Lambda \; \forall t \in \mathbf{R}$; 2) Λ is compact; 3) Λ has the property of the sensitive dependence on initial conditions; 4) Λ is topologically transitive [**9**]

It means that the property of the sensitive dependence in a compact manifold with geodesic congruence leads to the chaotic behaviour.

2. The generalized local instability criterion of geodesic flows

To investigate the sensitivity with respect to initial conditions in the case of a geodesic concruence in a Riemannian or (pseudoriemannian) space $(\mathcal{M}, g)$ it is convenient to consider the geodesic deviation equation. This equation determines the deviation of normal vector n. This vector is normal to a geodesic.

We can write down the geodesic deviation equation in an analogous form to the Newtonian equation of motion

$$\nabla_u \nabla_u n \equiv \frac{D^2 n}{ds^2} = -\mathrm{grad}_n V_u(n) \tag{2.1}$$

where the potential is

$$V_u(n) = \frac{1}{2} R(u, n, u, n) = \frac{1}{2} K_{u;n} g(u, u) g(n, n) \tag{2.2}$$

and ∇_u means the covariant differentiation with respect to the vector u tangent to a geodesic; s is a natural parameter along a geodesic; $[\mathrm{grad}_n]^i = g^{ij} \frac{\partial}{\partial n^j}$; $g^{ij} g_{jk} = \delta^i_k$; The Riemannian tensor $R(u, n, u, n) = g(u, R(u, n)n)$ and the sectional curvature $K_{u:n}$ in two-plane determined by vectors u and n are defined by

$$R(u,n,u,n) = K_{u;n}[g(n,n)g(u,u) - g(n,u)g(u,n)]. \tag{2.3}$$

We can look upon the quantity $V_u(n)$ as a potential whose gradient determines the right-hand side of the geodesic deviation equation. The norm of the normal vector deviation $\|n\|^2 = g(n,n)$ satisfies the following equation

$$\begin{aligned} \frac{D^2\|n\|^2}{ds^2} &= 2[g(n, R(u,n)u) + g(\nabla_u n, \nabla_u n)] = -2R(n,u,n,u) + 2g(\nabla_u n, \nabla_u n)] \\ &= -2K_{u;n}\|n\|^2 g(u,u) + 2g(\nabla_u n, \nabla_u n)] \end{aligned} \tag{2.4}$$

Here we distinguish two cases

1. If the metric g is positive definite, a geodesic congruence is defined on a Riemannian manifold.

2. If the metric g is indefinite and has the Lorentzian signature $(-,+,+,+)$, the geodesic congruence is defined on a pseudoriemannian manifold.

We can use the classical Maupertuis-Jacobi principle to reduce a Hamiltonian system to a geodesic flow on a Riemannian manifold with boundary. The systems which obey this procedure are called classical simple mechanical systems (with a positive definite kinetic energy form).

Non-classical simple mechanical systems are, for example, Hamiltonian systems (with an indefinite kinetic energy form) for which we can use the generalized Maupertuis-Jacobi principle to reduce them to a geodesic flow on a pseudoriemannian manifold. The second example is the geodesic motion of a test particle and of photons in spacetimes of general relativity.

In the non-classical case we cannot, in general, draw any conclusion about a sign of the norm $\|n\|^2$ from the sign of $g(u,u)$. However in two situations described below we can determine the sign of $g(u,u,)$. Let a plus sign in the metric signature corresponds to a spacelike direction and a minus sign — to a timelike direction.

Thus, only in the 2-dimensional case, we find that if a tangent vector to a geodesic is timelike $(g(u,u) < 0)$ then a normal vector is spacelike $(g(n,n) > 0)$, and *vice versa.*

Because of orthogonality of vectors u and n, we have $g(u,n) = 0$ and $g(u, \nabla_u n) = 0$. Consequently, if a tangent vector to a geodesic is timelike, then $\nabla_u n$ is spacelike, i.e., $g(\nabla_u n, \nabla_u n) > 0$.

Therefore, if the geodesic motion takes place in 2-dimensional pseudoriemannian manifold or in a pseudoriemannian one with timelike geodesics and if the Riemannian curvature is negative or, equivalently, the sectional curvature is positive, then we have

$$\frac{D^2\|n\|^2}{ds^2} > 0. \tag{2.5}$$

Inequality (2.5) suggests that the normal distance between nearby geodesics is a convex function of the natural parameter s along the geodesic (an arc length in the classical case). Relationship (2.5) also holds for classical simple mechanical systems provided that the sectional curvature is negative.

Additionally, if the Riemannian curvature is bounded or if the sectional curvatures in (u,n)-twoplanes at a point x of a manifold $\mathcal{M}$ satisfy conditions

$$K_{(u;n)}(x) < \quad -c^2 < 0 \qquad \text{for the Riemannian case}$$

$$K_{(u;n)}(x) > \quad c^2 > 0 \qquad \text{for the pseudoriemannian case,} \tag{2.6}$$

then nearby geodesics diverge exponentially near $x \in \mathcal{M}$

$$\|n\|^2 \geq \|n\|(0)\exp(c|s|). \tag{2.7}$$

This defines the property of the local (exponential) instability of geodesic flows.

3. Global Properties of Geodesic Motion

Let us consider manifolds of negative curvature. The Lobatchevsky-Hadamar theorem states that a geodesic flow on a closed Riemannian manifold of negative curvature is a C-flow.

Heuristically, C-flow is a flow of exponentially diverging geodesics with respect to both the past and the future. It is sufficient for a dynamical system to be a K-system if a manifold is closed ($E - V > 0$) and the sectional curvature is negative (a manifold is said to have the negative Riemannian curvature if all its sectional curvatures multiplied by $g(u, u)$ and $g(n, n)$ are everywhere negative).

However, to solve the problem of motion we have to work with the geodesic equation in a Riemannian space with boundary ($E - V = 0$) on which the Jacobi metric degenerates (if the metric g is indefinite, as it is the case in general relativity, we have to work with the pseudoriemannian manifold with the singular boundary set). The problem is such that usually in physically interesting cases (such as classical simple mechanical systems) the boundary of motion ∂D is not empty and we cannot directly apply the Lobatchevski-Hadamar theorem.

In the pseudoriemannian case there is no counterpart of the Lobatchevsky-Hadamar theorem. However, the Anosov theorem states that any locally unstable geodesic flow on a closed manifold is a K-system. Therefore, we need a generalized criterion of the local instability of geodesic flows on a *closed pseudoriemannian* manifold.

Some theorems on the global behaviour of the geodesic flow on pseudoriemannian manifolds can be found in the book by Beem and Ehrlich [**10**].

The Cartan-Hadamar theorem is here very important. This theorem says that if spacetime $(\mathcal{M}, g)$ has everywhere non-negative timelike sectional curvature then the spacelike geodesics have no conjugate points. This implies that the so-called Lyapunov exponents (which measure the average separation rate of nearby trajectories) are different from zero, and also something more, namely, that, with some additional assumptions, Pesin's theory applies here. Another important tool for drawing global conclusions is the so-called timelike diameter which can be estimated from the timelike sectional curvature. If this curvature is positive, the divergence rate of nearby geodesics is larger than that in the Minkowski spacetime.

To decide whether a geodesic flow is locally unstable one can follow one of the two strategies.

3.1. The Average Local Instability Criterion. Instead of solving geodesic deviation equation (2.1) we apply the averaging procedure to it [**6**], and then we obtain the Ricci scalar curvature as an indicator of the *average* local instability of a geodesic flow.

The geodesic deviation potential $V_u(n)$ is averaged with respect to randomly chosen vectors u and n. Now, the deviation equation assumes the form

$$\frac{D^2 n}{ds^2} = -\text{grad}_n \bar{V}_u(n) = -\text{grad}\frac{1}{2N(N-1)} Rg(n,n)g(u,u) = -\frac{Rg(u,u)}{N(N-1)} n, \tag{3.1}$$

where $N = \dim \mathcal{M}$.

And the sufficient conditions of the geodesic flow instability are

$$\begin{aligned} R &< 0 \quad \text{for the Riemannian case,} \\ Rg(u,u) &< 0 \quad \text{for the pseudoriemannian case.} \end{aligned} \tag{3.2}$$

We can go a step further and introduce the Fermi base along a randomly chosen geodesic as follows: $\{E_a,\ a = 1, \dots, N-1,\ E_N = u\}$, such that $g(E_a, E_b) = \delta_{ab}$ and $g(E_a, E_N) = 0$. Now we have

$$\frac{d^2 n^\alpha}{ds^2} = -\frac{Rg(u,u)}{N(N-1)} n^\alpha, \tag{3.3}$$

where n^α (such that $n = n^\alpha E_\alpha$) are components of the normal separation vector of nearby geodesics.

Equation (3.3) can be regarded as an equation of a linear non-autonomous harmonic oscillator with the potential $\tilde{V}(n^\alpha) = \frac{k}{2}\sum_{\alpha=1}^{N-1}(n^\alpha)^2$ in the generalized coordinates n^α. System (3.3) is locally unstable if its potential energy $\tilde{V}(n^\alpha)$ with $k = \frac{Rg(u,u)}{N(N-1)}$ is negative. If a system is locally unstable then nearby geodesics diverge in the same way as balls slipping down from a top of a hill. This happens if $\tilde{V}(n^\alpha) < 0$ (or $k < 0$) or $[\text{grad}\tilde{V}(n)]^i < 0$. Now, we can rewrite conditions (3.2)

$$\begin{aligned} R(W(q(s))) &< 0 \quad \text{for the Riemannian case,} \\ R(W(q(s)))\text{sgn}(E-V) &< 0 \quad \text{for the pseudoriemannian case.} \end{aligned} \tag{3.4}$$

Let us notice that if potential V of a hamiltonian system is such that $E - V < 0$ then the vector tangent to a trajectory is timelike because $\|u\|^2 = \text{sgn}(E-V)$ and, consequently, a normal separation vector is spacelike. It also implies $R > 0$ as the instability criterion. It is worth mentioning that the local instability of a geodesic flow depends not only on a positiveness of the Ricci scalar but also on its value.

3.2. Sectional Curvature for Dynamics. The knowledge of the sectional curvature in a twodirection (or a Riemann tensor) enables us to apply theorems on global properties of a system. To calculate the curvature tensor of $(\mathcal{M}, g)$ it is convenient to introduce the following tensor [6]

$$C_{ij} = -\frac{N-2}{2W^2}(-\nabla_i\nabla_j W + 3\nabla_i W \nabla_j W - \frac{1}{2} g_{ij} N_{W^2}) \tag{3.5}$$

where $\hat{g}_{ij} = 2W g_{ij}$ is a metric of space $(\mathcal{M}, \hat{g})$ conformally related to g_{ij}; $i, j = 1, \dots, N$ and the wedge denotes quantities calculated in the conformally related metric; ∇_i is the symbol of covariant differentiation with respect to the metric g; $N_{W^2} = g^{ij}\nabla_i W \nabla_j W$.

The components of the Riemannian tensor have the following form

$$\hat{R}_{ijkl} = \hat{g}_{ir} R^r_{jkl} + \frac{e^{2\phi}}{N-2}[g_{il}C_{jk} - g_{ik}C_{jl} + g_{kj}C_{il} - g_{lj}C_{ik}] \tag{3.6}$$

where $2W = e^{2\phi}$.

Since the fundamental tensor C is symmetric at each point $x \in \mathcal{M}$ it can be identified with the self-adjoint linear operator

$$\tilde{C}(x) : T_x\mathcal{M} \to T_x\mathcal{M}.$$

Let π be any two-plane spanned by orthonormal vectors A, B (with respect to the metric g). The sectional curvature is

$$\hat{K}_{A;B} = -\frac{1}{2(N-2)W}[C(A,A) + C(B,B)] \tag{3.7}$$

where $C(A,B) = C_{ij}A^iB^j = C(B,A)$.

The Ricci scalar $\hat{R} = \hat{g}^{ij}\hat{R}_{ij} = e^{-2\phi}g^{ij}R_{ij}$ is given by

$$\hat{R} = \frac{R}{2W} + \frac{N-1}{8W^3}[-4W\Delta W - (N-6)N_{W^2}] \tag{3.8}$$

where the Laplacian is $\Delta W = \nabla_i \nabla_j W$.

From the self-adjoint operator $\tilde{C}$ we can construct the new rescaled self-adjoint operator Ω defined as

$$\Omega \equiv -\frac{\tilde{C}}{2(N-2)W} \tag{3.9}$$

Let us consider the eigenvalue problem of this operator

$$\Omega_q e_a = \lambda_a(q) e_a, \qquad e_a \in T_q\mathcal{M}, \qquad a = 1, \dots, N. \tag{3.10}$$

Since the operator $\Omega_q : T_q\mathcal{M} \to T_q\mathcal{M}$ is self-adjoint, its eigenvectors e_a corresponding to different eigenvalues are orthogonal (in the sense of the metric g). Now, we can choose an orthonormal base e_a in $T_q\mathcal{M} : g(e_a, e_b) = \delta_{ab}$. Then the sectional curvature in the twoplane π spanned by orthonormal vectors e_{a_1}, e_{a_2} is

$$\hat{K}_A(q) = \hat{K}_{(e_{a_1}, e_{a_2})}(q) = \lambda_{a_1}(q) + \lambda_{a_2}(q), \tag{3.11}$$

where the index $A = (a_1, a_2)$ represents the pair of indices corresponding to a twodirection determined by the eigenvectors (e_{a_1}, e_{a_2}).

It is clear that bivectors $(e_{a_1} \wedge e_{a_2})$, $a_1, a_2 = 1, \dots, N$ form a basis in the space of all twodirections (at the point $q \in \mathcal{M}$).

Let us consider the special case of a conformally flat space with $R^i_{jkl} = 0$. In this case, equations (3.6) and (3.11) imply

$$\hat{R}_{AB} = \hat{K}_A \hat{G}_{AB}, \tag{3.12}$$

where $\hat{G}_{AB}$ plays the role of the fundamental metric tensor in the bivector space.

Finally in the base of eigenvectors $e_{a_1} \wedge e_{a_2}$ the operator $\hat{R}_A^{\ B} = \hat{R}_{AC}\hat{G}^{CB}$ is diagonal, $\hat{R}_A^{\ B} = \hat{K}_A \delta_A^B$, and the eigenvectors of this operator are equal to the sectional curvatures in the twoplane spanned by the eigenvectors of the operator Ω.

It is also important that the characteristic equation for the operator Ω in the bivector base $E_A = e_{a_1} \wedge e_{a_2}$ can be written in terms of symmetric curvature polynomials in the following way

$$\det(\hat{R}_A^{\ B} - \lambda \delta_A^{\ B}) = (-1)^{\binom{N}{2}} \prod_{1 \leq a_1 < a_2 \leq N} [\lambda - (\lambda_{a_1} + \lambda_{a_2})] \tag{3.13}$$

where the polynomials $Q(z)$, given by

$$Q(z) = \prod_{1 \leq a_1 < a_2 \leq N} [z - (\lambda_{a_1} + \lambda_{a_2})] = z^M - q_1(\lambda_1, \dots, \lambda_N) z^{M-1} + \cdots$$

$$+ (-1)^k q_k(\lambda_1, \dots, \lambda_N) z^{M-k} + \cdots + (-1)^M q_M(\lambda_1, \dots, \lambda_N); \quad M = \binom{N}{2}, \tag{3.14}$$

define the polynomials $q_k(\lambda_1, \dots, \lambda_N)$ which are symmetric with respect to $(\lambda_1, \dots, \lambda_N)$. They are called symmetric curvature polynomials for the operator Ω.

The following theorem [**7, 11**] characterizes the non-positiveness and non-negativeness of the sectional curvature

THEOREM 3.1. *If* $R^i_{jkl} = 0$ *and* $q_k(\Omega(q))$ *are curvature polynomials for the operator* $\Omega(q)$, *then*

$$\hat{K}_{X,Y}(q) \leq 0 \Leftrightarrow \begin{cases} q_k(\Omega(q)) \leq 0 & \text{for } k \text{ odd} \\ q_k(\Omega(q)) \geq 0 & \text{for } k \text{ even} \end{cases}$$

and

$$\hat{K}_{X,Y}(q) \geq 0 \Leftrightarrow q_k(\Omega(q)) \geq 0 \quad \text{for } i = 1, \dots, M.$$

There is a simple algorithm for constructing $\binom{N}{2}$ symmetric curvature polynomials in any dimension N which gives us a geometric picture (maps of signs of sectional curvature) of the behaviour of the sectional curvature of the space $\mathcal{M}$ (configuration space) in an invariant way. Only the first polynomial is proportional to the Ricci scalar

$$q_1 = (N-1)\hat{R}. \tag{3.15}$$

References

[1] Y.B. Pesin 1977, Usp. Math. Nauk 32, 55.

[2] V.I. Arnold, Mathematical Methods in Classical Mechanics, (Springer-Verlag, New York 1989); V.I. Arnold and A. Avez, Ergodic Problems of Classical Mechanics (Benjamin, New York 1968); D.V. Anosov and Y.G. Sinai 1967, Russ. Math. Surv. 22, 103.

[3] A. Wintner, The Analytical Foundations of Celestial Mechanics (Princeton University Press, Princeton 1941).

[4] V.V. Kozlov 1976, Appl. Mat. Mech. 40, 399; V.V. Kozlov 1977, Vestnik Mosk. Inst. Mat. Mech. 5, 118; V.V. Kozlov 1985, Usp. Mat. Nauk 40, 33.

[5] I.A. Tajmanov 1992, Usp. Mat. Nauk 47, 143; S.V. Bolotin 1978, Vestnik Mosk. Inst. Mat. Mech. 6, 72.

[6] M. Szydlowski and J. Szczesny, Phys. Rev. D 50 (1994) 819; M. Szydlowski, J. Math. Phys. 35 (1994) 1850; M. Szydlowski and A. Krawiec, Appl. Mech. Rev. 46 (1993) 427; J. Szczesny, M. Szydlowski and A. Krawiec, Chaos, Solitons & Fractals 3 (1993) 249; M. Szydlowski, J. Szczesny and M. Biesiada, Chaos, Solitons & Fractals 1 (1991) 233.

[7] M. Szydlowski, Phys. Lett. A 201 (1995) 19; M. Szydlowski, Phys. Lett. A 176 (1993) 22; M. Szydlowski and A. Krawiec, Phys. Rev. D 47 (1993) 5323; M. Szydlowski and M. Biesiada, Phys. Rev. D 44 (1991) 2369; M. Szydlowski and A. Lapeta, Phys. Lett. A 148 (1990) 239;

[8] M. Szydlowski, in: L.O. Chua (ed.), Chaos and Nonlinear Mechanics (World Scientific, Singapur 1994); M. Szydlowski, in: J. Seimeis (ed.) Hamiltonian Mechanics: Integrability and Chaotic Behaviour, NATO ASI Series Physics B331 (1994); M. Szydlowski, Acta Cosmologica XVIII (1992) 85; M. Biesiada and M. Szydlowski, Phys. Lett. A 160 (1991) 123; M. Szydlowski and M. Biesiada, Phys. Lett. B 220 (1989) 32.

[9] S. Wiggins, Introduction to Applied Nonlinear Dynamical Systems and Chaos (Springer-Verlag, New York 1990).

[10] J. Beem and P. Ehrlich, Global Lorentzian Geometry (Marcel Dekker, New York 1981).

[11] Chong Pin Ong, Adv. Math. 15 (1975) 269.

Astronomical Observatory, Jagellonian University, Orla 171, 30-244 Krakow, Poland
E-mail address: `uoszydlo@cyf-kr.edu.pl`